Statistics for Engineering and Physical Science

Akaike/Kitagawa: The Practice of Time Series Analysis.

Hawkins/Olwell: Cumulative Sum Charts and Charting for Quality Improvement.

Statistics for Engineering
and Physical Science

Series Editors
P. Green, J.F. Lawless, V. Nair

Springer
New York
Berlin
Heidelberg
Barcelona
Hong Kong
London
Milan
Paris
Singapore
Tokyo

Hirotugu Akaike
Genshiro Kitagawa

Editors

The Practice of
Time Series Analysis

Springer

Hirotugu Akaike
Genshiro Kitagawa
The Institute of Statistical Mathematics
Minato-ku, Tokyo
Japan

Series Editors
P. Green
Department of Mathematics
University of Bristol
Bristol B58 1TW
England

J.F. Lawless
Department of Statistics
University of Waterloo
Waterloo, Ontario, N2L 3G1
Canada

V. Nair
Department of Statistics
University of Michigan
Ann Arbor, MI 48109
USA

Library of Congress Cataloging-in-Publication Data
The practice of time series analysis / Hirotugu Akaike, Genshiro
    Kitagawa, editors.
        p.   cm. — (Statistics for engineering and physical science)
    Includes bibliographical references and index.

    ISBN 978-1-4612-7439-1   ISBN 978-1-4612-2162-3 (e-Book)
    DOI 10.1007/978-1-4612-2162-3

    1. Engineering—Statistical methods.   2. Time-series analysis.
    I. Akaike, Hirotugu, 1927–   . II. Kitagawa, G. (Genshiro), 1948–
    . III. Series.
    TA340.P72   1998
    519.5´5—dc21   98-31331

Printed on acid-free paper.

Production managed by Lesley Poliner; manufacturing supervised by Jacqui Ashri.
Camera-ready copy prepared by the authors using LaTeX.

9 8 7 6 5 4 3 2 1

Preface to the English Version

This book is aimed at presenting examples of the successful application of time series analysis and control in various fields such as engineering, earth science, medical science, biology, and economics. The development of statistical science seriously depends on the accumulation of successful experiences of the solution of concrete real problems. Once a statistical method finds a new application it often encourages further applications in other areas of research.

The examples included in this book are related to various stages of the practical application of the time series analysis. We hope that the examples will be helpful to the readers in finding solutions to their own problems.

The original version of this book was published in Japanese, in two volumes in 1994 and 1995, respectively, by Asakura Publishing Company, to whom we are grateful for giving permission to publish this English version. Thanks are also due to Mr. Toru Nishikawa, representative of Technical Translator's Agency, for preparing the first draft of the English translations of most of the chapters and to Mrs. Setsuko Ono for her efforts in inputting, proof reading and laying out the text in LATEX.

Hirotugu Akaike
Genshiro Kitagawa
Tokyo, Japan
July 1998

Preface

Due to the introduction of the information criterion AIC and development of practical use of Bayesian modeling, the method of time analysis is now showing remarkable progress. In attempting the study of a new field the actual phenomenon is rarely so simple as to allow direct applications of existing methods of analysis or models. The real thrill of the statistical analysis lies in the process of developing a new model depending on the purpose and the characteristics of the object of the research. The purpose of this book is to introduce the readers to successful applications of the methods of time series analysis in a variety of fields, such as engineering, earth science, medical science, biology, and economics.

The editors have been aware of the importance of cooperative research in statistical science and carried out various cooperative research projects in the area of time series analysis. The Institute of Statistical Mathematics was reorganized as an inter-university research institute in 1985 and the activities of the Institute have been organized to promote the cooperative researches as its central activity. This book is composed of the outcomes of cooperative researches developed within this environment and contains the results ranging from the pioneering realizations of statistical control to the latest consequences of time series modeling.

The initial process of the research of the control of cement kilns discussed in this book revealed the crucial importance of the model selection and lead to the introduction of the information criterion AIC. The method of analysis and control of dynamic systems established by this research found applications in various fields. The expansion of the fields of application encouraged the development of new statistical methods and lead to the practical application of Bayesian models. Several results obtained by the utilization of Bayesian models are included in this book.

Results at various stages of the development of the time series analysis are shown to enable the readers to closely observe an ideal form of the development of statistical science where the theory and application help each other. It is hoped that the readers will learn how the authors of this book have solved the problems by developing their own ideas and will properly use the acquired knowledge as an aid in undertaking researches in further wider areas of applications.

The publication of this book was planned as one of the activities to celebrate the 50th anniversary of the Institute of Statistical Mathematics. We are very grateful to professor Kunio Tanabe, chairman of the publication committee, who enthusiastically encouraged the publication of this book. We are also deeply thankful to Professor Takashi Nakamura of the institute for his help in preparing the style file by LaTeX. Special thanks are due to Mrs. Setsuko Ono who efficiently performed the work of inputting and editing of some of the material in LaTeX. It is by the efforts of these people that the publication was made possible within a very short period of time of about one year after the initial planning of the book.

Hirotugu Akaike
Genshiro Kitagawa
Tokyo, Japan
March 1994

The Structure of This Book
and General References

The contents of the chapters of this book can be roughly classified by the subject and the model used as follows.

	Control	Engineering	Earth science	Medical biological	Economics
AR model	1, 4, 12	6, 13, 14		2, 15, 16	17
Bayes model		7	19, 20	9	
State space model			11, 21, 22	10	3, 18
Other	8		5		

(Numbers denote chapters)

The readers are supposed to have basic knowledge on the time series analysis. The references explicitly cited within the text are shown in the references of each chapter. The basic general references of this book are as follows:

1. Akaike, H. and Nakagawa, T. (1988), *Statistical Analysis and Control of Dynamic Systems*, Kluwer Academic Pub., Dordrecht. (Original Japanese version was published in 1972 by Saiensu-Sha, Tokyo).

2. Kitagawa, G. and Gersch, W. (1996), *Smoothness Priors Analysis of Time Series*, Springer-Verlag, New York.

3. Sakamoto, Y. Ishiguro, M. and Kitagawa, G. (1986), *Akaike Information Criterion Statistics*, D. Reidel Pub., Dordrecht.

Most of the necessary computer software are published in the following TIMSAC (Time Series Analysis and Control Program Package) series with source code and numerical examples.

4. TIMSAC. (Included in the above reference 1.)

5. TIMSAC-74, TIMSAC-78, TIMSAC-84: *Computer Science Monograph,* The Institute of Statistical Mathematics, Nos. 5 (1975) & 6 (1976), No. 11 (1979), Nos. 22 & 23 (1985).

Contents

12 Statistical Control of Cement Process   193

Yoshitaka Yagihara

13 Analysis of a Human/2-Wheeled-Vehicle System by ARdock   209

Makio Ishiguro and Takao Oya

14 Vibration Data Analysis of Automobiles   229

Shinzi Yamakawa

15 Auto-Regressive Spectral Analysis of RR-Interval Time
Series in Healthy Fetus and Newborn Infants   247

Teruyuki Ogawa

20 Analysis of Earth Tides Data

Yoshiaki Tamura

21 Detection of Groundwater Level Changes Related to Earthquakes

Norio Matsumoto

22 Processing of Missing Observations and Outliers in Time Series

Genshiro Kitagawa

23 Mental Preparation for Time Series Analysis

Hirotugu Akaike

Contributors

Hirotugu Akaike   1-7-14-204 Toride, Toride, 302 Japan.

Kosei Fukuda   Economic Planning Agency, 3-1-1 Kasumigaseki, Chiyoda-ku, Tokyo 100–8970 Japan.

Kohyu Fukunishi   Advanced Research laboratory, Hitachi, Ltd., Hatoyama, Saitama 350-0395 Japan. (Present address: Department of Electronic Engineering, Graduate School of Engineering, Osaka University, Yamada-Oka 2-1, Suita 565-0871 Japan. E-mail: fukunisi@ele.eng.osaka-u.ac.jp)

Tomoyuki Higuchi   The Institute of Statistical Mathematics, 4-6-7 Minami-Azabu, Minato-ku, Tokyo 106-8569 Japan. E-mail: higuchi@ism.ac.jp.

Toshio Iseki   Tokyo University of Mercantile Marine, 2-1-6 Etchujima, Koto-ku, Tokyo 135-8533 Japan. E-mail: iseki@ipc.tosho-u.ac.jp.

Makio Ishiguro   The Institute of Statistical Mathematics, 4-6-7 Minami-Azabu, Minato-ku, Tokyo 106-8569 Japan. E-mail: ishiguro@ism.ac.jp.

Xing-Qi Jiang   Department of Economics, Asahikawa University, 3-jo 23-Chome, Nagayama, Asahikawa, Hokkaido 079–8501 Japan. E-mail: jiang@asahikawa-u.ac.jp.

Genshiro Kitagawa   The Institute of Statistical Mathematics, 4-6-7 Minami-Azabu, Minato-ku, Tokyo 106-8569 Japan. E-mail: kitagawa@ism.ac.jp.

Fumiyasu Komaki   Department of Mathematical Engineering and Information Physics, School of Engineering, University of Tokyo, 7-3-1 Hongo, Bunkyo-ku, Tokyo 113-8656 Japan. E-mail: komaki@stat.t.u-tokyo.ac.jp.

Norio Matsumoto   Geological Survey of Japan. 1-1-3 Higashi, Tsukuba, Ibaraki 305-8567 Japan. E-mail: norio@gsj.go.jp

Hideo Nakamura   6-18-6 Minamiga-oka, Ohnojo-shi, Fukuoka-ken 816-0964 Japan.

Sadao Naniwa   Kansai University of International Studies, 1-18 Aoyama, Shijimi-cho, Miki-shi, Hyogo Japan. E-mail: naniwa@kuins.ac.jp.

Teruyuki Ogawa   Professor Emiritus of Pediatics, Oita Medical University, School of Medicine, Midorigaoka 5–13–7, Oita 870–1172 Japan. http://www.lukaster@oec-net.or.jp.

Kohei Ohtsu   Tokyo University of Mercantile Marine, 2-1-6 Etyujima, Koto-ku, Tokyo 135-8533 Japan. E-mail: ohtsu@ipc.tosho-u.ac.jp.

Takio Oya   Meiji University, 1-1-1 Higasi-Mita, Tama-ku, Kawasaki-shi 214-0034 Japan.

Akinori Shimazaki   691-2 Kokubu, Ueda-shi, Nagano-ken 386-0016 Japan.

Hitoshi Soma   Japan Automobile Research Institute, 2530 Karima, Tsukuba, Ibaraki 305-0822 Japan. E-mail: hsoma@jari.or.jp

Tetsuo Takanami   Institute of Seismological and Volcanology, Graduate School of Science, Hokkaido University, Kita 10-jyou, Nishi 8-choume, Kita-ku, Sapporo 060-0810 Japan. E-mail: ttaka@eos.hokudai.ac.jp.

Yoshiaki Tamura   National Astronomical Observatory, Division of Earth Rotation, Mizusawa Astrogeodynamics Observatory, Mizusawa, Iwate-ken 023-0861 Japan. E-mail: tamura@miz.nao.ac.jp.

Hiroshi Tsuda   NLI Research Institute, 1-1-1 Yurakucho, Chiyoda-ku, Tokyo 100 Japan. E-mail: tsuda@nli-research.co.jp

Akifumi Yafune   Bio-Iatric Center, The Kitasato Institute, 5-9-1, Shirokane Minato-ku, Tokyo 108-8642 Japan.

Yoshitaka Yagihara   System Sogo Kaihatsu Co. Ltd., 4-8-17 Hongo, Bunkyo-ku, Tokyo 113-0033 Japan.

Shinzi Yamakawa   Kogakuin University, 1-24-2 Nishi-Shinjyuku, Shinjyuku-ku, Tokyo 160-0023 Japan. E-mail: yamakawa@cc.kogakuin.ac.jp.

Takao Wada   Former director of the Inaki Hospital. (Dr. Wada passed away on November 7, 1997.)

Chapter 1

Control of Boilers for Thermoelectric Power Plants by Means of a Statistical Model

Hideo Nakamura
6-18-6 Minamiga-oka, Ohnojo-shi, Fukuoka-ken 816-0964, Japan

1.1 Introduction

This chapter introduces a successful application of the optimal regulator to steam temperature control of thermal electric power plants, in which the state equation for the control system design is derived based on a statistical model.

The present author had been with an electric power utility and its subsidiary for years, where he was engaged in the control of thermal electric power plants (hereafter referred to as "thermal power plant" or simply "power plants").

From his long-year experience the author has an opinion that, in the so-called multivariable control system in which multiple input-and-output variables mutually interact within the process, improving control performance is rather difficult with the conventional control system consisting of P(proportional), I(Integral), and D(derivative) control elements. In other words, introducing a system that utilizes prediction of state variables is indispensable to obtain desirable control performance.

The first candidate to utilize prediction of state variables that would substitute for the PID controller is the optimal control system designed on the basis of the state-space representation of the object system.

As well known, the first step to apply an optimal control system to an actual plant is to derive a state equation, a mathematical expression, describing the dynamics of the object system. However, it is not so easy to obtain a proper state equation that expresses the dynamic relationship among the process variables through physical or chemical approaches, because actual plants are usually very complex and include nonlinearity within the process.

In such a complex system, it would also be difficult to derive the state equation experimentally through field tests, because many unmeasurable process noises usually exist in actual plants.

As a matter of fact, in the late 1960s, when the author attempted to apply the optimal regulator to actual plants, he tried several times to obtain the state equation of the object plant through experimental approaches, and finally failed in the trial:

1

such difficulties made the author almost nearly give up the application of the state-
space method to the power plant control.

Fortunately at that time, a method based on a statistical model was advocated
by Akaike (Akaike and Nakagawa 1972, 1988) and its application to the control of a
cement rotary kiln was published (Otomo, Nakagawa and Akaike 1972).

Encouraged by this successful example, the author was convinced that the statistical method proposed by Akaike was the most practically effective way of imple-
menting the optimal control at thermal power plants. Then, under the conduct of
Prof. Akaike, the author collaborated with the researchers and engineers from Kyushu
Electric Power Company, Central Research Institute of Electric Power Industries in
Japan (CRIEPI), Kyushu University, manufacturers of the boiler, computer and control systems until successful implementation of the optimal regulator at an actual
power plant was finally achieved in 1978.

In this chapter, the difficulty of controlling a multivariable system, the motivation
that led the author to the study, is described in the first place. Then, derivation of
the state equation and analysis of the system based on a statistical model, practical
procedures for controller design, control performance of the optimal regulator are
described with some examples, followed by a list of the plants now being in operation
under the proposed optimal regulator.

Of course, the methodology introduced in this chapter can be applied to general
industrial processes other than thermal power plants: some implementation examples
by means of the same kind of approach have been reported in Japan.

1.2   Problems in Controlling Multivariable System

At first, difficulties in controlling a multivariable system, which led the author to
introduce an optimal regulator, is explained using an example of the control system
for a once-through boiler which is commonly used in modern electric power generation
plants.

Figure 1.1 shows the conceptual view of the structure of a large-capacity supercritical once-through boiler for electric power generation. In this type of boiler, the
mainsteam temperature or the superheater outlet steam temperature is controlled
by adjusting the flow rates of the fuel and the superheater spray (a portion of the
feedwater injected into the attemperator or temperature adjuster located between the
primary and the final superheaters): on the other hand, the reheat steam temperature is controlled by adjusting the opening of the gasdamper which regulates the
distribution of the fluegas flowing along the heat conducting surfaces of the primary
superheater and the reheater located at the rear path of the boiler.

As can be understood from this control mechanism, the operation of the flue-gas
damper affects both the mainsteam temperature and the reheat steam temperature
in the opposite direction, causing a mutual interaction between these two process
variables.

In the electric power system of Japan, in which thermal power generation plays
the principal role in the frequency control of the system, thermal power plants are
required to vary their power generation as quickly as possible responding to the load
command (MWD:Megawatt demand) issued to the plants from the power system's
dispatch center. Even for such load changes power plants must keep the variations
of their process variables within specified ranges in order to maintain plant efficiency

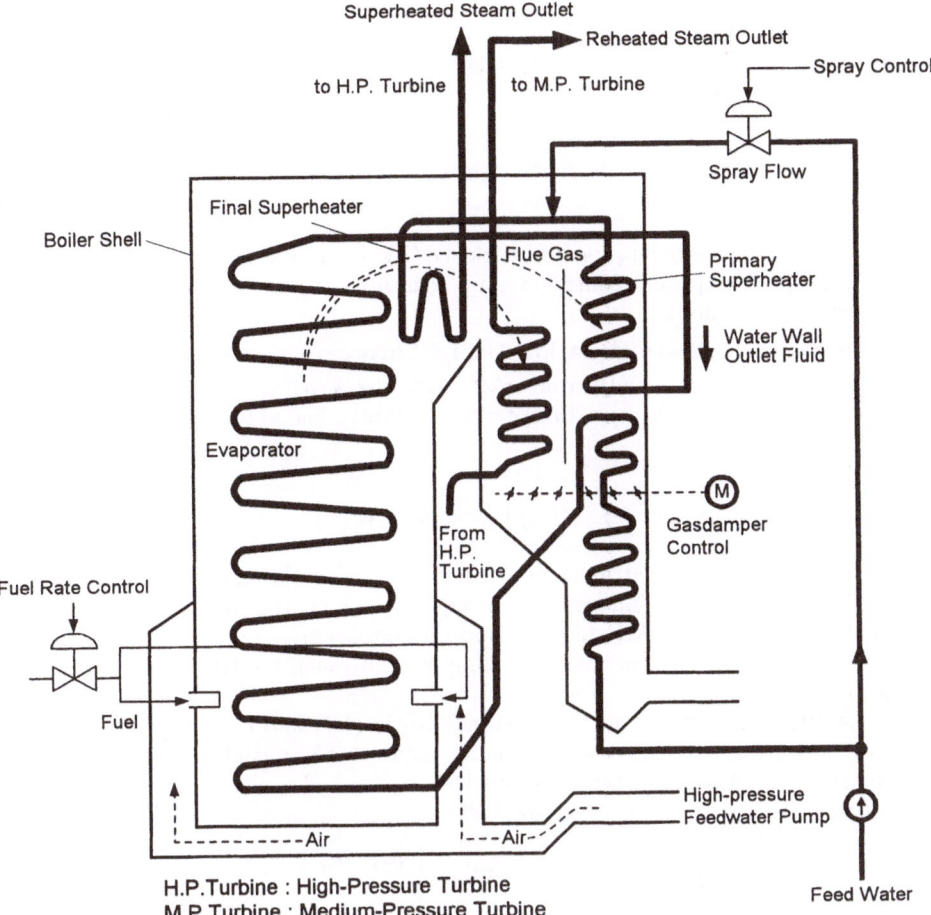

Figure 1.1 Conceptual view of F-W type supercritical boiler

and also to ensure security of the boiler and the turbine.

To satisfy these requirements the plant control system adjusts the flow rates of feedwater, fuel, air, and the opening of the flue-gas damper to their proper values responding immediately to the changes in MWD. Because such manipulation of control variables causes transient deviations of the controlled variables from their set point values, feedback control loops re-adjust process inputs so as to cancel such deviations. However, since these feedback control loops interact each other within the boiler process in the complicated ways, the whole system constitutes a typical multivariable system.

Therefore, with the conventional PID control theory which was developed substantially to deal with SISO (single-input single-output) systems, it is quite difficult to control many input variables properly and compensate for such interactions among the control loops and, as a result, to keep many controlled variables to their specified values during transient period. This was the principal factor that restricted the amplitude and the rate of changes of the power output of thermal power plants. The optimal control system described in this chapter was planned and implemented to solve this problem.

1.3 System Analysis and Control by Means of Statistical Model

In this chapter, the method proposed by Akaike (Akaike and Nakagawa 1972, 1988), the basic concept for optimal controller implementation by means of a statistical model, is briefly introduced.

1.3.1 Derivation of State Equation Using Autoregressive (AR) Model

Suppose that we fitted a multivariate AR model to $X(n)$, $n = 1, 2, \cdots, N$, a k-dimensional vector time series consisting of plant variables, and obtained the following expression:

$$X(n) = \sum_{m=1}^{M} A(m)X(n-m) + W(n),   \tag{1.1}$$

where $\boldsymbol{A}(m)$ are coefficient matrices of the AR model, M is the model order, $\boldsymbol{W}(n)$ is the residual vector called innovation. In equation (1.1), the model order M is determined to minimize the AIC (Akaike Information Criterion) or the FPE (Final Prediction Error) criterion, which is approximately equivalent to AIC for stationary Gaussian processes.

Then, we divide the system variable vector $\boldsymbol{X}(n)$ in (1.1) into r-dimensional state vector $\boldsymbol{x}(n)$ and l-dimensional control variable vector $\boldsymbol{u}(n)$, and at the same time, divide the matrices $A(m)$ into two sub matrices A_m and B_m, which correspond to $\boldsymbol{x}(m)$ and $\boldsymbol{u}(m)$ respectively, as shown in (1.2)

$$X(n) = \begin{bmatrix} \boldsymbol{x}(n) \\ \boldsymbol{u}(n) \end{bmatrix}, \quad A(m) = \begin{bmatrix} a_m & b_m \\ * & * \end{bmatrix},   \tag{1.2}$$

where $*$ means sub matrices irrelevant to the prediction of state variables.

Using (1.2), we can write the terms corresponding to $\boldsymbol{x}(n)$ as follows:

$$\boldsymbol{x}(n) = \sum_{m=1}^{M} a_m \boldsymbol{x}(n-m) + \sum_{m=1}^{M} b_m \boldsymbol{u}(n-m) + \boldsymbol{w}(n),   \tag{1.3}$$

where $\boldsymbol{w}(n)$ is the vector expressing modeling errors, or the innovation vector.

In (1.3), if we put

$$\boldsymbol{x}_0(n) = \boldsymbol{x}(n)   \tag{1.4}$$

$$\boldsymbol{x}_k(n) = \sum_{m=k+1}^{M} [A_m \boldsymbol{x}(n+k-m) + B_m \boldsymbol{u}(n+k-m)], \quad k = 1, 2, \cdots, M-1$$

then we get the state equation (1.5), which is well known as the observable canonical form:

$$\begin{aligned} \boldsymbol{Z}(n) &= \boldsymbol{A}\boldsymbol{Z}(n-1) + \boldsymbol{B}\boldsymbol{u}(n-1) + \boldsymbol{w}(n) \\ \boldsymbol{Y}(n) &= \boldsymbol{C}\boldsymbol{Z}(n) \end{aligned}   \tag{1.5}$$

where

$$
\boldsymbol{Z}(n) = \begin{bmatrix} \boldsymbol{x}_0(n) \\ \boldsymbol{x}_1(n) \\ \vdots \\ \boldsymbol{x}_{M-2}(n) \\ \boldsymbol{x}_{M-1}(n) \end{bmatrix}, \quad
\boldsymbol{A} = \begin{bmatrix} \boldsymbol{a}_1 & I & 0 & \cdots & 0 \\ \boldsymbol{a}_2 & 0 & I & \cdots & 0 \\ \vdots & \vdots & & \ddots & \\ \boldsymbol{a}_{M-1} & 0 & 0 & \cdots & I \\ \boldsymbol{a}_M & 0 & 0 & \cdots & 0 \end{bmatrix},
$$

$$
\boldsymbol{B} = \begin{bmatrix} \boldsymbol{b}_1 \\ \boldsymbol{b}_2 \\ \vdots \\ \boldsymbol{b}_{M-1} \\ \boldsymbol{b}_M \end{bmatrix}, \quad
\boldsymbol{W}(n) = \begin{bmatrix} \boldsymbol{w}(n) \\ 0 \\ \vdots \\ 0 \\ 0 \end{bmatrix}, \quad
\boldsymbol{C} = [I \ \ 0 \ \cdots \ 0].
$$

Figure 1.2 shows the block diagram which visually describes the relationship among the system variables in (1.5).

1.3.2 Design of Optimal Controller

Once the state equation that describes system dynamics is derived, the optimal gain matrix $\boldsymbol{G}$ for state-vector feedback can be computed by the DP (Dynamic Programming) procedure so as to minimize the following quadratic equation:

$$
J_I = E \sum_{n=1}^{I} [\boldsymbol{Z}^T(n)\boldsymbol{Q}\boldsymbol{Z}(n) + \boldsymbol{u}^T(n-1)\boldsymbol{R}\boldsymbol{u}(n-1)], \tag{1.6}
$$

where T denotes transposition of a vector, Q and R are weighting matrices that adjust the amplitude of the state vector $\boldsymbol{Z}$ and the manipulated vector $\boldsymbol{u}$, respectively.

In the DP computation, the state feedback gain matrix at each control stage can be obtained backward from the final control stage to the first control stage, and when the DP span k, the total number of the control stages in (1.6), is chosen large enough, say 40 or 50 in this example, the elements of the feedback gain matrix $\boldsymbol{G}$ for the first control stage converge to certain constant values.

In our controller design, we use this stationary gain matrix, assuming that the power plant constitutes a linear stationary process for small load variations, and perform the state feedback control using the manipulated vector $\boldsymbol{u}$ calculated by

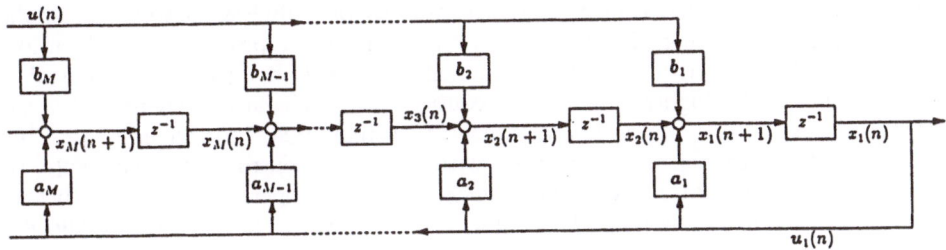

Figure 1.2 Block diagram expressing observable canonical form, equation (1.5)

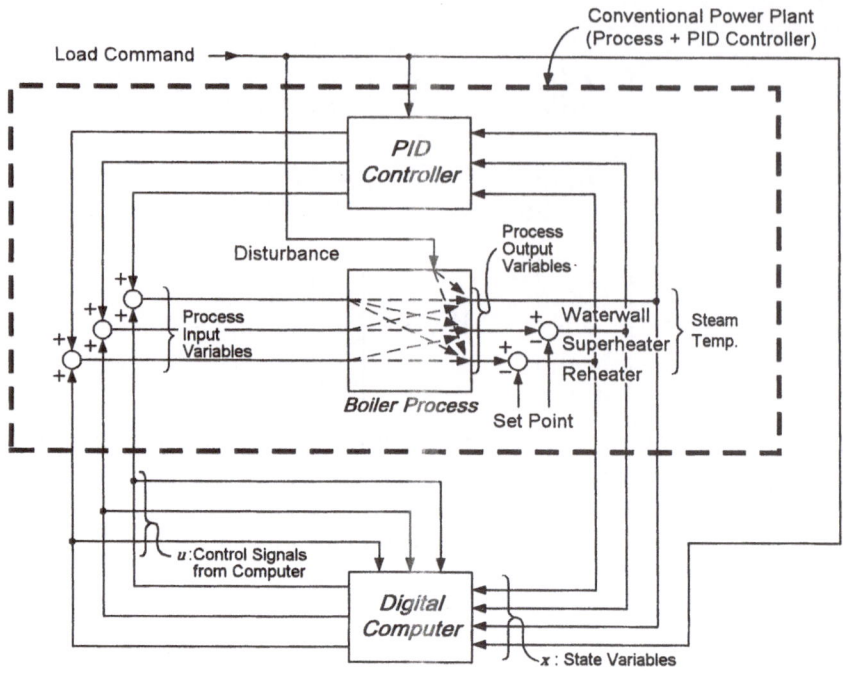

Figure 1.3 Configuration of proposed optimal control system

$u = -GZ(n)$ with a fixed G: in other words, considering optimal control for the control range from the present time-point to $I \cdot \Delta t$ (Δt is the control period) future time-point, we provide at every control time the optimal control signal using the state feedback gain matrix G calculated by DP for the first stage.

1.4   Practical Procedure for Optimal Controller Design

1.4.1   Composition of Control System

Figure 1.3 shows the composition of the optimal control system. As shown in the figure, the optimal control system composed of a digital controller is applied to the plant consisting of the combination of the boiler process and conventional PID controllers. In this system, the fundamental functions of the plant control are performed by the highly-reliable PID controllers, which have been refined with a long-year experience, and the optimal control system is used to cooperate with it to compensate for the mutual interactions within the PID control loops. Therefore, this system is suitable for the plants requiring extremely high operating reliability, because it not only assures integrity and robustness of the plant operation, but also allows an easy adjustment and maintenance work of the control system.

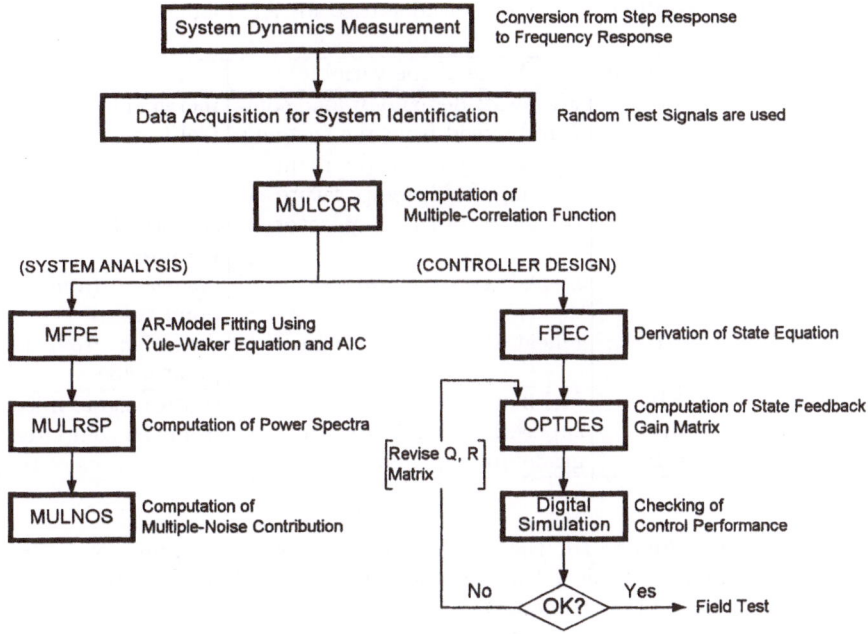

Figure 1.4 Procedure of optimal regulator implementation based on TIMSAC Library

1.4.2　Procedures for Optimal Controller Design

For the optimal controller design, the TIMSAC (Time Series Analysis and Control) program package developed by Akaike and his collaborators is used together with its some modifications. Figure 1.4 shows the flow diagram expressing the concept of TIMSAC Library.

Selection of system variables

Controlled variables such as the mainsteam and reheat steam temperatures, and the process variables that seem relevant to the controlled variables are chosen as the system variables for AR model building. In the case of the control system for a supercritical once-through boiler described in this chapter, seven system variables shown in Table 1.1 were employed.

In Table 1.1, the mainsteam temperature (SHT) and the reheat steam temperature (RHT) are the principal variables to be controlled. The WWT, the working fluid temperature at the outlet of the evaporator, is the variable useful as the preceding index to predict the behavior of the mainsteam temperature. The load demand or MWD, which the plant receives from the system's dispatch center, is virtually not a state variable. But since MWD is the largest disturbance to the plant that causes steam temperature deviations and is measurable at the plant, MWD is included in the state variable vector as a pseudo-state variable so as to predict its influence on the steam temperatures.

Table 1.1   System variable used for AR model building

	Symbol	Name of the variables	Reference
State Variable	MMD	Megawatt demand or load command issued from the system's dispatch center to the plant	increment at every control time
	WWT	working fluid temperature at the outlet of the evaporator	deviation from the values smoothed by a first-order lag filter
	SHT	mainsteam temperature	deviation from the set-point value
	RHT	reheat steam temperature	deviation from the set-point value
Manipulated Variable	FR	fuel flow rate	control signal injected
	SP	superheater spray flow rate	from the computer to the
	GD	gasdamper opening	manipulators of the plant

As shown in Figure 1.1, the fuel flow rate, superheater spray flow rate and gas-damper opening in Table 1.1 are the principal manipulated variables for the steam temperature control of the boiler.

Data collection for system identification

In the data collection for system identification, four independent pseudo-random signals are injected from the control signal computer in Figure 1.3 to the manipulators of MWD and three manipulated variables so as to stimulate the plant. Under this condition the data of all the system variables shown in Table 1.1 are collected at every regular time interval.

For this purpose, four pseudo-random binary signals, i.e. M-sequence (Maximum-period sequence) (Bendat and Piersol 1977) signals with different M and Δ (period of signal renewal), are produced from the computer and smoothed by second-order lag filters to attenuate undesirable higher frequency components, then injected to the plant. The amplitudes of these signals are adjusted based on the preliminary test so that the steam temperature excursions during the experiment remain within 3 to 4 ℃ in the standard deviations around their set-points. The sampling period and the data length are somewhat different depending upon the plant dynamics: in the case of thermal power plants, five to eight hours' data collected at the time interval of 20 to 40 seconds are practically sufficient for model building to be described in the following section.

By using the time series data thus obtained, the optimal state-feedback gain matrix is computed according to the Dynamic Programming procedure provided in TIMSAC Library shown in Figure 1.4.

AR model fitting to time series data

For the AR model fitting, we first compute the covariance matrices among the system variables of the multivariate time series. The coefficient matrices of the multivariate AR model and the residual innovation vectors are obtained by solving the

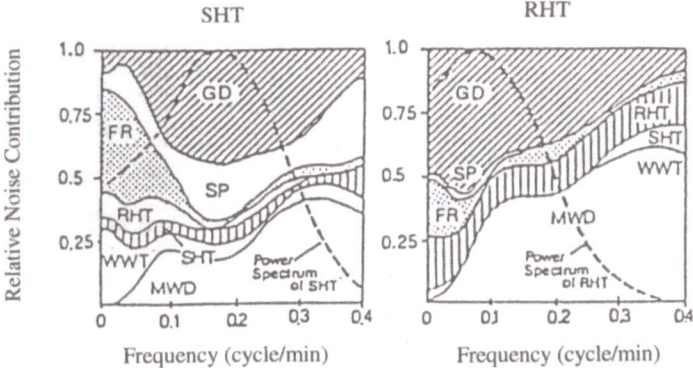

Figure 1.5 Relative noise contribution to the power spectra of SHT and RHT

Yule-Walker equation which is derived from the covariance matrices. Levinson's algorithm utilizing the systematic property of block Toeplitz matrix is useful to save time and expense for computation, in which the Yule-Walker equation is solved sequentially, starting from model order 1 and increasing it one by one, while computing the AIC criterion corresponding to each model order. The model order which gives minimum AIC is chosen as the optimal one.

System analysis using AR model

The power spectral density function and the cross spectral density function for each system variable can be obtained from the coefficient matrix of the AR model and the covariance matrix of the innovation obtained as the residual of model fitting. Using these results and assuming the independence among the innovations, we can estimate the contribution of system variables to a specified variable.

Figure 1.5 shows an example of such estimation, in which the relative power contribution, the contribution of system variables to the power spectra of the mainsteam and the reheat steam temperatures is illustrated. In Figure 1.5, the abscissa shows frequency in cycle per minute and the ordinate shows the relative power contribution. The hatched portion in Figure 1.5 shows the variation of the specified variable itself which seems irrelevant to other system variables.

The relative power contribution analysis is useful to determine the variables to be included in the AR model. As a matter of fact, the seven variables shown in Table 1.1 were selected through the power contribution analysis: in other words, noise power contribution analysis was performed for each of the AR models including several combinations of variables measured along the boiler tube, and by comparing the results of the analysis the variables in Table 1.1 were finally determined.

As can be seen in Figure 1.5, out of the seven variables in Table 1.1, the contribution of MWD to the mainsteam and the reheat steam temperatures is dominated, and it was found by simulation study that the performance of the optimal control system, designed based on the AR model built with the six variables excluding MWD, is not as good as the one designed with the seven variables shown in Table 1.1.

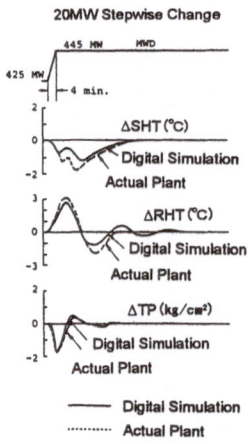

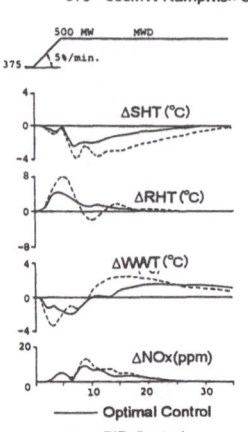

Figure1.6 Checking of the appropriateness of the state equation

Figure1.7 State feedback gain tuning by means of digital simulation

Verification of the appropriateness of state equation

The appropriateness of the state equation can be verified by means of a digital simulation by checking how exactly the state equation expresses system dynamics. The simulation is made in the following way: for instance, we apply a certain change in load command, such as a stepwise increase, to the element of MWD in the state vector and make the state transition, while providing the MWD with the specified value at every control period and keeping the manipulated variables from the optimal controllers zero. Then the records of the steam temperatures during this transition process express their fluctuations against the specified MWD change when the plant is in operation under the PID controller.

If this simulation results show a good agreement with the records obtained at the actual plant against the same MWD change under the same operating condition, the state equation can be considered practically appropriate.

Figure 1.6 shows an example of this kind of simulation in which the simulation results are compared with the records obtained at a 500MW supercritical variable-pressure boiler. In this example, although some disagreement is recognized in the amplitudes of the state variables, the simulation and the actual plant record show relatively a good agreement in the behavior of the responses. It is said that, if the similarity between the simulation and the actual plant record is to this extent, the optimal control system designed on the basis of this state equation would work as effectively as expected.

Computation of state-feedback gain matrix

Using the state equation verified in the above-mentioned way, we compute the state-feedback gain matrix G that minimizes the quadratic criterion function (equation (1.6)) through DP (Dynamic Programming) procedure.

In the case of thermal power plants, when the control period is chosen, for instance, as 30 seconds, the gain matrix G converges to nearly a constant value, if the span of

DP computation, I in (1.6), is set to 40 to 50. So, we use this G for state-feedback control. The control system obtained through this procedure is called an LQ (Linear Quadratic) regulator.

The values of the elements in the gain matrix G are dependent on the magnitudes of the elements in the weighting matrix Q and R, and how to determine the weights in Q and R appropriately is the crucial factor in the optimal regulator design.

In our system, all the elements of Q are put to zero except for the diagonal elements corresponding to the sub vector $x_0(n)$ in the state equation. The element corresponding to MWD in the sub vector $x_0(n)$ is also put to zero so as to allow free movements of MWD in the DP computation. The size of R, the weighting matrix for the manipulated variables, is 3×3, the same as that of the manipulated-variable vector.

In the digital simulation described in the previous section, if we make the state transition at every control time using the manipulation vector $u(n-1)$ obtained by multiplying G by the state-variable vector, then we can estimate the performance of the optimal control for the proposed G. The solid lines in Figure 1.7 is an example of this kind of simulation in which the optimal control is added to the PID control system having the characteristics shown with the broken lines in Figure 1.7.

Tuning of the weights in the matrices Q and R is performed repeatedly through this kind of digital simulation by starting from the unit matrix and gradually adjusting the magnitudes of their diagonal elements until desired performance is obtained on the simulation basis. The performance of the candidates of the gain matrix thus obtained is checked at the actual plant and the one that shows most desirable performance in view of the variance of both the controlled and manipulated variables is finally adopted.

Optimal control algorithm

Figure 1.8 shows the algorithm of the optimal control, in which the portion enclosed by the broken line is the part for computing optimal control signals. As shown in the figure, the block for control signal computation consists of the parts for state-variable vector computation and control signal computation, where the state equation and the state-feedback gain matrix obtained beforehand by off-line computation are stored.

The computation of control signals is performed in the following way: at the time point $(n-1)$ the prediction of the state vector at the next time point n is computed based on the values of the state vector and the manipulated vector at time point $(n-1)$. At the time point n, when the actual values of the state variables $y(n)$ are measured, the top most sub vector $x_0(n)$ of $Z(n)$, which provides the prediction of the state vector at the time point n, is replaced with the vector of the actual values $y(n)$: the control signal $u(n)$ computed as the product of this state vector $Z(n)$ and the gain matrix G is immediately applied to the plant. After that, the optimal control signal computation algorithm changes the time index from n to $n-1$ and computes the prediction of the state variable vector so as to prepare for the next control time. This procedure is repeated at every control time.

Measures to compensate for the process nonlinearity

Originally, the LQ (Linear Quadratic) regulator is a control system developed for the linear system. On the other hand, the steady-state and dynamic characteristics of thermal power plants vary considerably depending upon the magnitude of the plant

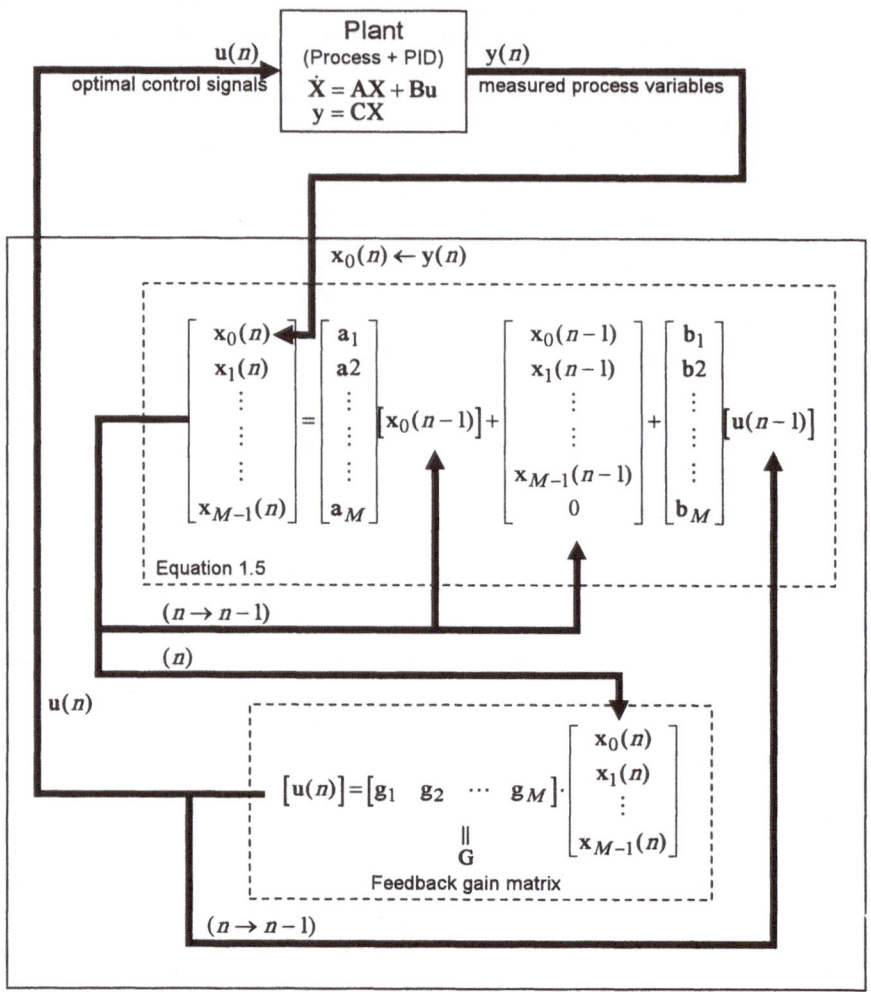

Figure 1.8   Optimal control algorithm

load. To cope with such system nonlinearities, a method which we call "load-adaptive parameter adjustment" is employed.

In this method, the time series data for system identification are collected at three or more load regions, say high, medium, and low load regions. Using these data, we compute the state equation and the optimal state-feedback gain matrix for each load region and store the elements of these matrices in the specified memory areas of the controller.

In the actual control, the control parameters are adjusted by "gain scheduling": in other words, the parameters of the state equation and the state-feedback gain matrix are adjusted at every control time by linear interpolation according to the magnitude

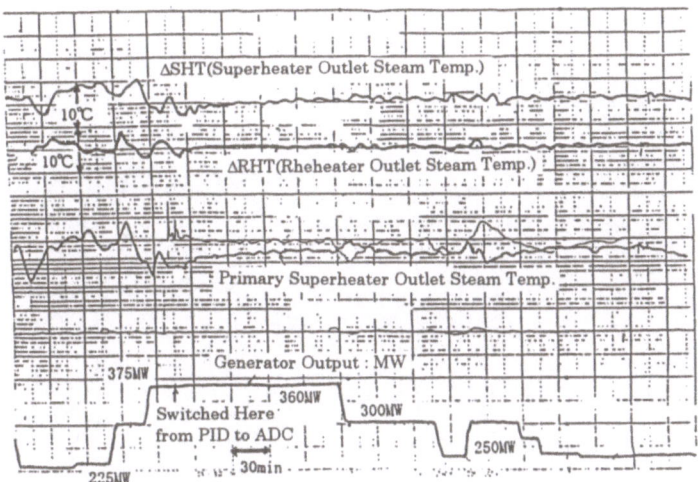

Figure 1.9 Control performance comparison of PID and ADC (Advanced Control)

of the plant load so as to meet the current plant dynamics.

In the digital simulation result illustrated previously in Figure 1.7, the gain scheduling method mentioned above was employed. Apart from the theoretical validity of the gain scheduling method, the experiment carried out at the CRIEPI (Central Research Institute of Electric Power Industries of Japan) by means of the precise power plant simulator proved the effectiveness of the method. In this experiment, the control performance of the system with the gain scheduling was compared with the one whose control parameters were fixed: it was found that, for the same large amount of rampwise increase and decrease of the plant load, the amplitudes of both the controlled variables and the manipulated variables were considerably smaller in the former case than in the latter one, exemplifying the improvement of the control performance brought about by the introduction of the gain scheduling method.

1.5   Application Results at Actual Plants

In this section, some results demonstrating the performance of the proposed optimal control system obtained at actual power plants are introduced.

Figure 1.9 shows a portion of the chart recorded at a 500MW oil-fired supercritical thermal power plant in routine operation. In this figure, the optimal control signals from the computer are added to the conventional PID control system at the time points shown with the arrow marks. As can be seen in the figure, the control performance of the steam temperatures at the outlets of the superheater, the reheater, and also of the primary superheater is remarkably improved by applying the optimal control system.

Figure 1.10 shows the behavior of the manipulated variables recorded at a 600MW

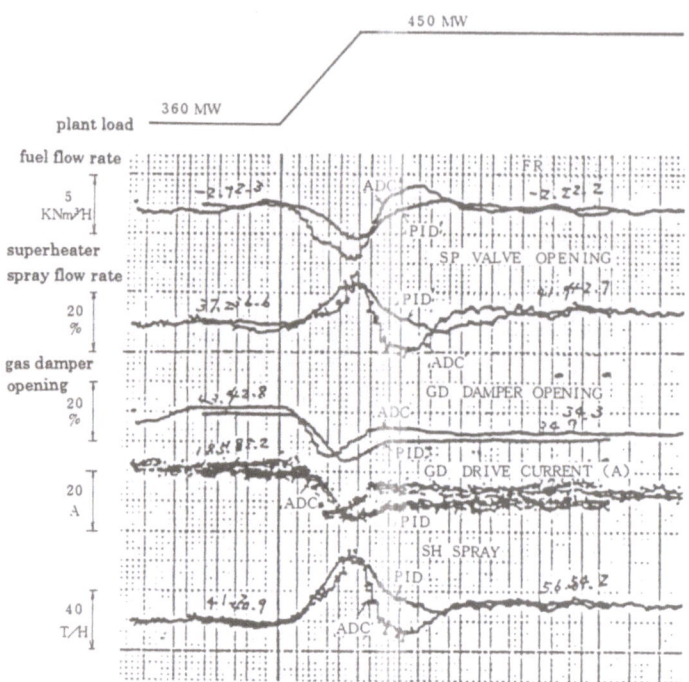

Figure 1.10   Comparison of the response of the identifical rampwise change

supercritical plant against the rampwise load increase from 360MW to 450MW, in which the optimal control system (indicated by ADC or Advance Control in the figure) and the conventional PID control system are compared for the same load increase. As shown in this figure, control signals move faster in the optimal control ADC than in the PID control during the load increase, and after the load increase is completed, the control signals return to their steady state values sooner in the optimal control than in the PID control. As a result, in the optimal control, the deviations of the controlled variables are remarkably reduced and the overshoot of the state variables, which is commonly observed in the PID control, is eliminated. Such favorable control characteristics results from the use of the state prediction which is inherent to the system designed on the basis of the state space concept.

The application of optimal control to actual power plants described above had a large impact on the control strategy of thermal power plants. After the publication of the reports on these successful implementations, the same kinds of control systems have been applied by boiler and control system manufacturers to many thermal power plants both in Japan and abroad. Table 1.2 shows a list of such implementations which the author knows of. The total output power of these plants reaches 13,725MW as of 1995, which amounts to nearly 15% of the total capacity of thermal power plants owned by Japanese electric power utilities.

Table 1.2   List of power plants under optimal control implementation by TIMSAC

No.	Name of plant	Output	Fuel	date of implementation
1	Buzen N0.1	500MW	oil	1978
2	Shinkokura No.3	600MW	gas	1980
3	Shinkokura No.4	600MW	gas	1981
4	Buzen No.2	500MW	oil	1983
5	Hekinan No.1	600MW	oil	1984
6	Hekinan No.2	600MW	coal	1985
7	Sendai K. No.2	500MW	oil, gas	1985
8	Shajiao No.1 (China)	350MW	coal	1987
9	Shajiao No.2 (China)	350MW	coal	1987
10	Akoh No.2	500MW	oil	1988
11	Sendai T.No.3	175MW	coal	1989
12	Shinsendai T.No.3	375MW	oil	1990
13	Nanticoke No.7 (Canada)	500MW	coal	1991
14	Higashi-ohgishima No.2	1000MW	gas	1991
15	Niigata No.4	250MW	oil, gas	1991
16	Sendai T. No.2	175MW	coal	1991
17	Nanko No.3	600MW	gas	1992
18	Hekinan No.3	700MW	oil, gas	1992
19	Hirono No.4	1000MW	oil, gas	1992
20	Sendai T. No.2	175MW	coal	1993
21	Anegasaki No.4	600MW	oil,gas	1994
22	Noshiro No.2	600MW	coal	1994
23	Shiriuchi No.1	375MW	oil	1994
24	Kudamatsu No.3	700MW	oil	1995
25	Tomato-azuma No.2	700MW	coal	1995
26	Reihoku No.1	700MW	coal	1995

1.6   Closing Remarks

The outstanding features of the above-mentioned optimal controller design methodology based on a statistical model are the fact that this method provides field engineers, who are not always familiar with the control theories, with the practical means to implement the modern control theory to actual plants.

In other words, the controller design procedures, from system identification, system characteristics analysis to optimal controller design, are provided as a complete program package called TIMSAC Library in this method. Therefore, even a field engineer, who is not an expert of the control theory and industrial process modeling, can design the optimal control system through the dialogue with the computer, if he follows the procedures described in the program.

It should be emphasized that such consideration is indispensable for a new technology to be inherited by the next generation and take root in industrial fields as a useful means.

The implementation of the optimal regulator described in this chapter was originally carried out from the late 1970s to the 1980s, when most of the thermal power plants in Japan were oil-fired or gas-fired ones.

Recently, the situation of fuel procurement in Japan has remarkably changed and many coal-fueled power plants have been constructed which require more sophisticated control strategies than the oil- or gas-fired plants.

In the coal-fired plants, there are difficult problems such as (i) the variations of heat rate due to the various brands, quality and water content of the coal, as well as the variations of the blending ratio of the coal, (ii) clogging of the coal in the coal feeding line, (iii) the change in process dynamics caused by commissioning or shutdown of burners and coal mills, or changes in the gas/coal mixing ratio, and such. In addition, many plants are operated at a reduced mainsteam pressure aiming at the high efficiency in the low and medium load range.

Any of these factors is the difficult problem which cannot be solved by the ordinary linear control theory, and dealing with these items properly to make the application of the control theory effective is an important problem that faces field control engineers.

The following is an example of such measures carried out at a 500MW coal-fueled power plant in Canada (Braggeman 1992) listed in Table 1.2. In this plant the abrupt raise of mainsteam temperature caused by the ignition of burners is suppressed by the technology called gas tempering which utilizes burner ignition signals to change temporarily the distribution of the fluegas flow along the reheater surface: the disturbance of the boiler process that occurs as an aftermath of this action is quickly calmed by the system stabilizing effect of the optimal control system which is one of the outstanding features of an optimal control system.

For the switching of the coal bland, the measure, which applies the optimal controller after adjusting the PID controller to meet the current plant operating condition, would be effective.

In this chapter the author introduced the practical usefulness of a statistical model in the industrial fields. The successful implementation of the optimal control system at the power plants by means of a multivariate AR model had a large impact on both the industrial and academic fields.

As well known, the modern control theory was advocated by R. E. Kalman in the early 1960s, but its application to the industries, especially to the large-scale plants, has been very few, although the theory had been expected to open a new world of control: as a matter of fact, in the 1970s, the optimal control theory was criticized by some practical engineers as a theoretical hobby of mathematicians.

Such situation can be understood by the fact that the subjects of the panel discussions held by SICE (Society of Instrument and Control Engineers in Japan) were "Is the modern control theory practically useful?" at the early 1970s, and "What is the hindrance preventing the application of modern control theory to industrial fields" at the late 1970.

However, the situation completely changed in 1978, when the first optimal control system at a thermal power plant was reported: for instance, the subject of the panel discussion held by SICE in 1993 was "The way to apply the optimal control to industries".

Today it is a common sense that an optimal controller appropriately designed and implemented is effective to improve system performance. This fact should be noted as an important contribution of the statistical model approach to industrial fields.

References

Akaike, H. and Nakagawa, T. (1972), *Statistical Analysis and Control of Dynamic Systems*, Saiensu-sha, Tokyo (in Japanese, with a computer program package TIMSAC written in FORTRAN IV with English comments.)

Akaike, H. and Nakagawa, T. (1988), *Statistical Analysis and Control of Dynamic Systems*, Kluwer Scientific Publisher, Dordrecht/Boston/London.

Bendat, J. B. and Pierson, A. G. (1977), *Pandom data: Analysis and measurement procedures*, John Wiley.

Bruggeman (1992), "Acord nanticoke test results and implementation", June 1992 EPRI/ISA Power Symposium in Kansas City, June 4–5, 1992.

Nakamura, H. and Akaike, H. (1981), "Statistical identification for optimal control of supercritical thermal power plants," *Automatica*, Vol. 17, No. 1, 143–155.

Otomo, T., Nakagawa, T. and Akaike, H. (1972), "Statistical approach to computer control of cement rotary kilns," *Automatica*, Vol. 8, No. 1, 35–48.

Chapter 2

Feedback Analysis of a Living Body by a Multivariate Autoregressive Model

Takao Wada[1]

2.1 Introduction

In the fields of medical or biological science, time series analysis including the analysis of fluctuation become popular in recent years. In spite of the fact, the analysis of the univariate system is mostly used and the analysis of the multivariate system is very few. Of course, there are some handbooks for multivariate time series analysis for medical data. However even in these handbooks, only the coherency among mutual variables, i.e. correlation, although it is classified according to the frequency, is referred to as a pivotal factor and lacks the consideration on the effect of feedback. It should be recognized that such a situation is quite regrettable considering from medical researchers' viewpoint. Perhaps it is due to the fact that, in addition to the difficulty in taking in the multivariate data for time series analysis, the difficulty in the analysis or absence of methods usable for analysis of actual feedback systems.

The author of this chapter was interested in a multivariate autoregressive (AR) model when he came across a monograph entitled "Analysis and Control of a Dynamic System" (Akaike and Nakagawa 1988) about 10 years ago. No sooner had he noticed that the method suggested the approach to a feedback system of living bodies despite the fact that a system of quite a different kind of cement rotary kiln is dealt with, he was instinctively aware of the usefulness of the model.

The concept of the feedback was probably introduced in the field of engineering, especially in the fields of machinery or electricity. However when the world of nature is considered apart from such an artificial system, the living bodies are the most complicated systems composed of many feedback loops. This is a sophisticated mechanism built up by nature after the lapse of 3,500 million years in controlling the highly-advanced organic bodies called living bodies.

A living body, considered as an object of study, is an independent multivariate feedback system, and more than 5,000 million systems exist even when human is exclusively considered as research object. It might be understood how profoundly the

---

[1]**Editor's note**: Dr. Takao Wada, former director of the Inaki Hospital, passed away on November 7, 1997.

Table 2.1 Examples for applications of multivariate autoregressive modeling for analyzing medical or biological data

1. **Immunological systems**
   - CD4–CD8 regulation in rheumatism
   - Rejection phenomena in renal transplantation
   - Immunoglobulin production in hemodialysis patients
2. **Metabolic and endocrinological systems**
   - TSH-thyroxin regulation
   - Cortisol secretion
   - Serum albumin versus anemia in HD patients
   - Renin-angiotensin system
   - Glucose-insulin feedback
   - Sodium and water regulation
   - Acid-base regulation
   - K-Cl-bicarbonate relationship in plasma
   - Calcium-phosphate relationship in plasma
   - Cholesterol-triglyceride relationship
3. **Nephrotic syndrome**
   - Lipid metabolism
4. **Cardiovascular systems**
   - Hypertension in chronic renal insufficiency
   - Hypertension and anemia in hemodialysis patients
   - Heart rate variability
5. **Neuro-muscular systems**
   - Equilibrium and position control
   - Postural control in sports players

practical methods to analyze such a system are required. Table 2.1 shows a list of various kinds of feedback systems in a living body which the author has considered so far (Wada *et al.* 1988, 1990, 1993, Wada 1993, Wada 1989, 1990). Due to the development of medical electronics in recent years, it becomes very easy to obtain multivariate time series data and importance of the analysis of the multivariate system, especially that of the feedback system has been recognized. The objective of this chapter is to discuss the application of the method by Akaike and Nakagawa (1988) to the analysis of the feedback systems in living bodies.

2.2 Body Liquid Control and Feedback

In order to show the importance of the analysis of the feedback system of a living body, we will present an example of the analysis of a typical feedback system in

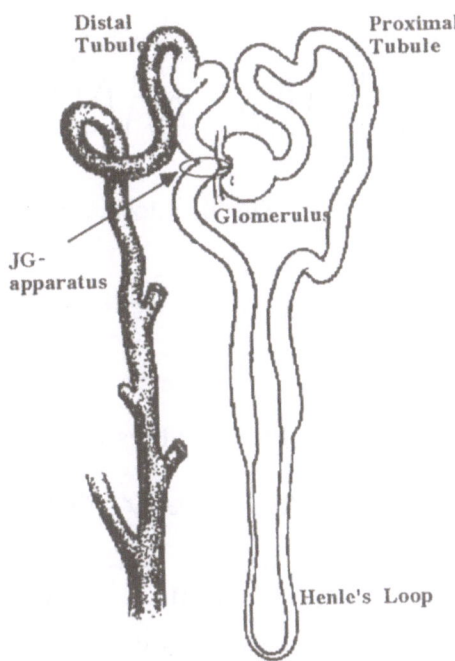

Figure 2.1   Simple nephron's running course and JG apparatus

medicine concerning how to establish control between glomerules and tubule.

A kidney is an organ to make urine. There exist 1,000,000 mini-organs called nephrons in a kidney, and each nephron produces urine. The nephron has a shape illustrated in Figure 2.1. The nephron's starting point has the spherical structure called glomerulus, in which capillaries are rolled in a style of thread balls. The liquid that comes out from the blood passing through the capillaries to extract blood corpuscles, protein molecules, etc. is filtered to produce crude urine. A long pipe following the glomerulus is tubule, where the component of the crude urine is adjusted.

The proximal tubule is once detached from the glomerulus and contacts again with the glomerulus as distal tubule in the rear side of the Henle's loop. The portion where the tubule contacts with becomes a special cell mass called macula densa and forms part of the JG (juxta-glomerular) apparatus. A macula densa theory is proposed among the academic circles in purport that the mass might perhaps perform feedback actions on that portion. That is to say, a possibility is repeatedly pointed out that the amount of the sodium or chloride filtered with the glomerulus is feedback-controlled with the JG apparatus.

Recently, Holstein-Rathlow and Marsh (1989) succeeded in accomplishing continuous measurement of the pressure in the proximal tubule and the concentration of the chloride. The values of the pressure and concentration are shown in Figure 2.2. The power spectra of the series are illustrated in Figure 2.3. The proximal pressure (P) and chloride concentration (Cl) have almost the same spectra. It is quite inter-

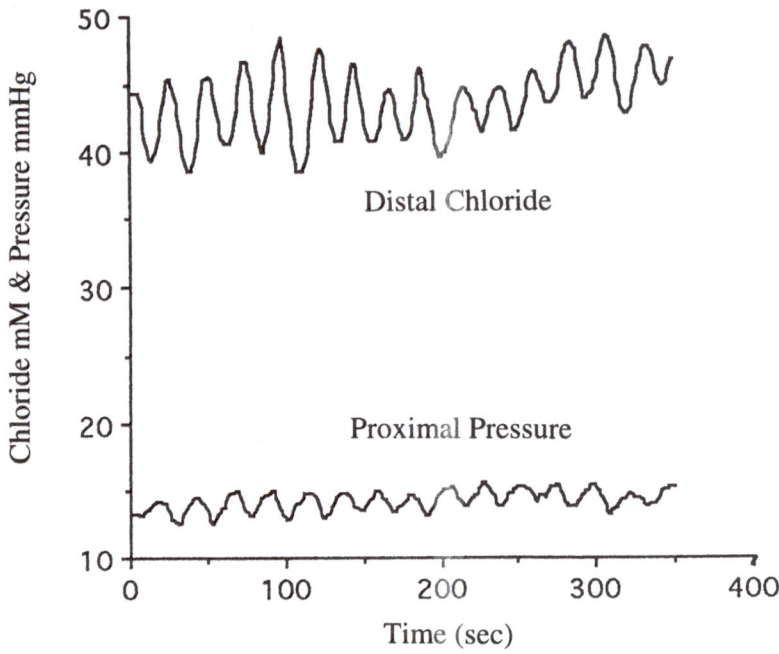

Figure 2.2 Pressure (P) of the proximal tubule and chloride concentration (Cl) of the distal tubule (traced from the figure by Holstein-Rathlow and Marsh (1989))

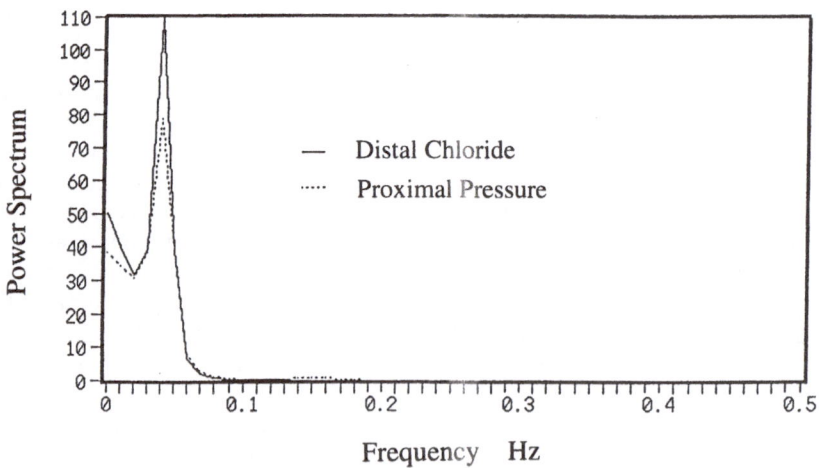

Figure 2.3 Power spectra of P and Cl

esting to note that especially the peaks in the vicinity of 0.04Hz entirely coincide. The authors of this books point out that some deviation is noticed with the original waveform between the 2 variables, and stated that this provides important ground to exhibit the existence of the feedback control.

In this chapter, we will analyze the behavior of the 2 variables by bivariate AR model. With mathematical meanings left to be referred to later, we first show how the feedback will be expressed by the AR model analysis. We employ mainly two methods: the relative power contribution and the impulse response. With the former, the feedback effect is detected viewed from the frequency domain. Meanwhile the latter is a method to describe the linkage between the variables viewed from the time domain.

In relation to the time series data measured with the above mentioned 2 variables, we start with fitting a bivariate AR model to the data. When the optimal AR model is determined, it is possible to obtain both of the relative power contribution and the impulse response. The graph of these quantities help grasp the style of the feedback of the system. The result is useful for researchers who are knowledgeable about the medical or biological meanings of the object system.

2.3 Example of the Relative Power Contribution and Impulse Response

Figure 2.4 shows the relative power contribution obtained from the 2-dimensional time series referred to above. The schematic diagram in the left side of the figure shows to what extent the chloride concentration (Cl) in the distal tubule is driven depending on the variation of the inner pressure (P) in the proximal tubule. The horizontal axis indicates the frequency, whereas the vertical axis shows the contribution of the individual variables to the power at each frequency. It is noticed that the contribution from the pressure of the proximal tubule is almost 45% in the vicinity of 0Hz, but the

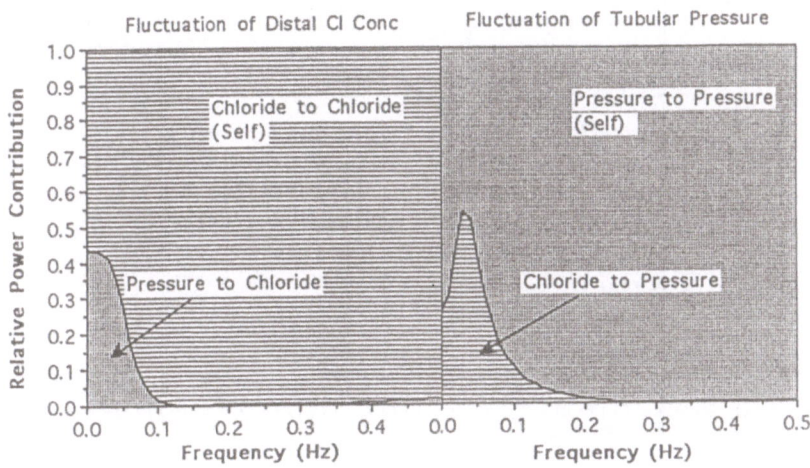

Figure 2.4 Power contribution showing the mutual relation of P and Cl

contribution is lowered in other frequency band. The right side of the figure shows the contribution of the variation caused by the Cl complying with the power of the P. The contribution of the Cl takes the maximum in the vicinity of 0.04Hz.

The frequency of 0.04Hz in question coincides with the peak of the power spectra of the P and Cl given in Figure 2.3. That is to say, the analytical result obtained from the power contribution reveals that despite the fact that both the 2 variables show mainly the oscillation with about 0.04Hz, it is not caused by P but is originated from the variation of the Cl. As seen above, it is comprehended that the information of the feedback that cannot be obtained from the ordinary power spectrum analysis can be obtained by using the power contribution.

However only the information in the frequency domain can be obtained from this. From this fact, it is understood that there exists an asymmetric feedback relation between P and Cl, and thus it is difficult to grasp in a concrete manner from the standpoint of medicine and biology. That is to say, it is uncertain in what a manner the Cl will react when the P is changed or how the P will be changed when the Cl is changed. The analysis of the impulse response is the most useful tool to answer such a problem.

The impulse response (of an open system) is shown in Figure 2.5. The mathematical meanings of the response will be referred to later, but we perform the simulation using the estimated AR model with the route of the feedback being cut. As shown in the upper half of the figure, P increases 0.48mmHg 7–8 min after the pulse-like

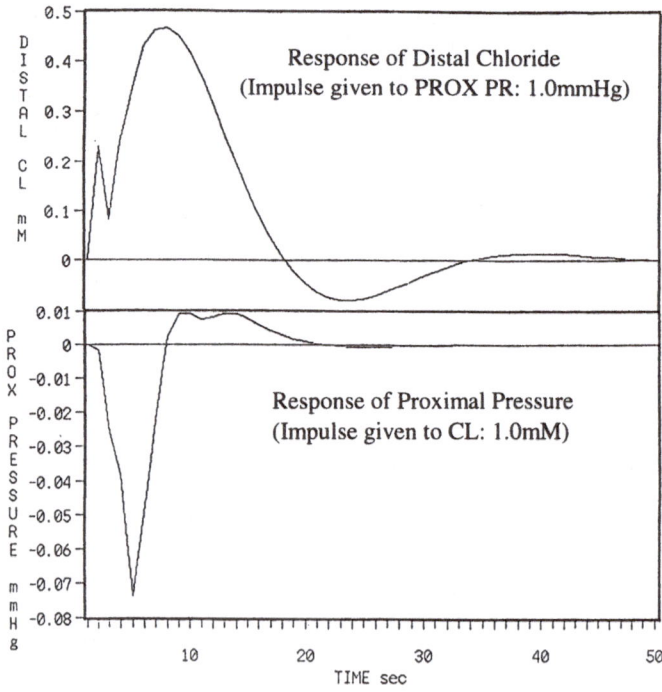

Figure 2.5   Open-loop impulse response between P and Cl

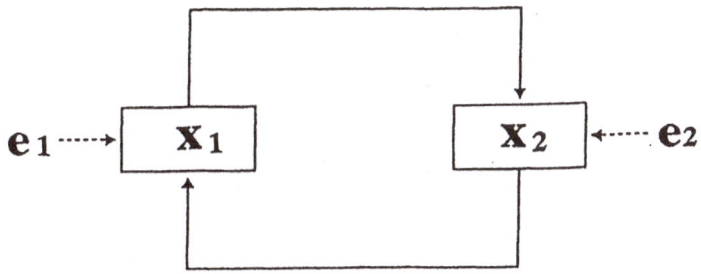

Figure 2.6   Simple feedback system with 2 variables of x_1 and x_2

stimulus of the size of 1.0mmol/ℓ is given to the Cl concentration for 1 sec. The lower half of the schematic diagram in the figure shows the response of the Cl concentration in case that the pulse-like stimulus of 1.0mmHg is given to the P, and the decreases of about 0.07mmol/ℓ is seen 5 sec later.

Glancing at the 2 types of the response from the controlling point of view, it is understood that a feedback loop is formed as follows. As the pressure of the proximal tubule is raised, the chloride concentration of the distal tubule is lowered resulting in lowering of the proximal tubule contrarily to the above. This is an interesting result supporting the Macula densa theory. The result was sent to Prof. Marsh who was interested in it and approved of the author to use his data in this chapter.

2.4   Using an Autoregressive Model for Feedback Analysis

A feature of the feedback system is fully embodied with the example shown above. If there exists a feedback relation between Cl and P, then the output from the Cl influences as input to the P. On the contrary, the output from the P influences on the Cl. Thus endless entanglement is started, and the output curves of the 2 variables become quite similar to each other. The Cl and P shown in Figure 2.2 are, in fact, quite similar in their oscillation, and the power spectra have almost the same shapes as shown in Figure 2.3.

Therefore the correlation coefficient between the Cl and P, for example, become considerably high at a certain time lag. Judging from this, it is reasonable to consider that there exists some causality between the both, but it is utterly uncertain which controls which. The power contribution and impulse response are important clue to judge the causal relation between these 2 variables.

To understand the principle of the feedback analysis method using a linear model, we consider a simple feedback system with two variables, x_1 and x_2, as shown in Figure 2.6. The variation of x_1 is transferred to x_2 in this system, and the variation of x_2 is transferred to x_1. Therefore, the variation of both is mixed to the variation of the x_1 and x_2, and the linear combinations of the past values of both variables form the present value.

If the noise inputs e_1 and e_2 causing the variation of x_1 and x_2 can be separated from the x_1 and x_2, it is quite effective to express how the both are controlled each other. That is to say, our original problem of estimating how x_1 controls x_2 (or how

x_2 controls x_1) is replaced with the one of estimating how e_1 controls x_2 (or how e_2 controls x_1). The autoregressive model

$$x_1(s) = \sum_{j=1}^{2}\sum_{m=1}^{M} a_{1j}(m)x_1(s-m) + e_1(s)$$

$$x_2(s) = \sum_{j=1}^{2}\sum_{m=1}^{M} a_{2j}(m)x_2(s-m) + e_2(s),$$

is used to compute e_1 and e_2 from x_1 and x_2, where $a_{1j}(m)$ and $a_{2j}(m)$ are autoregressive coefficients.

By generalizing the above model to k variate case, we obtain a general multivariate autoregressive model

$$x_i(s) = \sum_{j=1}^{k}\sum_{m=1}^{M} a_{ij}(m)x_j(s-m) + e_i(s), \tag{2.1}$$

where the value of each variable indicates the deviation from the mean value. M is the order of the model and is determined so that the value of the Akaike information criterion (AIC) will be minimized.

2.5   Obtaining the Power Contribution

To estimate the autoregressive model from the observed time series data, the Yule-Walker equation

$$\sum_{j=1}^{k}\sum_{m=1}^{M} a_{ij}(m)r_{jh}(s-m) = r_{ih}(s) \qquad (i,h = 1,2,\ldots,k), \tag{2.2}$$

is used, where $r_{jh}(s)$ is a crosscovariance function of the variables x_j and x_h, and is an autocovariance function for $j = h$. The AR coefficient $a_{ij}(m)$ is obtained by substituting the sample cross covariance $r_{jh}(s)$ obtained from the data into the above equation to solve the simultaneous equations.

In actual computation, Levinson-Durbin's algorithm is employed. To obtain the optimal order, the calculation is repeated by increasing the order, and the optimal order is determined by minimizing the AIC. The power contribution can be obtained from the AR coefficient obtained by this procedure. We shall hereunder briefly show the procedure by Akaike and Nakagawa (1988).

First of all, the covariance $Re_ie_j(n)$ of the innovations $e_i(t+n)$ and $e_i(t+n)$ in (2.1) is given by

$$Re_ie_j(n) = \sum_{\ell=0}^{M}\sum_{m=0}^{M}\sum_{r=1}^{k}\sum_{s=1}^{k} a_{ir}(\ell)a_{js}(m)R_{rs}(n-\ell+m),$$

where $R_{rs}(n-\ell+m)$ is the covariance of $x_r(t+n-\ell)$ and $x_s(t-m)$.

Fourier transform of the above equation yields

$$s_{ij} = \sum_{r=1}^{k}\sum_{s=1}^{k} a_{ir}(f)\overline{a_{js}(f)}p_{rs}(f),$$

where $a_{js}(f) = \sum\limits_{m=0}^{M} a_{js}(m) \exp(-i2\pi fm)$ and $p_{rs}(f)$ is the cross spectral density between x_r and x_s and $s_{ij} = Re_i e_j(0)$. With matrix notations it can be expressed as

$$S = A(f)P(f)A(f)^T,$$

where

$$A(f) = \sum\limits_{m=0}^{M} A(m) \exp(-i2\pi fm) = \sum\limits_{m=1}^{M} A(m) \exp(-i2\pi fm) - I,$$

and i denotes the imaginary number. $A(m)$ is an AR coefficient matrix whose elements are $a_{ij}(m)$. Therefore the $k \times k$ cross spectral matrix $P(f)$ is given by

$$P(f) = A(f)^{-1}SA(f)^{-T}, \tag{2.3}$$

where $A(f)^T$ denotes the conjugate transposed matrix of $A(f)$.

Assuming that the innovations $e_i(s)$ of the different variables are independent of each other, $s_{ij} = 0$ ($i \neq j$) and S becomes a diagonal matrix with diagonal element s_{ii}. Accordingly $p_{ii}(f)$ is expressed as

$$p_{ii}(f) = \sum\limits_{j=1}^{k} q_{ij}(f), \tag{2.4}$$

where

$$q_{ij}(f) = \left|(A(f)^{-1})_{ij}\right|^2 s_{jj}.$$

This means that the power spectrum $p_{ii}(f)$ of the variable x_i is expressed as the sum of the contributions of the innovation of the variable x_j. Akaike defined the relative power contribution $r_{ij}(f)$ as

$$r_{ij}(f) = \frac{q_{ij}(f)}{p_{ii}(f)}. \tag{2.5}$$

In the definition of the power contribution, it is essential to assume that the correlations of the innovations are zero. The calculation referred to above can be done by the program MULNOS in the TIMSAC package (Akaike and Nakagawa 1988), where the correlation of the noise can also be checked.

2.6   State Equation and Impulse Response

The impulse response function is used as a tool for visualizing the dynamic characteristics of the feedback system. A linear system having the variable x as input and the variable y as output has a response waveform of the output y for the impulse input of a unit magnitude 1. This is called the impulse response function. Since the output y continues to respond to each of the impulsive noise inputs, the actual output is given as the weighted average of impulse response function with weight determined by the inputs.

The impulse response transfers the impulsive input to the output as the time lapses. The AR coefficient $a_{ij}(m)$ is the response of the variable x_i, and can be considered as an element exhibiting the impulse response. Since a variable expresses, however, a closed system including the influence of the feedback via itself or other variables, it cannot be true that a single $a_{ij}(m)$ determines a response relation between the i component and the j component. The impulse response of the closed system is evaluated by the whole alignment of the AR coefficients. To express the impulse response in this closed system, a simulation is executed using the AR coefficients as shown below.

Assuming that the order of the model is 2 in a bivariate system, we have

$$
\begin{aligned}
x_1(s) &= a_{11}(1)x_1(s-1) + a_{12}(1)x_2(s-1) \\
&\quad + a_{11}(2)x_1(s-2) + a_{12}(2)x_2(s-2) + e_1(s) \\
x_2(s) &= a_{21}(1)x_1(s-1) + a_{22}(1)x_2(s-1) \\
&\quad + a_{21}(2)x_1(s-2) + a_{22}(2)x_2(s-2) + e_2(s).
\end{aligned}
$$

Arranging the above two equations in a vector-matrix form gives

$$
\begin{bmatrix} x_1(s) \\ x_2(s) \end{bmatrix} = \sum_{m=1}^{2} \begin{bmatrix} a_{11}(m) & a_{12}(m) \\ a_{21}(m) & a_{22}(m) \end{bmatrix} \begin{bmatrix} x_1(s-m) \\ x_2(s-m) \end{bmatrix} + \begin{bmatrix} e_1(s) \\ e_2(s) \end{bmatrix}. \tag{2.6}
$$

Such matrix expression can be generalized to a model for k-dimensional vector $X(s)$ with order M

$$
X(s) = \sum_{m=1}^{M} A(m)X(s-m) + E(s), \tag{2.7}
$$

where both X and E are k-dimensional vectors, and A denotes a $k \times k$ matrix.

Define the $k \times M$-dimensional state vector $Z(s)$ by

$$
Z(s) = \begin{bmatrix} X(s) \\ X(s-1) \\ \vdots \\ X(s-M+1) \end{bmatrix}. \tag{2.8}
$$

Then the model (2.7) can be expressed in the state space form

$$
Z(s) = \Psi Z(s-1) + V(s),
$$

where Ψ is called a transition matrix and is given as

$$
\Psi = \begin{bmatrix} A(1) & A(2) & \cdots & A(M-1) & A(M) \\ I & O & \cdots & O & O \\ O & I & \cdots & O & O \\ \vdots & \vdots & \ddots & \vdots & \vdots \\ O & O & \cdots & I & O \end{bmatrix},
$$

and $V(s)$ is the $k \times M$-vector defined by

$$
V(s) = \begin{bmatrix} E(s) \\ O \\ \vdots \\ O \end{bmatrix}.
$$

Once the model is identified and the AR coefficients are obtained, it becomes possible to obtain the response of the individual variables when a unit impulse is added to a noise term of a variable. This is called impulse response of the closed system. The method of obtaining the impulse response function is exemplified with a bivariate AR model with order 2 in Figure 2.7.

2.7 Impulse Response of the Closed and Open Systems

By expressing the variable x_1 as the sum of the influence of the past values of x_2 and a noise u_1, it is possible to define the impulse response of the variables x_1 to x_2. Plot of the $\alpha_{ij}(m)$ obtained from the autoregressive coefficient, $a_{ij}(s)$

$$\alpha_{ij}(1) = a_{ij}(1)$$

$$\alpha_{ij}(m) = a_{ij}(m) + \sum_{k=1}^{m-1} a_{ii}(k)\alpha_{ij}(k-m) \qquad (m = 2, 3, \ldots, M)$$

$$\alpha_{ij}(m) = \sum_{k=1}^{M} a_{ii}(k)\alpha_{ij}(k-m) \qquad (m = M+1, M+2, \ldots)$$

for $m = 1, 2, 3, \ldots$ shows the impulse response of the open system. It remains to be seen which type of the response will be easily comprehensible with the open system or with the closed system referred to earlier.

Therefore in the computer program TISMAC modified by the author, both the open loop and closed loop impulse response functions are given. Figure 2.8 shows an example of the outputs. This is a result of the analysis to examine how the sodium blood disease in a patient of inappropriate secretion of antidiuretic hormone (ADH) is generated. The uppermost row of the figure indicates the increase of the blood sodium serum value in the open system (dotted line) and the closed system (solid line) in case that daily observations are obtained after impulse input is added to sodium balance.

Glancing at the response of the closed system, it is seen that inputting the impulse of 100mEq/L to the sodium balance helps increase the blood sodium serum approximately 3mEq/L (refer to the upper row) in the first example (the left side). However after the 8th day, the blood serum sodium is deflected in a minus direction owing to the repulsion of the increase of the sodium. This tendency is more remarkable in the second example, and the increase of the sodium balance is steered in a slightly biased direction of the sodium expediting the lowering of the blood serum values. This might stem from the reason that thirsty compels the patient to drink water resulting in changing the water balance into plus. In fact, no deflection in a minus direction can be seen with the response of the closed system. This implies that the response of the closed system reflects not only the direct relations between the variables but also the indirect response via the movement of the other variables.

Meanwhile as seen in the middle row, in the second example the increase in taking in hydration 1000ml by the impulse input help lower the blood serum sodium 3mEq/L. In the lower row, it is revealed that gain in weight by 2kg causes lowering of the blood serum sodium of 1 or 2mEq/L. In this occasion, a problem arises as to what degree of the impulse should properly be inputted. Although no absolute criterion is established with this matter, the magnitude once or twice as large as the standard deviation of the innovation (residual) might be adequate. However by this rule, the magnitude of the input changes data by data, since the standard deviations are different.

[1] Let $Z(0)$ be a zero matrix

$$Z(0) = \left[\begin{array}{c} X(0) \\ X(-1) \end{array} \right] = \left[\begin{array}{c} x_1(0) \\ x_2(0) \\ x_1(-1) \\ x_2(-1) \end{array} \right] = \left[\begin{array}{c} 0 \\ 0 \\ 0 \\ 0 \end{array} \right].$$

[2] Compute $Z(1)$ by using an impulsive noise to $V(1)$,

$$\begin{aligned} Z(1) &= \Psi Z(0) + V(1) \\ &= \left[\begin{array}{cc} A(1) & A(2) \\ I & 0 \end{array} \right] Z(0) + \left[\begin{array}{c} N(1) \\ 0 \end{array} \right] \\ &= \left[\begin{array}{cccc} a_{11}(1) & a_{12}(1) & a_{11}(2) & a_{12}(2) \\ a_{21}(1) & a_{22}(1) & a_{21}(2) & a_{22}(2) \\ 1 & 0 & 0 & 0 \\ 0 & 1 & 0 & 0 \end{array} \right] \left[\begin{array}{c} 0 \\ 0 \\ 0 \\ 0 \end{array} \right] + \left[\begin{array}{c} 1 \\ 0 \\ 0 \\ 0 \end{array} \right] \\ &= \left[\begin{array}{c} 0 \\ 0 \\ 0 \\ 0 \end{array} \right] + \left[\begin{array}{c} 1 \\ 0 \\ 0 \\ 0 \end{array} \right] = \left[\begin{array}{c} 1 \\ 0 \\ 0 \\ 0 \end{array} \right]. \end{aligned}$$

[3] Compute $Z(2)$ from $Z(1)$ assuming the noise $V(2)$ is zero,

$$\begin{aligned} Z(2) &= \left[\begin{array}{cccc} a_{11}(1) & a_{12}(1) & a_{11}(2) & a_{12}(2) \\ a_{21}(1) & a_{22}(1) & a_{21}(2) & a_{22}(2) \\ 1 & 0 & 0 & 0 \\ 0 & 1 & 0 & 0 \end{array} \right] \left[\begin{array}{c} 1 \\ 0 \\ 0 \\ 0 \end{array} \right] \\ &= \left[\begin{array}{c} a_{11}(1) \\ a_{21}(1) \\ 1 \\ 0 \end{array} \right]. \end{aligned}$$

[4] Repeat the computation for $Z(3)$, $Z(4)$, ... similarly to the step [3].

Figure 2.7 The procedure to obtain the impulse response in the closed system (the order 2 model with 2 variables)

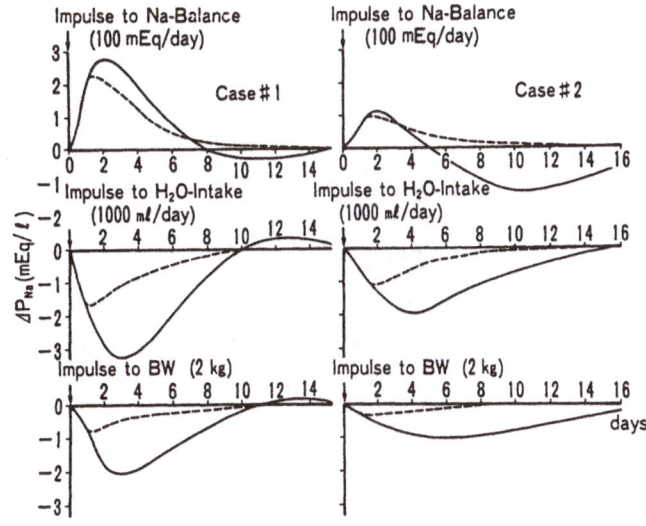

Figure 2.8 Impulse response of a patient inappropriate secretion of ADH (the solid line indicates, the closed system, whereas the dotted line denotes the open system)

In such a case, we also use the impulse input having a fixed value. Despite the above, the impulse with adequate magnitude is not necessarily chosen depending on how the units of the individual variables should be taken. From such a reason, in the above case input is set to approximately twice as large as the standard deviation of the innovation. However since the response is estimated by a linear model, i.e., an AR model, attention should be paid to the fact that the output, provided that the values in the output is multiplied 10 times, will be multiplied 10 times even if it is far from reality. In that sense, it is necessary to investigate by selecting the physiologically meaningful values in a region close enough to the activity in reality.

2.8 Confirmation by a Virtual Feedback System

The author of this chapter has been engaged in various types of medical study by using a method referred to above. The proposed method can cover a considerably wide range as listed in Table 2.1, and the author is under the impression that the usefulness of the method has almost been established. However some researchers might be skeptical as to whether the method really extract an essential information from the feedback system. To dispel such skepticism, assume the structure of a system and analyze the data generated from the system.

Now, we consider three variables with feedback as shown in Figure 2.9. That is to say, A controls B, whereas B controls A. Furthermore C is unidirectionally controlled by A. In short, A is expressed as the linear combinations of the values of B in the past and noise, whereas B is expressed as the linear combinations of the values of A

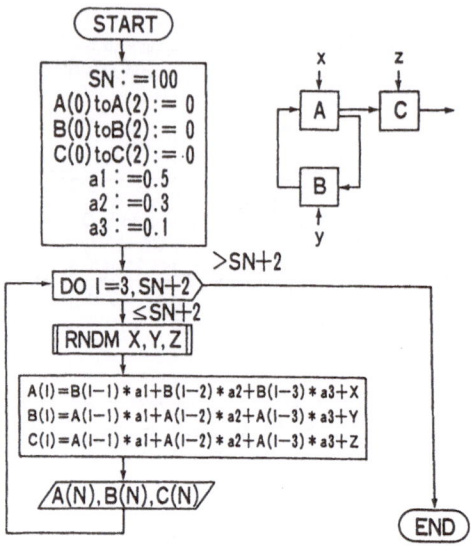

Figure 2.9 Virtual feedback system composed of 3 variables of A, B and C and its program flowchart

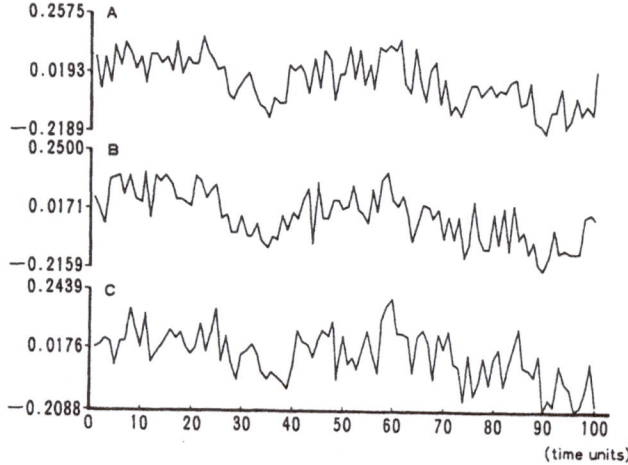

Figure 2.10 Output of the 3 variables from a virtual feedback system

in the past and noise (innovation). C is also expressed as the linear combinations of the values of A in the past and noise. However the past of B is included in the values of A, from which it is understood that C is indirectly influenced by the past of B. Here, for the sake of brevity it is assumed that the noise is white.

Figure 2.10 shows realizations of the 3 variables output from such a system. Tracing the movement of the 3 variables at that time, it is seen that the variables have low-

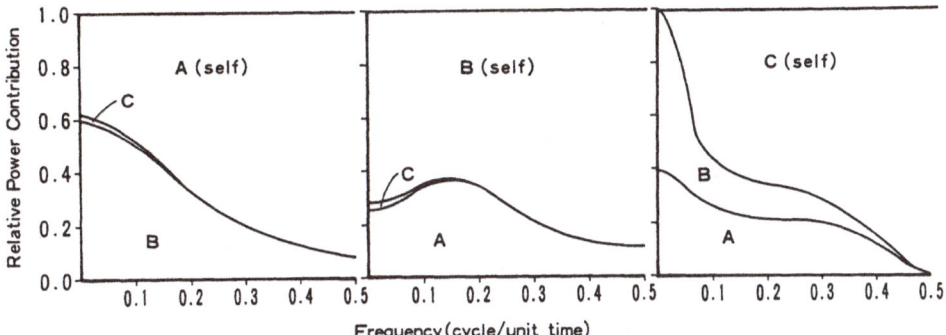

Figure 2.11 Power contribution showing the mutual relation of the virtual feedback system

frequency components considerably similar to each other. Thus it is incomprehensible at a glance which variable controls which. First, the power contributions are shown in Figure 2.11. As can be indicated from these, the innovation of B contributes to a considerable part of the power spectrum of A. Thus it is shown that almost all of the contribution of the innovation of A itself accounts for the remaining part of the above. It seems likely that the innovation of C makes slight contribution to the above, but the magnitude of the contribution remains very small and is within the range of calculation error. On the other hand, the innovation of A contributes to the power spectrum of B, and the innovation of B itself accounts for the remaining part of the above. At this stage as well, the contribution from C is minor and also remains in a range of the calculation error. In other words, it can be said that although A and B are mutually controlled between them, they are never controlled by C.

In connection with C, it becomes clear from the low-frequency part that almost all are controlled by A and B. Thus it is noted that the innovation of C itself comes to be added in the part of higher frequency. In short, C is controlled directly by A, but is indirectly controlled by B as well. On the other hand, C never contributes to the power of A and B. Therefore it might be stated that C is uni-directionally controlled by A and B. These analysis correctly reflect the matter how the control is maintained among the three variables in A, B, and C given in Figure 2.10.

Figure 2.12 shows the estimated impulse response. Here, the solid line indicates the impulse response of the closed system, whereas the dotted line denotes the response of the open system. First of all, let attention be paid to the solid line. When the impulse input is applied to A, the influence of the impulse is transferred to B with a delay. Furthermore the influence is transferred to C. Also when the impulse input is applied to B, the influence of the impulse is transferred to A and furthermore to C. However even if the impulse input is applied to C, it is noticed at a glance that the influence of the impulse is never transferred to A or B.

At this stage if we pay attention to the impulse response of the open system, the convergence of the wave is expedited owing to the fact that direct response is entirely omitted. Furthermore glancing at the response of C when the impulse input is applied to B, it is seen that the response has almost become zero different from the case of the solid line. That is to say, the response is lost because of the fact that the response

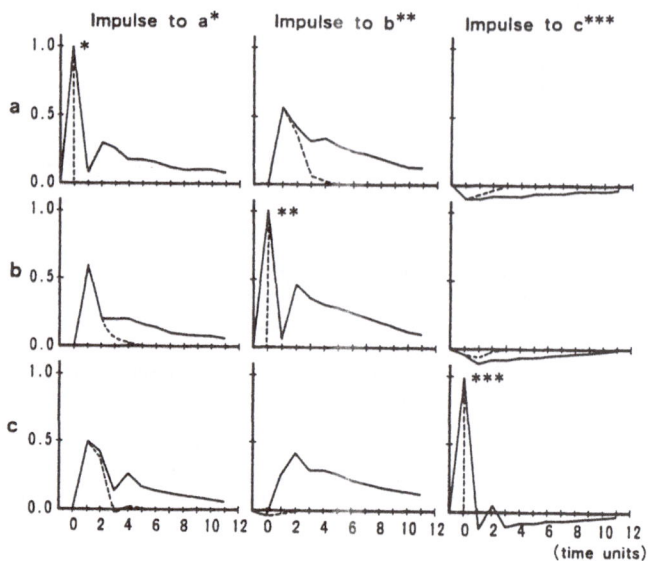

Figure 2.12 Impulse response showing the mutual relation in the virtual feed-back system (the solid line indicates the closed system, the dotted line denotes the open system)

of C has directly no route to C except just the indirect response via A when impulse input is applied to B. As seen above, it is confirmed that the multivariate AR model can give a tool to analyze a linear feedback system.

2.9   Conclusions

In this chapter, we discussed the feedback analysis using a multivariate AR model, but several of problematic points are left untouched with the method. They are non-stationarity and non-linearity of the data. With the problems in question, researches are underway in a cooperative research project at the Institute of Statistical Mathematics. However exceedingly difficult matters are pointed out as untouched subjects with the multivariate system.

The author appreciates the copious learning of Dr. Akaike who suggested to promote the study on the application of this method anticipating the consequences, although he also admits himself the existence of these problematic points. Needless to say, nothing can be better than the fact that accuracy of the researches is high enough. However a proverb has it that the remedy is worse than the evil. It is quite doubtful if the study has arrived at such a high stage as this, should we have been scrupulous in minute subjects. The author wishes to solve the remaining problems with the help of statisticians.

References

Akaike, H. and Nakagawa, T. (1988), *Statistical Analysis and Control of Dynamic Systems*, Kluwer Academic Publishers, Dordrecht. (Original Japanese version was published in 1972 from Saiensu-sha.)

Holstein-Rathlow, N.H. and Marsh, D.J. (1989), "Oscillations of tubular pressure, flow, and distal chloride concentration in rats," *Amer. J. Physiol.*, Vol. 256, F1007–F1014.

Wada, T., (1989), "Analysis of clinical data by multivariate autoregressive model," *Medical Informatics*, Vol. 9, 263–272 (in Japanese).

Wada, T., (1990), "Analysis of clinical data by multivariate autoregressive model (Akaike model)," *Morbid Pathology*, Vol. 9, 984–990 (Japanese).

Wada, T. (1994), "Multivariate autoregressive modeling for analysis of biomedical systems with feedback," *Proceedings of the First US/Japan Conference on Frontiers of Statistical Modeling*, Kluwer Academic Publishers, 293–317.

Wada, T., Akaike, H., Yamada, H. et al. (1988), "Application of multivariate autoregressive modeling for analysis of immunologic networks in man," *Comput. Math. Appl.*, Vol. 15, 713–722.

Wada, T., Jinnouchi, M. and Matsumura, Y. (1988), "Application of autoregressive modeling for the analysis of clinical and other biological data," *Ann. Inst. Statist. Math.*, Vol. 40, 211–227.

Wada, T., Kojima, F., Aoyagi, T. and Umezawa, H. (1988), "Feedback analysis of renin-angiotensin system under the effect of angiotensin converting enzyme inhibitors," *Biotech. Appl. Biochem.*, Vol. 10, 435–446.

Wada, T., Yamada, H., Inoue, H., Iso, T., Udagawa, E. and Kuroda, S. (1990), "Clinical usefulness of multivariate autoregressive modeling as a tool for analyzing T-lymphocyte subset fluctuations," *Math. Comput. Model*, Vol. 14, 610–613.

Wada, T., Sato, S. and Matuo, N. (1993), "Application of multivariate autoregressive modeling for analyzing chloride-potassium-bicarbonate relationship in the body," *Med. Biol. Eng. Comput.*, Vol. 31, 99–107.

Chapter 3

Factor Decomposition of Economic Time Series Fluctuations – Economic and statistical models in harmony –

Kosei Fukuda
Economic Planning Agency
3-1-1 Kasumigaseki, Chiyoda-ku, Tokyo,100–8970 Japan

3.1 Introduction

In analyzing factors of economic time series fluctuations, it is important to compare, first of all, the characteristics between a macro-econometric model and a time series model in order to explain to what extent an economic theory should be utilized or how deeply a statistical theory should be taken in. With the macro-econometric model, we make much of the fact that movement of a variety of factors mainly concerning business fluctuations should be described deterministically with the aid of the economic theory. At the same time, the macro-econometric model is often used for the procedures including simulation for the purpose of measuring the effect of the fiscal and monetary policies, and no attention is paid to the time series structure of economic variables. On the other hand, with a general time series model, we need none of economic theories to analyze the time series structure of the data. Therefore the model is described stochastically and we make much of the prediction performance.[1] Needless to say, there might actually be neither an econometric model ignoring the time series characteristics nor economic time series analysis neglecting the economic theories. The purpose of this chapter resides in the improvement of the time series model by taking in the economic knowledge expressed by an econometric model in analyzing economic data.

Under recognition as described above, we try to make a model which expresses adequately the fluctuations of the real GDP as a representative macro-economic variable. The fluctuations of the real GDP can be grasped by classifying them into a long-term viewpoint and a short-term one. With the former, it is pointed out that the potential growth rate of the Japanese economy made drastic change in the first half of 1970's. On the other hand, the short-term fluctuations of the real GDP are divided into autonomous business fluctuations and economic policy oriented ones such as fiscal and monetary policy. We need a model expressing adequately such points as descried above. Accordingly the model in question should satisfy the condition shown below:

1) to express adequately the trend as the expression of the potential of the Japanese economy.
2) to express adequately the business fluctuations.
3) to measure the effect of the macro-economic policies such as discount rate operations.
4) to possess prediction ability.

In this chapter, in the beginning, we express the fluctuations of the real GDP by using a time series model, and when we can not interpret the obtained results, we use economic theory in order to improve the model. To grasp the fluctuations of the time series by dividing them into a long-term viewpoint and a short-term one, a state-space model proposed by Kitagawa and Gersch (1984) is to be utilized.

3.2   Model 1 (Model with Stochastic Components Only)

3.2.1   Outline of Model 1

Let Y_t be the real GDP (seasonally adjusted quarterly data, with the index taking up 1985 as 100, ranging from the first quarter of 1965 to the fourth quarter of 1991). Also let y_t be natural logarithm of Y_t.[2]

Considering that the real GDP is in stationary variation expressed as an autoregressive process around the stochastic trend, Model 1 is expressed as

$$y_t = T_t + p_t + u_t, \tag{3.1}$$

where T_t, p_t, and u_t represent the trend, autoregressive process, and observation noise, respectively.

Here, the trend component T_t is formulated by the model

$$\Delta^m T_t = v_{T_t}, \qquad v_{T_t} \sim N(0, \tau_T^2). \tag{3.2}$$

On the other hand, the AR component p_t is formulated by

$$p_t = a_1 p_{t-1} + \cdots + a_\ell p_{t-\ell} + v_{p_t}, \qquad v_{p_t} \sim N(0, \tau_p^2). \tag{3.3}$$

At that time, (3.1), (3.2) and (3.3) can be represented, as expressed in Kitagawa and Gersch (1984), with a state-space model

$$
\begin{aligned}
x_t &= F x_{t-1} + G v_t, \\
y_t &= H x_t + u_t,
\end{aligned}
\tag{3.4}
$$

where the state vector x_t is defined as

$$x_t = (T_t, T_{t-1}, \ldots, T_{t-m+1}, p_t, p_{t-1}, \ldots, p_{t-\ell+1})^T.$$

Furthermore, F, G and H are given, for example in case that $m = 2$, $\ell = 3$, as shown below:

$$
F = \left[\begin{array}{cc|ccc}
2 & -1 & & & \\
1 & 0 & & & \\
\hline
 & & a_1 & a_2 & a_3 \\
 & & 1 & 0 & 0 \\
 & & 0 & 1 & 0
\end{array}\right], \qquad
G = \left[\begin{array}{cc}
1 & 0 \\
0 & 0 \\
0 & 1 \\
0 & 0 \\
0 & 0
\end{array}\right],
$$

$$H = [1 \; 0 \; 1 \; 0 \; 0].$$

Meanwhile the noise components v_t, u_t are defined as

$$\begin{bmatrix} v_t \\ u_t \end{bmatrix} \sim N\left(\begin{bmatrix} 0 \\ 0 \end{bmatrix}, \begin{bmatrix} Q & 0 \\ 0 & \sigma^2 \end{bmatrix} \right), \quad v_t = \begin{bmatrix} v_{Tt} \\ v_{pt} \end{bmatrix}, \quad Q = \begin{bmatrix} \tau_T^2 & 0 \\ 0 & \tau_p^2 \end{bmatrix}.$$

Although the seasonal component is included at the same time in Kitagawa and Gersch (1984), let it hereby be assumed for brevity of the discussion that the seasonally adjusted data are used although there are some problems. On the other hand, the data period is long enough to be 27 years, so that there may be economic structural change during the period. Thus we need to take the possibility into the model. In this chapter, such structural change is expressed as a kink of the trend. To be concrete with this, let two sets $(\tau_{T1}^2, \tau_{T2}^2)$ of the variance τ_T^2 of the system noise component be established. τ_{T2}^2 corresponds to the drastic change of the economic structure as seen in the beginning of the floating rate system. Let τ_{T1}^2 $(< \tau_{T2}^2)$ also be τ_T^2 at the ordinary period except that time.

By calculating the log likelihood with the application of the Kalman filter algorithm to the state space model referred to above, the hyper-parameter $\theta = (\sigma^2, \tau_{T1}^2, \tau_{T2}^2, \tau_p^2, a_1, \ldots, a_\ell)^T$ can be estimated with the aid of the numerical optimization. Furthermore when smoothing is made using the estimated hyper-parameter, the estimate of the state vector x_t is obtained and the trend component T_t and AR component p_t are estimated.

3.2.2 Analysis Procedure

In the process of this modeling, the following are cited as the manipulable variables:

1) The initial value of hyper-parameters except $a_1, \ldots, a_\ell$.
2) Orders of the trend and AR, and initial value of $a_1, \ldots, a_\ell$.
3) The location of the kink of the trend.

All the factors in 1) through 3) may influence each other, and therefore we had better consider all possible combinations of these factors. However considering the computational cost, it is decided in actual analysis that the optimal value of the individual hyper-parameters are to be obtained sequentially staring with 1) using AIC.

With respect to 1), we try to use some values such as 10^{-4} referring to the variation of the real GDP from previous quarter. Also with 2), we restrict the maxiimum orders of both the trend and AR components to be 3, so 9 models was tried. At that time, restriction of the stationarity condition was to be established even in the process of the optimization using the initial value of a_i satisfying the stationary condition. At the final stage 3), detection was made keeping Takeuchi (1991) in mind on the assumption that the structural change occurred once in the period ranging from the first quarter of 1970 to the fourth quarter of 1974.

3.2.3 Estimation Results

Decomposition was made as shown in Figures 3.1 and 3.2. Also the results of the estimation are summarized in Table 3.1.

From the comparison with the model having no kink, it is perceived that AIC having the kink is smaller. Also with the trend component, a kink is made with the first quarter of 1973 as a boundary, and the variance of the system noise at the ordinary period is exceedingly small. The yearly growth rate of the trend component is rapidly reduced in speed from 8.2% at the high economic growth period to 4.0%

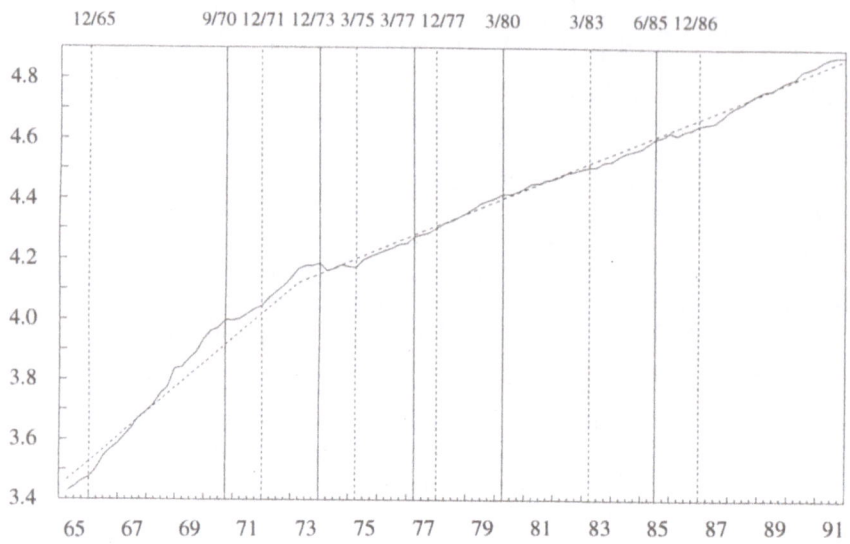

Figure 3.1 Real GDP and the trend component (the vertical solid lines indicate the peak of the business cycles dated by Economic Planning Agency,
whereas the vertical dotted lines denote the trough. For example, recession is
noted in the period ranging from the second quarter of 1985 to the fourth quarter
of 1986).

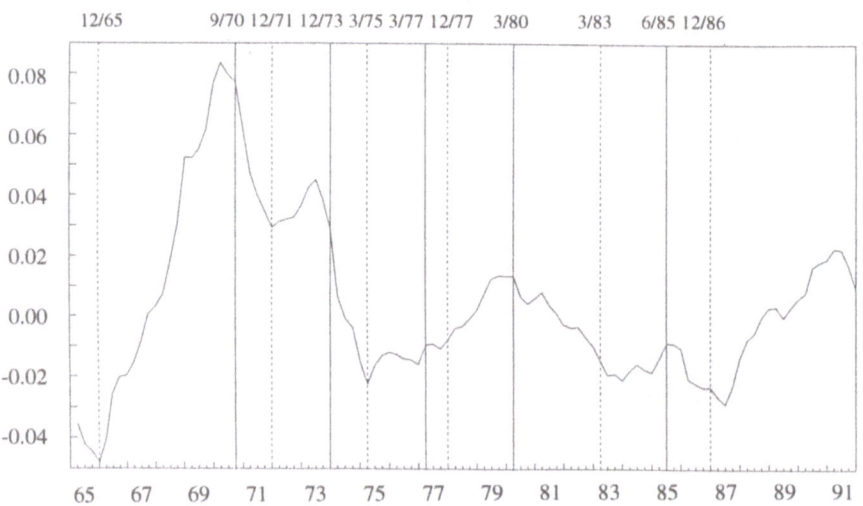

Figure 3.2 AR component

Table 3.1 Estimation results of each model

hyper-parameter		Model 1	Model 2	Model 3
AIC		-672.0	-808.8	-802.7
order of the trend	m	2		
order of AR	ℓ	2	2	2
order of TFR	$\dot{m}$		2	2
order of labor share	n		2	2
order of TVCEX(1)	k_1			2
order of TVCEX(2)	k_2			2
coefficient of AR(1)		1.397	1.575	1.524
coefficient of AR(2)		-0.403	-0.696	-0.720
variance of obs.noise	σ^2	0.176×10^{-4}	0.478×10^{-16}	0.329×10^{-16}
VSN of trend 1	τ^2_{t1}	0.139×10^{-15}		
VSN of trend 2	τ^2_{t2}	0.122×10^{-3}		
VSN of TFR	τ^2_t		0.171×10^{-5}	0.233×10^{-5}
VSN of labor share	τ^2_w		0.490×10^{-5}	0.154×10^{-5}
VSN of TVCEX(1)	τ^2_{b1}			0.248×10^{-6}
VSN of TVCEX(2)	τ^2_{b2}			0.158×10^{-8}
VSN of AR(X)	τ^2_a	0.507×10^{-4}	0.124×10^{-4}	0.119×10^{-4}

Remark:

1) TVCEX(1): the time varying coefficient of the exogeneous variable (1)
   VSN: variance of system noise

2) Owing to the difference among the factors including the formulation of the model, no simple comparison is prohibited among the models of AIC.

of the stable economic growth period. On the other hand, the AR component almost corresponds to the peak and trough of the business cycle in the past.

When we pay attention to a relation between the variation of the trend and AR component of the real GDP and the economic theory, the problematic points of this model are:

1) It can be considered that the growth rate of the trend component obtained is equivalent to the growth rate corresponding to the potential of the Japanese economy, and the factor equivalent to business fluctuations corresponds to the AR component. However it is difficult to make satisfactory explanation with no aid of the economic theory.

2) The growth rate of the trend component reveals wide reduction in speed with the first quarter of 1973 as a boundary, but the results that the potential of the Japanese economy drastically changes at a time is far from the actual economy. This is because Japanese economy is composed of a varieties of activities such as consumption or investment and it takes much time for the Japanese economy to finish structual change.

3.3 Model 2 (the Model Including Deterministic Components)

3.3.1 Outline of the Model

To solve the problematic points referred to in the previous section, we introduce deterministic trend component in this section by using an economic theory. Production function is adopted as an economic model to make an explanation with the real GDP. By so doing, it becomes possible to divide the growth rate of the trend component into the growth rates of labor, capital, and total factor productivity. Thus our model can be more closely related to the actual economy.

For that purpose, let the assumption shown below be made with Model 2.

The real GDP makes stochastic and stationary variation expressed in an autoregressive process around the trend determined by the production function composed of the total factor productivity (stochastic variation) and the input of the labor and capital stock. At that time, the labor share makes stochastic variation.

From a viewpoint of the economic theory, the trend component of the real GDP can be regarded as the potential of the Japanese economy, that is to say, the sum of the input of the labor and capital stock in average and total factor productivity (technological progress). Here, Cobb-Douglas type production function is introduced. The equation of the production function is

$$\log Y_t \approx \log TFP_t + W_t \log L_t + (1 - W_t) \log K_t, \tag{3.5}$$

where TFP_t is the total factor productivity, W_t is the labor share, and L_t is the labor input (1985 = 100). Meanwhile K_t is the capital stock (1985 = 100).[3] In connection with the problematic points concerning the total factor productivity and production function, refer to Kuroda (1984).

If the right side of the above equation is considered to be the trend component of the logarithmic value of the real GDP, then owing to the fact that business fluctuations are noted actually around the trend component, the model is given by

$$\log Y_t = \log TFP_t + W_t \log L_t + (1 - W_t) \log K_t + AR_t + N_t, \tag{3.6}$$

where AR_t is the AR process corresponding to the business fluctuations, and N_t is the observation noise component expressed by the white noise. In actual estimation, the above equation is modified as

$$\log \frac{Y_t}{K_t} = \log TFP_t + W_t \log \frac{L_t}{K_t} + AR_t + N_t. \tag{3.7}$$

Here we should take attention to the following point. While (the number of the workers) × (the total labor hours) should be used for the labor input with the estimation of the ordinary production function, (the private enterprise capital stock) × (equipment operation rate) should be used for the capital stock. This is originally for the purpose of estimating the total factor activity in which the business fluctuations are included in addition to the trend component, but the purpose of this section is to estimate the trend component. We therefore eliminate the parts of the business fluctuations in the individual elements. In the labor input, the following equation is formed.

(Labor force population) × (1 − Unemployment rate) = Number of workers

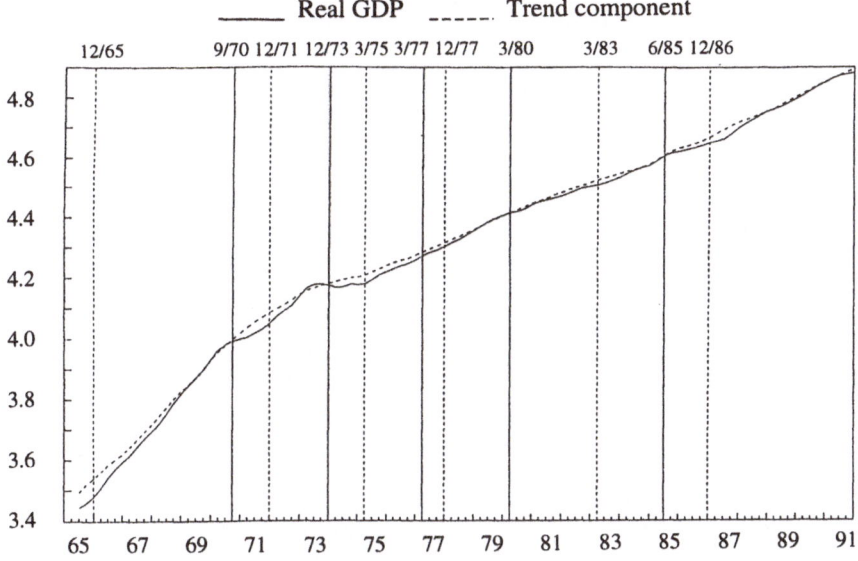

Figure 3.3   Real GDP and the trend component

Furthermore the business fluctuation part fundamentally corresponds to the part of the unemployment rate, so the labor force population is used.[4] Likewise with respect to the capital stock, the equipment operation rate is not taken in.

TFP_t and W_t are stochastically in variation based on the equation the same as (3.2) of Model 1. AR_t and N_t are dealt with as p_t, u_t of Model 1, respectively. The state space representation is given by changing the state vector $x_t = (T_t, \ldots, T_{t-m+1}, p_t, \ldots, p_{t-\ell+1})^T$ into $X_t = (T_t, \ldots, T_{t-m+1}, W_t, \ldots, W_{t-n+1}, p_t, \ldots, p_{t-\ell+1})^T$ to alter H in (3.4) of Model 1 as

$$H_t = (\underbrace{1 \ 0 \ \cdots \ 0}_{m} \quad \underbrace{\log L_t/K_t \ 0 \ \cdots \ 0}_{n} \quad \underbrace{1 \ 0 \ \cdots \ 0}_{\ell}).$$

With the state space model referred to above, the hyper-parameter $\theta = (\sigma^2, \tau_T^2, \tau_W^2, \tau_p^2, a_1, \ldots, a_\ell)^T$ is obtained by the minimization of AIC.

3.3.2   Analysis Procedure

Analysis procedure is almost the same as the case of Model 1, except for setting up the initial value of the state variable W_t. For this, the labor share (employer income/national income) in 1965 was adopted based on the theory of the ordinary production function.

3.3.3   Estimation Results

Decomposition was made as shown in Figures 3.3–3.4. The AR component expresses the business fluctuations, and the value in the first quarter in 1970 is, for example, approximately 0.03, but what the value means is that the real GDP at that time exceeds the level predicted by the production function by 3%. This implies econominically that the relation between the demand and supply is very tight in the goods or service market. However as with the case of Model 1, estimation results were quite deferent from actual economy. For example, the level of the AR component in

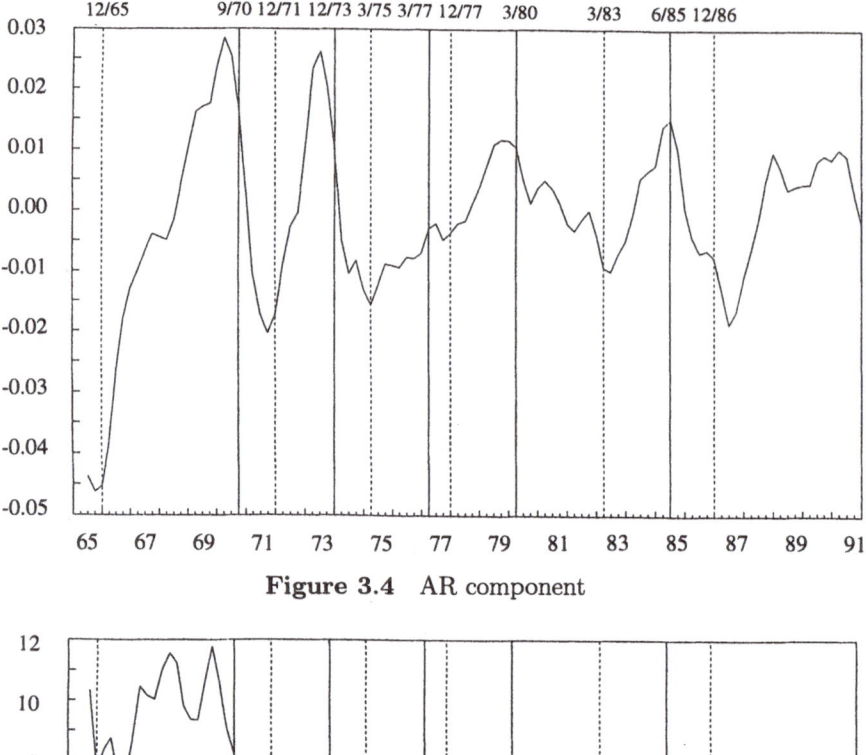

Figure 3.4   AR component

Figure 3.5   Growth rate of the trend component

1965 is slightly low, and that in the period ranging from 1988 to 1990 (so called "bubble economy") is almost the same as the level in 1985, etc. With the trend component, the yearly growth rate shows, different from the case of Model 1, the movement in conjunction with the business fluctuations (Figure 3.5). Therefore a relation between the variation of the trend component that was nothing but a black box with Model 1 and the actual economy is made clear. For example, it is understood that the high trend growth rate is brought about as a result of the extremely high labor input and capital stock. Furthermore, the yearly growth rate of the total factor productivity of model 2 is almost filled to those calculated by ordinary production function except for some minor points (Figure 3.6).[5]

Figure 3.7 shows details of the contribution decomposed into each factor of the

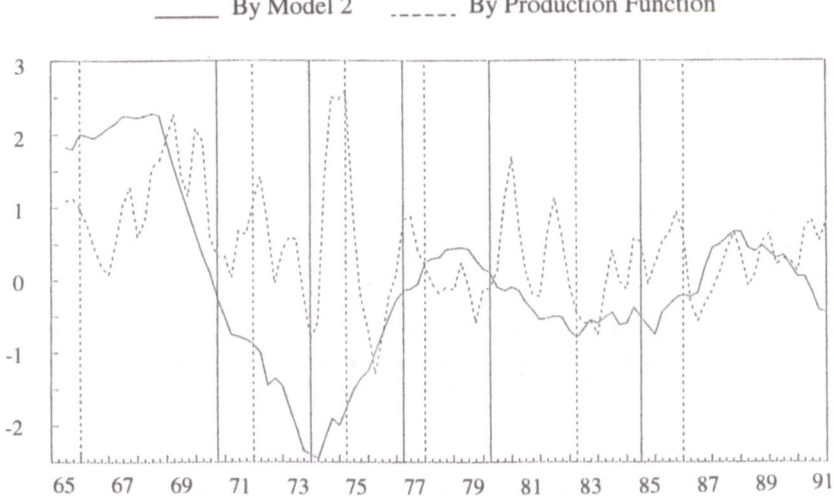

Figure 3.6 Yearly growth rate of the total factor productivity

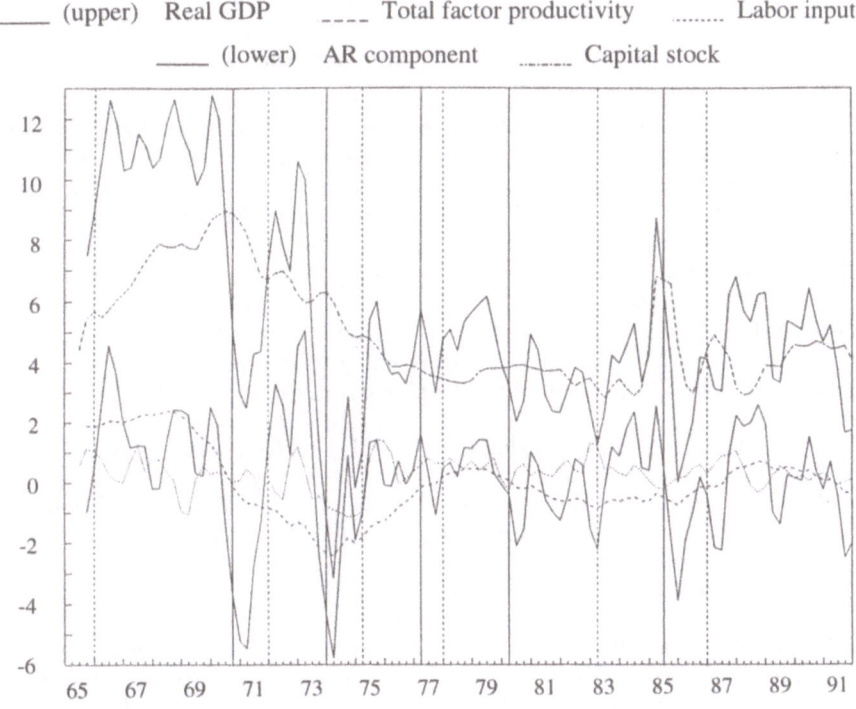

Figure 3.7 Contribution of each factor to yearly growth rate of the real GDP

yearly growth rate of the real GDP based on (3.6). From this it is explained that the short term fluctuations is brought about by the AR component corresponding to the business cycles. It is also explained that the trend of the growth rate is given by that of the capital stock, the labor input and the total factor productivity. Among them, contribution of the capital stock is especially great.

3.4 Model 3 (the Model by Which Macroeconomic Policy Effect can Also be Measured)

3.4.1 Outline of the Model

By applying an economic theory as seen in the previous section, the trend component has become more consistent with the actual economy. However in this stage, no model that can estimate the effect of the economic policy in the condition 3) in 3.1 is obtained yet.

Therefore assumption is furthermore made with Model 3 as shown below:

> The real GDP makes stochastic and stationary variation represented by an autoregressive process around the trend explained by the labor input and capital stock (deterministic variation with a production function) and total factor productivity (stochastic variation) under the influence of some economic policy variables such as a discount rate from the outside of the system. At that time, the degree of the influence from the outside of the system is variable according to the economic condition. On the other hand, the labor share is in stochastic variation.

The business fluctuations are expressed by the AR process with Model 2, but such formulation is insufficient as a model to express the actual economy because the factor bringing about business fluctuations is nothing but the system noise of AR process. In the Japanese economy financial and monetary policies have been taken corresponding to the business fluctuations, and therefore it is hoped that the influence of such a variable will be formulated explicitly and separately from the noise. Accordingly it is necessary to take into the model the exogenous variables (policy variables) exercising influence on the AR process. At that time, it might be natural to think that the degree of influence in question are changing corresponding to the business fluctuations.

The model obtained by formulating the above is given as

$$\log Y_t = \log TFP_t + W_t \log L_t + (1 - W_t) \log K_t + \text{ARX}_t + N_t, \qquad (3.8)$$

where TFP_t and W_t are the same as with the case of Model 2, and ARX_t means the autoregressive process having exogenous variables (ARX), and is formulated by

$$p_t = a_1 p_{t-1} + \cdots + a_\ell p_{t-\ell} + b_t E_t + v_{p_t}, \qquad v_{p_t} \sim N(0, \tau_p^2), \qquad (3.9)$$

where b_t is the time-varying coefficient, and it is formulated by the same model (3.2) as the trend component of Model 1. Meanwhile E_t is actually a bivariate time series expressing the real public investment and discount rate, but we use only one variable for the sake of simple exposition.

From the procedure shown above, a state-space model is to be constructed by changing the system model in Model 2 as shown below (in case that the order of the individual components is 2):

$$
\begin{bmatrix} T_t \\ T_{t-1} \\ \hline W_t \\ W_{t-1} \\ \hline p_t \\ p_{t-1} \\ \hline b_t \\ b_{t-1} \end{bmatrix}
=
\begin{bmatrix} 2 & -1 & & & & & & \\ 1 & 0 & & & & & & \\ \hline & & 2 & -1 & & & & \\ & & 1 & 0 & & & & \\ \hline & & & & a_1 & a_2 & E_t & \\ & & & & 1 & 0 & & \\ \hline & & & & & & 2 & -1 \\ & & & & & & 1 & 0 \end{bmatrix}
\begin{bmatrix} T_{t-1} \\ T_{t-2} \\ \hline W_{t-1} \\ W_{t-2} \\ \hline p_{t-1} \\ p_{t-2} \\ \hline b_{t-1} \\ b_{t-2} \end{bmatrix}
+
\begin{bmatrix} 1 & 0 & 0 & 0 \\ 0 & 0 & 0 & 0 \\ 0 & 1 & 0 & 0 \\ 0 & 0 & 0 & 0 \\ 0 & 0 & 1 & 0 \\ 0 & 0 & 0 & 0 \\ 0 & 0 & 0 & 1 \\ 0 & 0 & 0 & 0 \end{bmatrix}
\begin{bmatrix} v_{Tt} \\ v_{Wt} \\ v_{pt} \\ v_{bt} \end{bmatrix}.
$$

With the state-space model referred to above, the value of the hyper-parameter $\theta = (\sigma^2, \tau_T^2, \tau_W^2, \tau_p^2, \tau_{b1}^2, \tau_{b2}^2, a_1, \ldots, a_\ell)^T$ is obtained by the AIC minimization.[6]

3.4.2 Analysis Procedure

First of all, a problem at hand is stationarization of the exogenous variables. No trouble is fundamentally found with the discount rate, but it is necessary for the trend elimination to be made in the real public investment.[7] Although it might also be advisable to extract the AR component by the same method as Model 1, calculation might be a hard task because of the fact that the influence of the structural change is great and there will be several kink. At this stage, time trend with kinks was used as the second-best disposition although strictness is rather sacrificed.[8] As seen from the result shown in Figure 3.8, the structural change of the real public investment occurred in the second quarter of 1972 and in the second quarter of 1979. The former

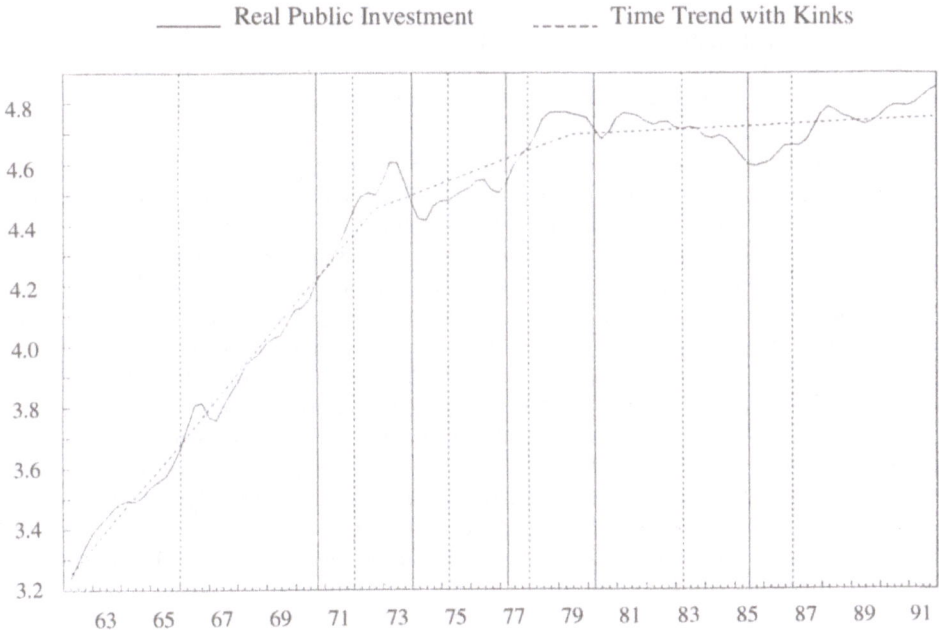

Figure 3.8 Real public investment and kinked time trend

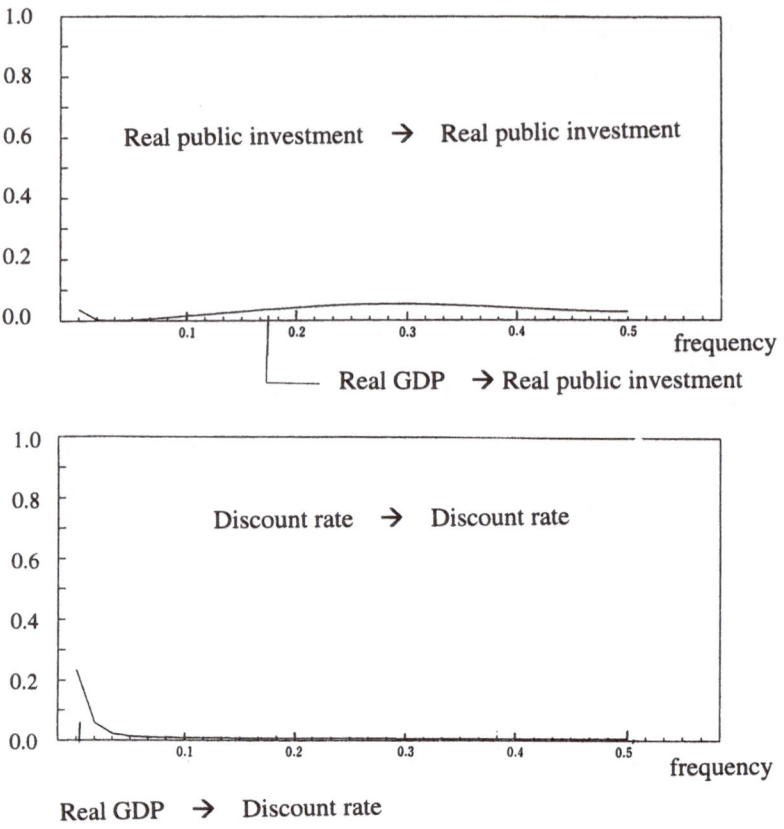

Figure 3.9   Relative power contribution

is the period of the financial tightening at the end of the high economic growth period, whereas the latter is the starting period of the financial reform. It can be said that the results almost correspond with the actual economy.

On the other hand, investigation of the exogenousness of these exogenous variables is also necessary. Several methods are available with detection of the said properties, but relative power contribution was employed in this occasion (TIMSAC was used for analysis software). From Figure 3.9, it is ensured that the real public investment and discount rate are hardly subjected to the influence of the real GDP (the AR component of Model 2) in either of their frequency domains. Also to settle the initial value of the time varying coefficient of the exogenous variables, regression analysis was made taking up the real GDP obtained in Model 2 with logged values up to second quarter and taking up 2 exogenous variables as explanatory variables. The result of the analysis is as shown below. In this connection, the 4-lag model was best fitted in both of the exogenous variables (E_1 means the public investment, whereas E_2 means the discount rate)

$$Y_t = \underset{(23.4)}{1.543Y_{t-1}} - \underset{(-11.4)}{0.703Y_{t-2}} + \underset{(1.0)}{0.00581E_{1,t-4}} - \underset{(-1.0)}{0.000205E_{2,t-4}} \, ,$$

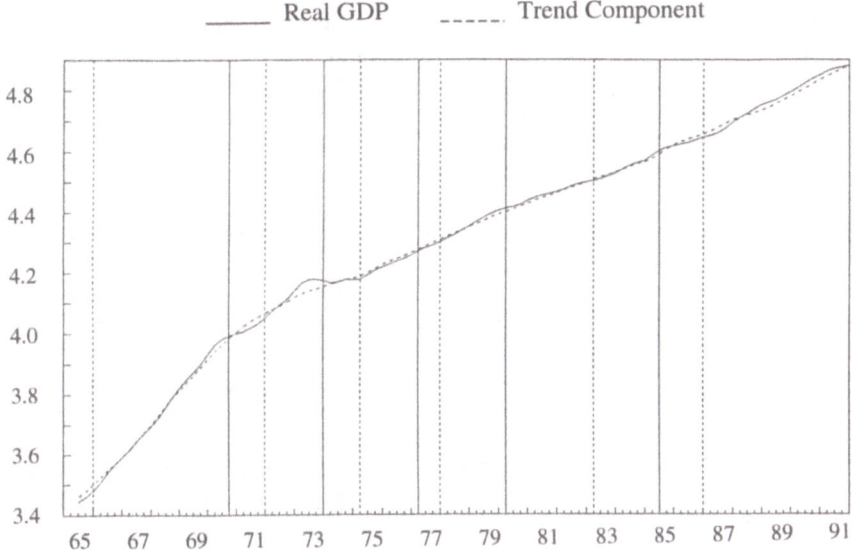

Figure 3.10   Real GDP and the trend component

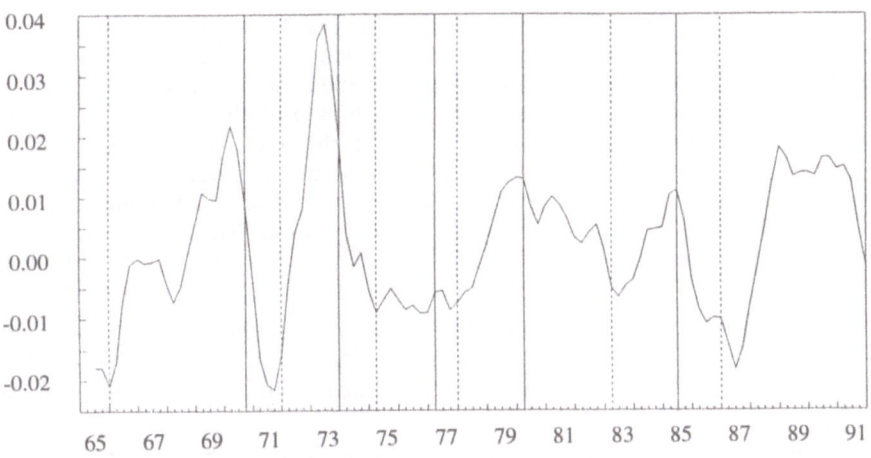

Figure 3.11   ARX component

where the values in the parentheses represent t-values. As long as the judgment from the correlation coefficient of $E_{1,t-4}$ and $E_{2,t-4}$ is concerned, the multi-collinearity is not so large. However t-values of these coefficients are low. This suggests that it depends on the business condition whether the economic policy is effective or not. Therefore it is necessary to adopt the time varying coefficient model.

3.4.3   Estimation Results

The results are shown in Figures 3.10 and 3.11. In connection with the ARX component, the results are in general good except for the fact that the 1965 depression is too serious and the depression after first oil crisis is too light.

Figure 3.12 shows the time varying coefficients of the real public investment and discount rate. The value of the first quarter of 1987 is, for example, 0.02 with respect to the real public investment. This means that when the real public investment is added 10% in the first quarter of 1986, the ARX component of the real GDP is raised 0.2% as the initial effect. Meanwhile with the discount rate, the fact that the first quarter of 1987 is, for example, −0.003 implies that the ARX component of the real GDP is raised 0.3% as the initial effect when the discount rate is lowered 1%. These values correspond to the multiplier of the macro-econometric model, but there are some movements different from the actual economy in each case. Especially the coefficients before 1975 fluctuate too much. This might happen from the reason that the model is unstabilized owing to the increase of the number of the hyper-parameters in comparison with Model 2. To stabilize the performance, a model with a constant term, for example, should be introduced. This is the subject to be examined in the future prospect.

As seen above, a model making it possible to express adequately the variation of the real GDP has successfully been obtained. More detailed discussion will be made in connection with the conditions 3) and 4) in the section 1.

3.5   Has Fine Tuning been Successful?

Has the so-called Keynesian policy levelized Japan's business fluctuations? Also what is the optimal opportunity to enforce the policy? It is not easy to provide a satisfactory solution with this problem. Shinpo (1984) has assessed the contribution of the macroeconomic policy variables to growth rate of the real GDP every time business fluctuations occur, and points out that the macroeconomic policy in the 1970s does not make levelization of the business fluctuations but amplify them.

By comparing the movement of the policy variables of Model 3 (Figure 3.13) with the movement of those time varying coefficients (Figure 3.12), the optimal period of the policy enforcement can be detected (take care that the 4-lag model is applied in each case).

1) In connection with the public investment, the time varying coefficient is in general heightened at the same time of the trough (ARX component of Model 3), and the optimal period for the policy enforcement is seen in general one year ahead of the trough.

2) Excluding 1983 in which identification is difficult, the policy enforcement is delayed in either case.

Description for the optimal policy period is listed in Table 3.2. In this connection, the optimal policy period is when the influence is most effectively exercised on the business fluctuations.

(1) **Trough at the fourth quarter of 1965**   Confronting the greatest depression, the government breaked itself of the balanced budget system, and the real Keynesian policy was introduced. However deducing from Figure 3.13, it is perceived that the public investment was, actually raised at the first quarter of 1966, and the policy was too lately taken and never contributed to quickening economic recovery, in consideration of the existence of the fourth quarter lag or of the movement to the

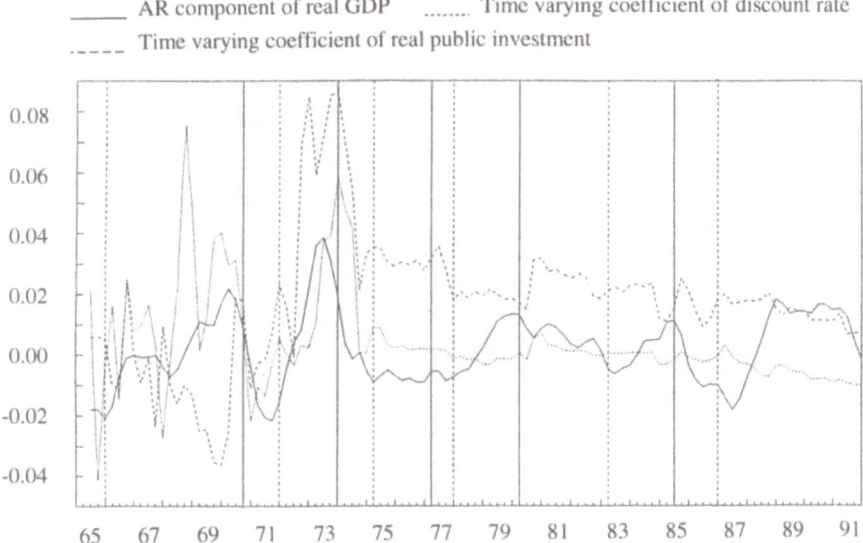

Figure 3.12 Time varying coefficients of the economic policy variables

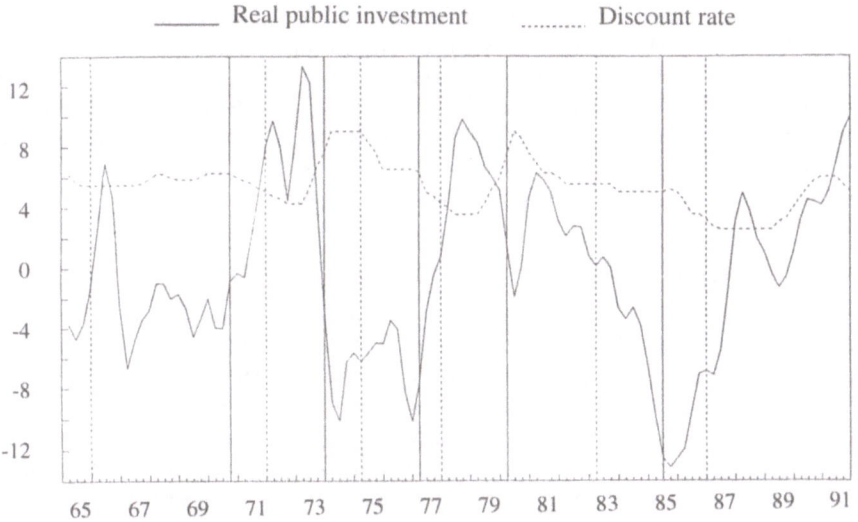

Figure 3.13 The economic policy variables

autonomous recovery of the ARX component.[9] Thus it only amplified the expansion after the recovery.

(2) Trough at the third quarter of 1971   From a viewpoint concerning the prevention of the excess liquidity inflation, it is conceived that the policy was perfectly in failure. Also judging from the time varying coefficient of the discount rate in 1973, it is comprehended that a tight money policy was needed. Since the

Table 3.2 Actual and optimal policy enforcement period

(Real public investment)

Date of trough	1965 IV	1971 III	1975 I	1983 II	1987 II
Actual period	1966 I	1971 II	1974 III	—	1987 III
Optimal period	1965 III	1970 IV	1974 I	1982 II	1986 I

(Discount rate)

Date of trough	1965 IV	1971 IV	1975 I	1983 II	1987 II
Actual period	1965 I	1970 IV	1975 II	1980 III	1986 I
Optimal period	1964 III	1969 IV	—	1981 III	1987 I

Remark:

1) Actual period is defined as day of the policy enforcement for the trough of the business cycles based on the ARX component (by Model 3) of the real GDP.

2) Optimal period is defined as day when the value of the time varying coefficient takes the maximum.

actual tightening was caused after 1973, it is understood that the tightening policy was late approximately one year.

(3) Trough in the second quarter of 1987   Steep rise of the yen appreciation was in progress starting with the beginning of 1985. Although business recovery was discussed because of the increase of real income by price going down, the business condition became worse. At that time, the discount rate policy was taken at the period too early. On the contrary, the fiscal policy such as the public investment was enforced one and half a year behind the optimal period. Should the fiscal policy have been carried out earlier one more year, the degree of the depression caused by the steep rise of the yen appreciation would have been quite a little one.

3.6   Is Prediction Ability Available?

To give investigation to the condition (4), 5 year prediction simulation for 4 cases was made by ARX model as listed in Table 3.3 (Figure 3.14).   With Model 3, the autonomous business fluctuations are shown in case that the exogenous variables of the ARX component or system noise are determined to be a given value (usually 0). Case 1 and 2 are the equivalent to the above. A higher growth rate is seen in case 2 than in case 1, which is due to the fact that in the process where the exogenous variable returns to 0 (trend level) (public investment $0.04 \rightarrow 0.0$, discount rate $0.48 \rightarrow 0.0$), the plus effect of the discount rate lowering was greater than the minus effect of public investment lowering. In any occasion, mild recovery is resulted the third quarter of 1992 as a trough.

On the other hand, case 3 and 4 show the business fluctuations when the macroeconomic policy was enforced. In this occasion, the business recovery happens in the second quarter of 1992.

It can safely be said that the prediction in question is quite appropriate to some extent. This is because it is certain for the real GDP to have exhibited mildly the movement of the recovery in the first quarter of 1993. Furthermore this might be

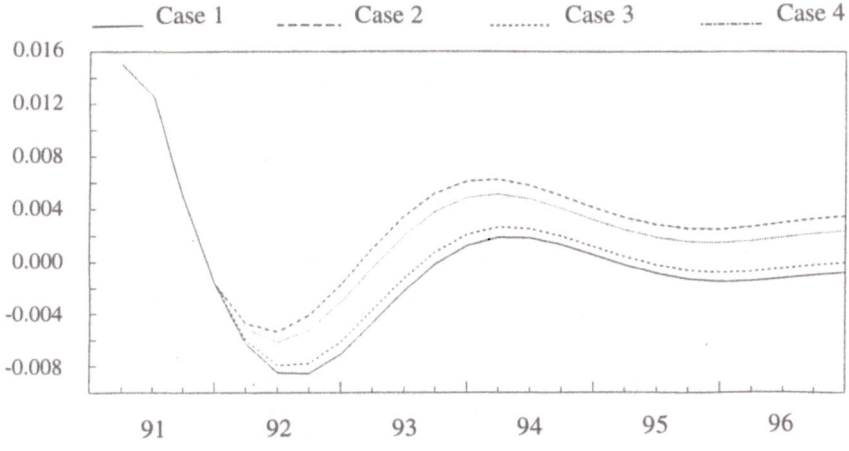

Figure 3.14 Extrapolated simulation

Table 3.3 The influence of the exogenous variables to ARX component

		value of elasticity	simulation (sustained)	period	initial effect
Case 3	public investment	0.0074	10% increase	1990IV	0.07%
Case 4	discount rate	−0.0010	1% decrease	1990IV	0.10%

Remark:
1) case 1: The values of the exogenous variables are constant equal to those of last estimation period.
   case 2: The values of the exogenous variables are equal to 0.
2) Values of elasticity are those of the time varying coefficients of last estimation period.

derived from the reason that adjustment of durable goods or capital stock was finished to some extent, and autonomous recovery factors of the business fluctuations were in operation. However owing to the exchange rate that is not adopted even in this model 3 or the influence of the cold summer, recession has again come to the Japanese economy.

Despite the above, there are essential problems with this simulation. Long term prediction simulation such as five years prediction, for example, is not appropriate because labor input and capital stock are not given and trend is variable.

3.7 Conclusions and Subjects in the Future

Business fluctuations have been produced by a variety of factors complicatedly exercising influence on each other. Under such a situation, the macro-econometric model has become large to be "a model that can make explanation with everything". However it is a well-known fact that attaining a larger model is not necessarily linked to the enhancement of the prediction ability.

In this paper, investigation has been promoted taking up a time series model (Model 1) expressing the variation of the real GDP as a starting point, and when it is necessary to apply the economic theory to the model, some economic models are taken in.

To make the trend component in stochastic variation (Model 1) correspond to the actual economy, the trend component was improved by taking in a production function as a deterministic economic model in Model 2. Furthermore, the policy variables as the external factors driving the stochastic AR component (Model 2) are explicitly formulated in Model 3, and estimation of the policy effect and prediction simulation were made. In this manner, the statistical model is improved properly with the aid of economic theory.

Finally, the necessity of the analysis of noise component is pointed out as a subject in the future. Let an ordinary AR model be considered. Although the source to drive this model is innovation, Gaussian white noise is usually kept in mind with this. If a turning point of the business fluctuations is predicted by using this model when the influence of large exogenous factors such as the oil crisis is dominant, the prediction ability of this model is remarkably degraded (Fukuda 1992). Different from the affairs in a field such as engineering, the movement of the noise is quite complicated in the economy, different from the assumption of the model. To solve such a problem, 2 types of approach seem to be available. One is to take in a non-Gaussian model for noise (Kitagawa 1987). The other is to adopt the change of the structure of the time series correlation of the noise, where prediction errors are deviated in a direction of either of plus or minus. Model 3 in this paper is the starting point for the next step.

Note 1: Documents on the detailed discussion is held in Bank of Japan (1981).

Note 2: With the real GDP, the X-11 (seasonal adjustment method) was used for adjusting seasonality in Model 1. However in Models 2 and 3, another version which exclude irregular variation is used. This is the disposition to prevent the solution from being unstable in consideration of the fact that Models 2 and 3 become complicated compared with Model 1.

Note 3: Since NTT was established in 1985 and JR in 1987, private capital stock of these period was adjusted.

Note 4: Strictly speaking, the labor force population is also subjected to the influence of the business fluctuations. For example with women's labor, offer for part-time job was increased owing to labor shortage in economic expansion. Thus housewives who had been categorized in nonlabor force population are reviewed as labor force population, resulting in increase of labor force population.

Note 5: Based on the Cob-Douglass-type production function, calculation was made as

$$\log Y_t = \log TFP_t + w \log L_t + (1 - w) \log K_t .$$

From the above,

$$\log TFP_t - \log TFP_{t-1} = \log Y_t - \log Y_{t-1}$$
$$+ w(\log L_t - \log L_{t-1}) + (1 - w)(\log K_t - \log K_{t-1}).$$

Note 6: τ_{b1}^2 is a variance of the system noise of a time varying coefficient of the real public investment (b_t in the state vector), whereas τ_{b2}^2 is that of the discount rate.

Note 7: Although unit root test is desirable to check the stationarity, this step is omitted.

Note 8: With the estimation method of the trend in question, $Y(P_1)$, $Y(P_2)$ are obtained so that sum of squared error will be minimized for arbitrary P_1, P_2 ($P_0 = 1$, $P_3 = 100$). By so doing, sum of squared error of $_{98}C_2$ (4753) cases is assessed to obtain P_1, P_2 of the minimum case. In this connection, the real public investment is given by the natural logarithm with the index taking up 1985 as 100.

Note 9: This is equivalent to the extrapolation simulation result of the ARX in case that the exogenous variable and system noise kept constant.

References

Bank of Japan (1981), "Panel discussion on the time series analysis", *Bank of Japan Monetary and Economic Studies*, (in Japanese).

Fukuda, K. (1992), "On the effectiveness of univariate time series models," *ESP* (edited by Economic Planning Agency), No. 247, 51–55 (in Japanese).

Gersch, W. and Kitagawa, G. (1983), "The prediction of time series with trends and seasonalities," *Journal of Business and Economic Statistics*, Vol. 1, No. 3, 253–264.

Harvey, A.C. (1981), *Time Series Models*, Philip Allan.

Hiromatsu, T. and Naniwa, S. (1990), *The economic time series analysis*, Asakura Shoten (in Japanese).

Kitagawa, G. (1986), "Decomposition of a nonstationary time series," *Proceedings of the Institute of Statistical Mathematics*, Vol. 34, No. 2, 255–271 (in Japanese).

Kitagawa, G. (1987), "Non Gaussian state space modeling of nonstationary time series," *Journal of American Statistical Association*, Vol. 82, No. 400, 1032–1063.

Kitagawa, G. (1993), *FORTRAMN 77 Programming for time series analysis*, Iwanami Shoten (in Japanese).

Kitagawa, G. and Gersch, W. (1984), "A smoothness priors state space modeling of time series with trend and seasonality," *Journal of American Statistical Association*, Vol. 79, No. 386, 378–389.

Kuroda, M. (1984), *Introduction to positive economic analysis*, Nippon Hyoron-sha (in Japanese).

Naniwa, S. (1985), "The trend estimation of economic time series," *Monetary Studies*, Vol. 4, No. 4, 60–93 (in Japanese).

Nelson, C. R. and Plosser, C. I. (1982), "Trends and random walks in macroeconomic time series," *Journal of Monetary Economics*, Vol. 10, 139–162.

Sinpo, S. (1984), "Was fine tuning successful?", in Microeconomic Policy in Japan, ed. National Institute for Research Advancement, 43–62, Toyo Keizai Shinposha (in Japanese).

Takeuchi, Y. (1991), "Trends and structural changes in macroeconomic time series," *Journal of Japan Statistical Society*, Vol. 21, No. 1, 13–25.

Yamamoto, T. (1988), *The economic time series analysis*, Sobunsha (in Japanese).

The author of this paper has been provided with precious advice and constructive suggestions by Dr. Hirotugu Akaike, former Director General of The Institute of statistical Mathematics, Professor Genshiro Kitagawa, and referees, for which the author wishes to express his sincere gratitude. Needless to say, the errors and mistakes, should there be any found in the paper, the blame is entirely attached to the author. The author has to add that the paper is drawn based exclusively on the author's personal opinion and is not necessarily in accordance with the view of the Economic Planning Agency.

Chapter 4

The Statistical Optimum Control of Ship Motion and a Marine Main Engine

Kohei Ohtsu
Tokyo University of Mercantile Marine
2-1-6 Etyujima, Koto-ku, Tokyo 135-8533 Japan
email: ohtsu@ipc.tosho-u.ac.jp

4.1 Introduction

In the identification of ship motion on ocean, it is especially required to adopt a statistical model, since external disturbances caused by wind and wave and furthermore the motion of the hulls itself responding to such an oceanic disturbances are intrinsically irregular. Moreover, the range of the strength of the external disturbances is very wide from a mirror-like quiet sea to a rough sea where violent storm is raging. Thus the ship motions under wide range of disturbances are so largely changed as can never be imagined in the case of the other transport facilities. A method of practical analysis of such irregular phenomena have earlier been established in a frequency domain (Isobe 1960) and then, the ship motions under disturbances has been dealt with as a stochastic process also in the field of ship-building engineering. Nowadays, it can be said that statistical methods to analyze the time series gained from model-tests in irregular waves or records of actual-sea tests in a frequency domain have been established (Yamanouchi 1961a, 1961b). However reliable modeling method in a time domain for irregular time series is quite difficult, and no attempt to use the obtained analytical results for control has been made.

On the other hand, recently in the field of statistics, it was revealed that the information criteria, FPE or AIC proposed by Akaike is quite effective to identify an autoregressive model to actual irregular time series (Akaike 1971, 1974; Akaike and Nakagawa 1972, 1988). As a result, also in the field of shipbuilding engineering, it was expected to identify the ship's system by applying this method and fitting a time series model in a time domain directly to the data observed on an actual ship navigating on irregular sea. Furthermore, it was also expected to use the identified model to the design of marine control systems. The authors of this chapter have been aware of the fact that the method referred to above is quite effective for the design

57

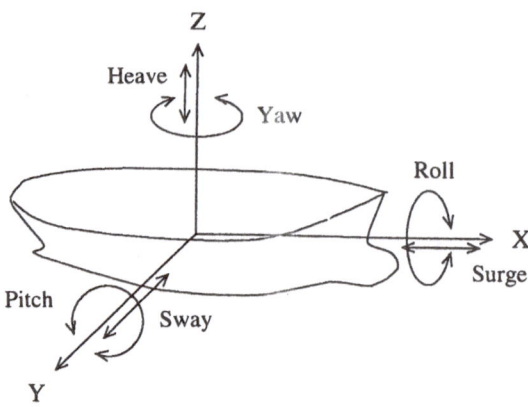

Figure 4.1 Ship motion

of the control system of the ships navigating on irregular sea, and they have exerted their endeavors to execution of the design of a variety of marine control systems.

In this chapter, two typical examples of the applications of the statistically identified autoregressive model are discussed. The first example is an application to design an autopilot system to control a ship's course and the second one is to design a ships governor to regulate the propeller revolution speed of the marine main engine. All examples taken up here involve the results of actual sea tests.

4.2   Outline of the Control of the Motion of the Hull and Main Engine

First of all, rough description is made with the motions of ship and main engine. The ship considered here is a conventional-type vessel, and such special vessels including the so-called high-speed launches are out of consideration. As shown in Figure 4.1, the ship navigating on sea can be expressed by the motion of 6 degrees-of-freedom.

Roll, pitch and heave are the motions having restoring force, whereas swaying, surging, and heading deviation (or yawing) are the motions having no restoring force. A rudder is a device to make it possible to control yawing and allow the ship to be directed to the desired course. The steering motions caused by the rudder can be classified into course-keeping motion to keep the desired course and course-changing motion to alter the course in a direction to another desired course. The former belongs to small motion, whereas the latter belongs to great motion. In this paper, mainly course-keeping motion is dealt with. Almost all of the ocean going ships are equipped with a control system (autopilot) for steering. As shown in Figure 4.2, the autopilot system forms a typical feedback control system.

On the other hand, propeller-driven methods are mainly used as the thrusting device for the ship's forward motion. Propellers, which have been conventionally provided with rotational motion by diesel engines or turbines, are mainly driven by diesel engines at present. However the propeller's rotation can never maintain a settled rotational frequency unless some regulator has been properly controlled. The device to allow the amount of the inlet fuel fed to the piston to be regulated so that the

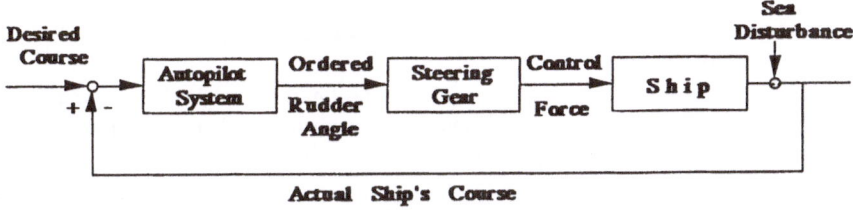

Figure 4.2   Ship's Autopilot system

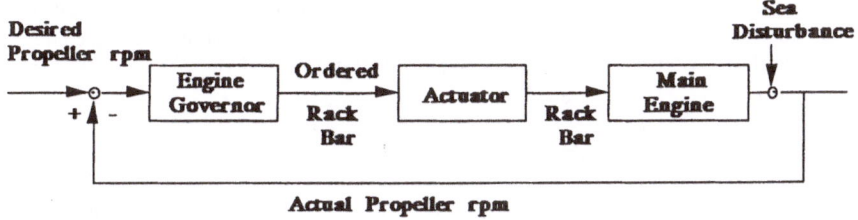

Figure 4.3   Propeller revolution control system (governor) of the marine main engine

propeller can maintain the desired rotational frequency is called an engine governor (Figure 4.3). Nowadays, centrifugal type of governors, which have been widely used as ship's governors, have been gradually replaced by electronic governors because of the rapid progress of electronic equipment and apparatus.

The data of the actual sea tests used in this chapter are obtained from the experiments executed on Shioji-Marus II and III, training ships of Tokyo University of Mercantile Marine. The principal dimensions and main engine specifications of both the ships are shown in Table 4.1.

Table 4.1   Principal dimensions of Shioji-Maru II, III

	Shioji-Maru II	Shioji-Maru III
Length	41.70m	49.93m
Breadth	8.00m	10.00m
Draft	2.575m	3.01m
Gross tonnage	331.37tons	425tons
Velocity (in voyage)	11.49knots	14.12knots
Engine type	diesel engine	diesel engine
Horse power	300PS×2	1400PS
Rated rotation frequency	1200rpm	700rpm

4.3   Statistical Model of Ship Motions and Its Control

4.3.1   Expression by a Control-type Autoregressive Model of the Ship's Control System

Suppose that the time series $x_1, \ldots, x_N$ of the heading angle driven by the rudder angle signals, $y_1, \ldots, y_N$, sampled at every Δt are observed from a ship in steering. Let us consider that an autoregressive model of x_n having y_n as input

$$x_n = \sum_{m=1}^{M} a_m x_{n-m} + \sum_{m=1}^{M} b_m y_{n-m} + u_n , \qquad (4.1)$$

is fitted to the observed time series. In (4.1), the heading deviation is expressed by a weighted sum of the past heading deviation and rudder angle, and an independent noise at time n. The model is a single-input single-output autoregressive model with output variable X_n and input one, Y_n. In general, the relation between the vector input Y_n and the vector output X_n is reasonably expressed by

$$X_n = \sum_{m=1}^{M} A_m X_{n-m} + \sum_{m=1}^{M} B_m Y_{n-m} + U_n . \qquad (4.2)$$

The model (4.2) shall hereafter be called a control-type autoregressive model. In this model, X_n is an r-dimensional control variable for the motion of the hull and main engine such as heading deviation, rolling, propeller revolution speed, etc. Meanwhile, Y_n is an ℓ-dimensional control variable, such as ordered rudder angle. U_n is assumed to be an r-dimensional Gaussian white noise.

The information criterion AIC for this model whose coefficients are estimated by the maximum likelihood method is given as

$$\mathrm{AIC}(M) = N \log |\Sigma_{r,M}| + 2r(r + \ell)M + r(r + 1), \qquad (4.3)$$

where N is the data length and $|\Sigma_{r,M}|$ the determinant of the least square estimate of the variance-covariance matrix of the residual U_n of the M-th order control-type autoregressive model. Also the second plus the third terms in the right-hand side are twice as large as the number of parameters of this model. According to the minimum AIC estimation (MAICE) method by Akaike, the model allowing AIC to be minimum is used as the best model (Akaike 1974).

However, a state space model is used as a standard formulation in the modern control theory whereas the models (4.1) or (4.2) is statistical model used in the time series analysis. To obtain the state space expression of a control-type autoregressive model, define a new variable $Z_{p,n}$ by

$$Z_{p,n} = \sum_{i=1}^{M-p} A_{p+i} X_{n-i} + \sum_{i=1}^{M-p} B_{p+i} Y_{n-i} \qquad (p = 1, \ldots, M-1) \qquad (4.4)$$

and then, the rM-dimensional state vector Z_n by

$$Z_n \equiv \begin{bmatrix} X_n \\ Z_{1,n} \\ \vdots \\ Z_{M-1,n} \end{bmatrix}. \qquad (4.5)$$

By this definition, the model (4.2) is expressed as (Akaike 1971; Akaike and Nakagawa 1988)

$$\begin{cases} Z_n = \Phi Z_{n-1} + \Gamma Y_{n-1} + W_n \\ X_n = H Z_n , \end{cases} \tag{4.6}$$

where Φ, Γ, H and W_n are defined by

$$\Phi = \begin{bmatrix} A_1 & I & 0 & \cdots & 0 \\ A_2 & 0 & I & \cdots & 0 \\ \vdots & \vdots & \vdots & \ddots & \vdots \\ A_{M-1} & 0 & 0 & \cdots & I \\ A_M & 0 & 0 & \cdots & 0 \end{bmatrix}, \quad \Gamma = \begin{bmatrix} B_1 \\ B_2 \\ \vdots \\ B_{M-1} \\ B_M \end{bmatrix}, \quad W_n = \begin{bmatrix} U_n \\ 0 \\ \vdots \\ 0 \\ 0 \end{bmatrix},$$

$$H = [\, I \ 0 \ \cdots \ 0 \,]. \tag{4.7}$$

Since this expression allows the latest data to be restored in a form to be used in future, the value of X_n in the next step can be predicted with a simple and small calculation when the newest data are obtained.

4.3.2 The Optimum Control Law

We present here a method of designing the optimum control law under an adequate performance index for a state space model (4.6) obtained in the previous section. As a performance index, the following two criteria that are widely used, are considered.

[1] A quadratic performance index evaluating the expected quadratic loss of a state variable and a control variable,

$$J_I = \mathrm{E}\left[\sum_{n=1}^{I} \left\{ Z_n^T Q Z_n + Y_{n-1}^T R Y_{n-1} \right\} \right]. \tag{4.8}$$

[2] A performance index which also penalizes the difference of the control input taking account of mechanical loss of the actuator,

$$J_I = \mathrm{E}\left[\sum_{n=1}^{I} \left\{ Z_n^T Q Z_n + Y_{n-1}^T R Y_{n-1} + (Y_{n-1} - Y_{n-2})^T T (Y_{n-1} - Y_{n-2}) \right\} \right], \tag{4.9}$$

where Q is an $rM \times rM$ positive semi-definite matrix, R is an $\ell \times \ell$ positive definite matrix, and T is an $\ell \times \ell$ positive semi-definite matrix. Since [1] is a special case of [2], an algorithm to obtain the optimum control law for [2] is hereunder shown (Ohtsu, Horigome, Kitagawa 1976, 1978).

Starting with

$$S_0 = Q, \quad R_0 = R, \quad P_0 = 0, \quad T_0 = T,$$

compute S_i, P_i and R_i, recursively for $i = 1, 2, \ldots$ by

$$\begin{aligned} S_i &= S_{i-1} + \Phi^T \left\{ S_{i-1} - (S_{i-1}^T \Gamma + P_{i-1})(\Gamma^T S_{i-1} \Gamma + P_{i-1}^T \Gamma + \Gamma^T P_{i-1} \right. \\ &\qquad \left. + R_{i-1} + T)^{-1}(\Gamma^T S_{i-1} + P_{i-1}^T) \right\} \Phi \tag{4.10} \\ P_i &= P + \Phi^T (S_{i-1} \Gamma + P_{i-1})(\Gamma^T S_{i-1} \Gamma + \Gamma^T P_{i-1} + P_{i-1}^T \Gamma + R_{i-1} + T)^{-1} T \\ R_i &= T + R - T^T (\Gamma^T S_{i-1} \Gamma + \Gamma^T P_{i-1} + P_{i-1} \Gamma^T + R_{i-1} + T)^{-1} T . \end{aligned}$$

The optimal control input Y_i is then given by the feedback control law

$$Y_i = G_i Z_i + F_i Y_{i-1} . \tag{4.11}$$

The control gains G_i and F_i are given by

$$
\begin{aligned}
G_i &= -(\Gamma^T S_{i-1} \Gamma + P_{i-1}^T \Gamma + \Gamma^T P_{i-1} + R_{i-1} + T)^{-1}(\Gamma^T S_{i-1} \Phi + P_{i-1}^T \Phi) \\
F_i &= (\Gamma^T S_{i-1} \Gamma + P_{i-1}^T \Gamma + \Gamma^T P_{i-1} + R_{i-1} + T)^{-1} T .
\end{aligned}
\tag{4.12}
$$

Here if the evaluation period I is taken long enough, then G_i and F_i respectively converge to G and F. Therefore, the optimum control law for a stationary state is determined as

$$Y_i = G Z_i + F Y_{i-1} . \tag{4.13}$$

The control law with constant gains (4.13) shall be used in the control system hereafter.

4.4   Design of Optimum Autopilot System Based on the Control-type Autoregressive Model

A feature of the method to obtain the optimum control law by using a state space expression of a control-type autoregressive model is that none of special experiment for a control objective is required and the model can be obtained from a time series of actual ship with controlled input as random as possible. Another feature is that the optimal control system can easily be realized by using of the existing system. These features are quite convenient for the ship system as a large-size system that is compelled to be in motion under strong disturbances. In this section, an example of the design of an autoregressive (AR) model based autopilot system by applying the method in the previous section (Ohtsu, Horigome, Kitagawa 1976, 1978).

4.4.1   Simulation Study

In this method, we start with observing of the time-series data obtained from a ship actually in sailing and modeling the causal relation of those time series by a control-type autoregressive model. It is desirable that the data of the heading deviation steered by rudder involve frequency components as wide as possible. For that purpose, it is recommended to steer manually to avoid possible dangerous problems caused by random steering. Figure 4.4 shows a part of the actual ships motion observed on a container ship navigating on a rough sea of wind force 9 or 10. From the top to the bottom the time series of the pitch, roll, yaw (heading deviation), sway acceleration and rudder angle are illustrated. Here, we first select important variables which should be used in designing a new autopilot system.

By the method of designing the optimum control system based on the control-type autoregressive model shown in the previous section, design of not only a single-input/single-output system but also a multiple-input/multiple-output system is also possible. At that stage, it is important to evaluate what influence the control input, the rudder, will give on the whole of the ship motion. Although, the analysis by a power contribution is effective for that purpose, a simulation study is hereby used so that the actual response of the state variable can be observed (Akaike and Nakagawa 1972, 1988).

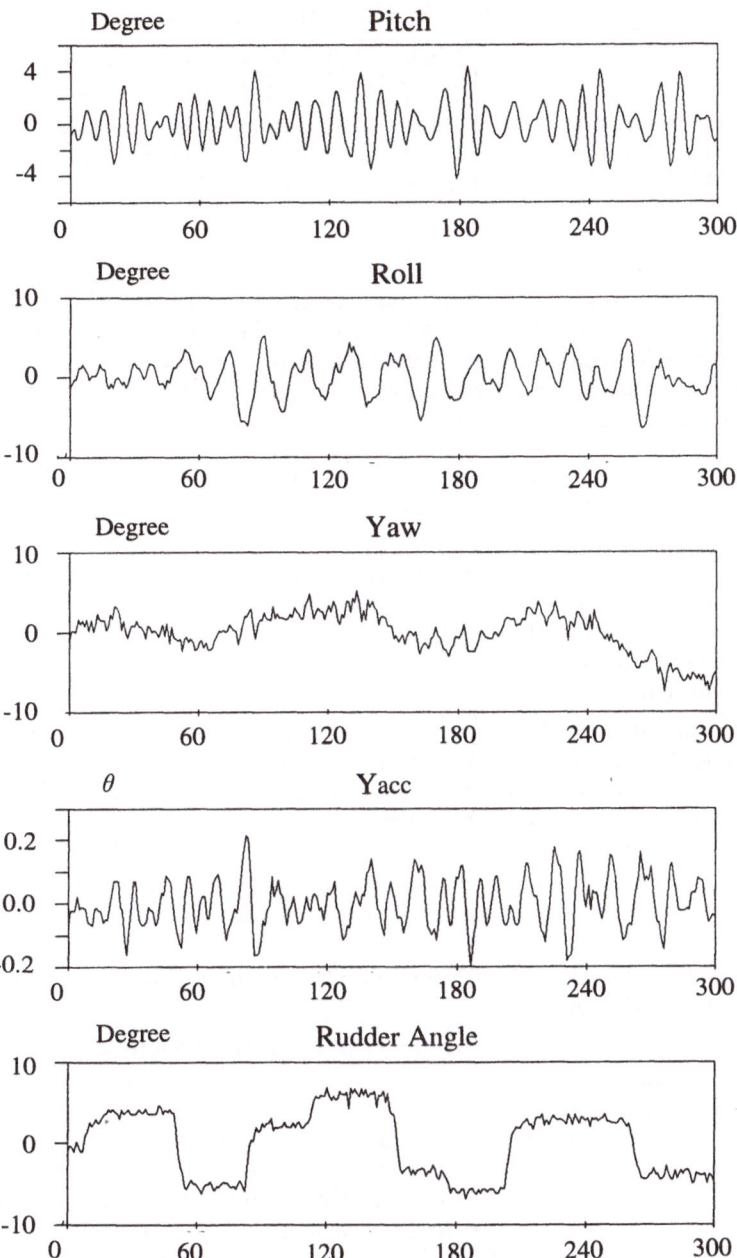

Figure 4.4   Actual data of a container ship

Table 4.2 Results of the simulation

Q		R	Pitch	Roll	Yaw	Y_{acc}	Rudder	Diff.
(1)	(2,7,35,1)	0.85	1.726	4.575	1.709	0.0063	18.72	4.546
(2)	(3.6,7,40,1.67)	1.015	1.444	4.915	1.674	0.0062	19.12	5.052
(3)	(0,6.33,57.8,0)	0.86	2.242	4.473	1.672	0.0068	18.66	4.931
(4)	MANUAL		2.771	6.557	7.781	0.0081	17.62	0.832

To comply with such a situation, a multi-dimensional control-type autoregressive model including all the variables is obtained by the MAICE method, and the optimal control gain is calculated for a predetermined performance index. After that, the simulation is performed by using feedback control gain and a realization of white noise process for the fitted control-type autoregressive model. That is to say, the simulation is made using the models

$$\begin{cases} Z_n &= \Phi Z_{n-1} + \Gamma Y_{n-1}^* + W_n \\ Y_{n-1}^* &= G Z_{n-1} + F Y_{n-2}^* \end{cases} \tag{4.14}$$

and the result is evaluated. Here, the variance covariance matrix of the residual of the fitted model is used as the variance covariance matrix of the white noise W_n. On the other hand, as the weighting matrices in the performance index, adequate values can almost automatically be selected by using the procedure shown below (Ohtsu, Horigome and Kitagawa 1979).

[1] Let the initial value of the matrixes Q, R and the integer K be properly set. For example, Q is a diagonal matrix whose diagonal elements are the inverse of the diagonal elements of the matrix $\Sigma_{r,M}$, and R is a diagonal matrix whose diagonal elements are the square of the reciprocal numbers of the tolerance limitation of the control amount.

[2] Let just one of the diagonal components $Q(i, i)$ of the matrix Q be multiplied by $1 + K^{-1}$ in order, and do the simulation and evaluate the result. Find the most effective direction j.

[3] Modify $Q(j, j) = (1 + K^{-1})$, and the value of K is adequately increased (for example, let $K = K + 1$).

Comparison is made between the variances of the optimal controller and that of the steering by the human operation. The optimal controller is designed by using the control-type autoregressive model fitted to the data shown in Figure 4.4 by the MAICE method and the weighting matrix Q, R, ($T = 0$) in the performance index (4.8) which are selected using a trial and error method or an automatic method shown above.

The result of the comparison is listed in Table 4.2 (Ohtsu, Horigome and Kitagawa 1979). Among the values in the result, the four variables of the pitch, roll, yaw (heading deviation), and Yacc (lateral acceleration) are selected as state variables in the case 1, whereas the rudder is selected as a control variable. The variances of the individual variables are obtained by repeating the simulation 10 times. Here, the parenthesized values of the weighting matrix Q in the left column of the table indicate the weight in the performance index for the corresponding variables. In the meanwhile, the values in the column of R indicate the weight for the rudder. The

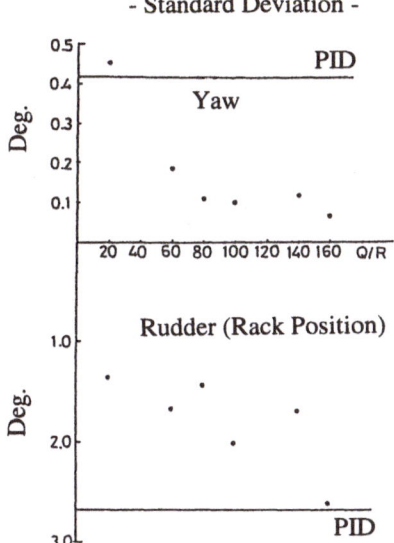

Figure 4.5 Relation between the gain ratio Q/R and the standard error of the heading deviation (right) and rudder angle amount (left) (actual sea test)

column having the heading of Diff. shows the variance of the changing speed of the operation variables (rudder). The case 2 gives the result by the gain automatically modified by the successive approximation method shown above from the gain of the case 1, whereas the case 4 provides the standard error around the desired course of the actual data. On the other hand, the case 3 brings about the simulated result obtained by designing a control gain aiming at exclusively the reduction of the yaw (heading deviation) and roll as a target without considering the pitch and transverse acceleration. These results suggest that not only the yaw (heading deviation) which is the main purpose of the conventional autopilot system but also the roll can be controlled by a new control system.

4.4.2 Actual Sea Test and Its Analysis

As a result of the repeated execution of the simulations, firm belief has been obtained that the proposed method is practical enough to be a design method for the new control system of a ship on an actual sea. First of all, we developed an optimum autopilot system based on a single-input/single-output control-type autoregressive model having a state variable of the control target as a heading deviation and the control input as a rudder angle should respectively. The ship used in this purpose is Shioji-Maru II, Tokyo Mercantile Marine University's training ship. The ship is equipped with a conventional control system. In addition to this, the system is instantly changeable to the computer-mode for experiments. Using the system, a time series of the heading deviation under the random steering by a human operator was measured for 600 points at every second and a control-type autoregressive model was fitted. This resulted in adoption of a model of order 7. Then simulation was made by using the model and the method shown in (4.14), and the results of the various

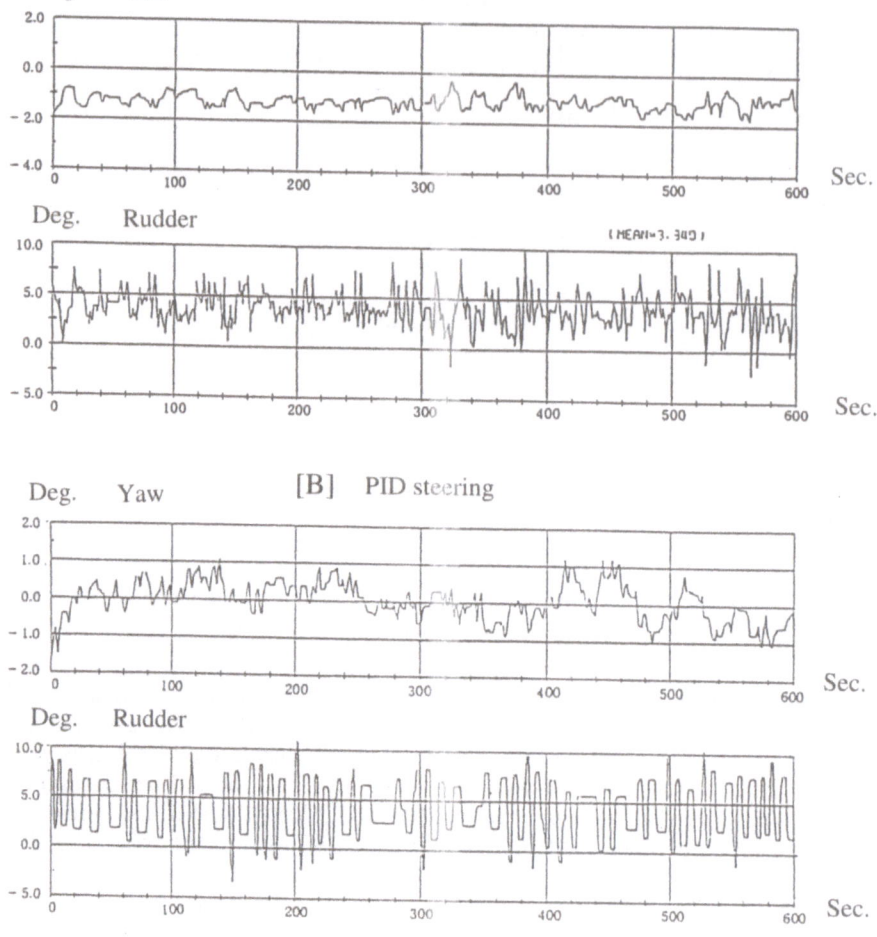

Figure 4.6 Examples of the actual sea tests [A]: The optimum type ($Q/R =$ 160) [B]: Conventional type)

gains obtained by changing the weight Q/R of a variety of performance indexes were compared (Ohtsu 1983). As a consequence of this procedure, it is found that an amount of the rudder angle is increased as Q/R is made greater, but the variation of the heading deviation remains almost unchanged at the stage of $Q/R =$ more than 60.

The results of the actual sea tests by Shioji-Maru II executed by changing the gain in various manners based on the outcomes obtained as above is evaluated by the standard deviation around the mean value of the individual variables (Figure 4.5). Also in Figure 4.6, a record of the AR type autopilot system with $Q/R = 160$ and an actual sea test of the conventional automatic steering system (which shall sometimes be expressed as PID) executed immediately after the last time series are shown. From

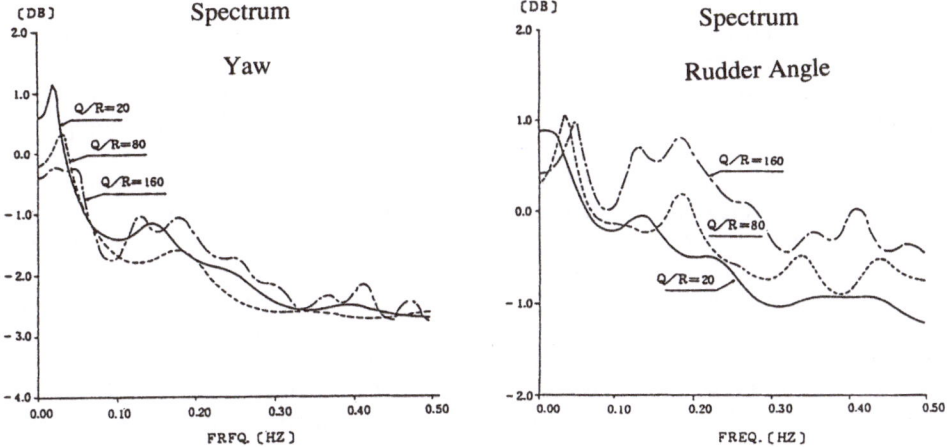

Figure 4.7 Comparison of the spectrum according to the change of the gain ratio Q/R (left: yaw (heading deviation), right: rudder angle amount) power spectrum directional angle rudder angle frequency

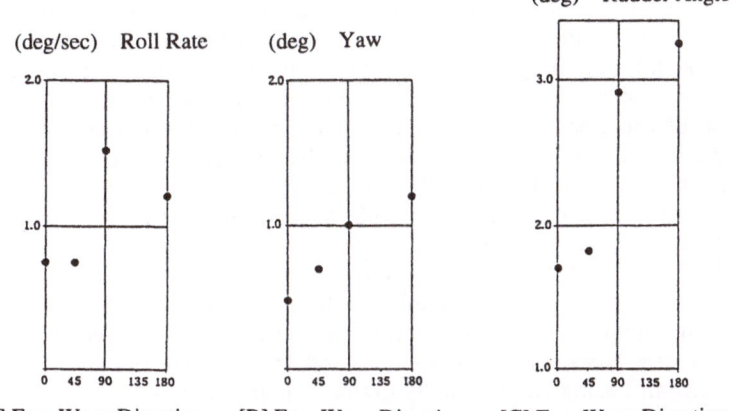

[A] Enc. Wave Direction   [B] Enc. Wave Direction   [C] Enc. Wave Direction

Figure 4.8 Change by the encounter angle with a wave (abscissa: from left (A) roll angular velocity, (B) yaw (heading deviation), (C) standard error of the rudder angle amount, ordinance: encounter angle)

these figures, (1) it is explained that the variation of the heading deviation in the AR type of autopilot is smaller than the angular variation of the heading deviation caused by the conventional autopilot except that $Q/R = 20$, and the ship is excellent in course-keeping ability. (2) It is also explained that, except for the case of $Q/R = 160$, just a smaller amount of the rudder angle is necessary compared with that of the conventional one.

Figure 4.7 illustrates the power spectra of the heading deviation and rudder angle when the gains are changed. It is explained that the rudder motion with high frequency is increased as Q/R is made greater, resulting in the decrease of the peak in the low frequency side. Also it is explained that the peak in the vicinity of 0.15 or 0.20Hz is rather increased.

As shown above, it is explained that the newly-developed system exhibits excellent performance. However to apply the system to practical use by actually equipping it onto a ship, a variety of performance is furthermore required. Several of the evaluation tests of the system in question are hereby shown (Ohtsu 1983). Figure 4.8 illustrates the evaluated result obtained by changing the ship's course in a short while in order to examine how the control characteristics change, depending on the difference of the encounter angle of the external wave disturbances. The best record is seen in case of the heading sea (encounter angle is 0 degree), and the records are worsened as the direction of waves become following sea to the experiment ship. This is because as the direction of waves become in following sea, especially in oblique sea, not only the ship motion induced by the waves becomes greater but also the steering effect becomes low owing to the decrease of the relative speed of the fluid to the rudder. However deterioration of the course-keeping performance is, judging from a point of view of practical use, considered to be in a tolerance range.

The most important role of autopilot system is to keep a course of the ship while she is on voyage at a desired course. However when the desired course is changed, the ship is required to change rapidly to a new course. Figure 4.9 gives a case when a ship altered her course to the 20 degree starboard side at the vicinity of 500sec and returned to the original course. It is understood that no overshoot is seen on the heading deviation and the course is changed to the new one very rapidly.

Another problem is the motion of the actuator to realize the control. A very rapid variation of the steering gear as an actuator of the steering system may destroy a steering gear. The weighting matrix T in the performance index (4.9) is introduced to avoid this problem. It is explained as a result that the system has course-keeping ability more excellent than the one of the conventional system (Ohtsu 1983). In addition to the above, a variety of actual sea tests were executed using the ship or other ships greater than the former. Those tests have revealed that the control system that is designed, keeping the sampling rate at around 1 sec, brings about a control system with high course-keeping ability and excellent course-changing performance for the vessels based on the data obtained by properly designed random steering experiment and using the optimally designed gain function.

4.5　Noise-Adaptive Control System

In the control-type AR autopilot system described in the former section, the fitted model is not changed through the experiments. Namely, the control gain was calculated beforehand and it was fixed. However although wind and wave can be regarded as stationary for a short time, it is necessary from a viewpoint of long-time observation to consider that the property of the stochastic process will be changed and be nonstationary. It is confirmed that the control-type AR autopilot system is robust enough to cope also with a variety of external disturbance to some extent. However, hereafter we will consider the development of a noise-adaptive control system based on a locally stationary AR model proposed by Ozaki and Tong (1978) (Ohtsu, Horigome and Kitagawa 1979).

For that purpose, it is desired to develop an autopilot system with an ability to

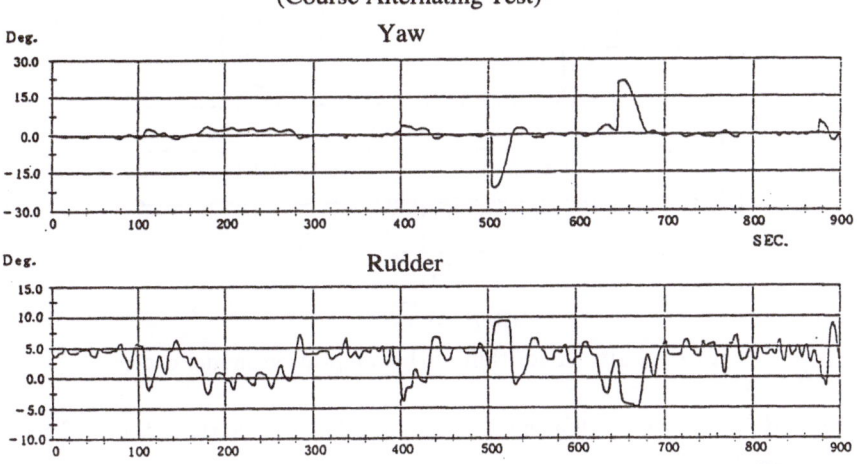

Figure 4.9 Course-changing ability (the new course is always taken as 0)

adapt to the change of disturbances of the model

$$x_n = \sum_{m=1}^{M} a_m x_{n-m} + \sum_{m=1}^{L} b_m y_{n-m} + u_n ,\qquad (4.15)$$

where x_n is the heading deviation, y_n the rudder input. Here u_n is an unknown noise process that expresses the influence of the external disturbances and is not necessarily the white noise but may be expressed by an autoregressive model

$$u_n = \sum_{\ell=1}^{K} c_\ell u_{n-\ell} + \varepsilon_n .\qquad (4.16)$$

Substituting (4.15) into (4.16) yields

$$x_n = \sum_{m=1}^{M+K} A_m x_{n-m} + \sum_{m=1}^{L+K} B_m y_{n-m} + \varepsilon_n .\qquad (4.17)$$

Here A_m and B_m are defined by

$$A_m = a_m - \sum_{j=1}^{m} c_j a_{m-j}, \qquad B_m = b_m - \sum_{j=1}^{m} c_j b_{m-j} ,\qquad (4.18)$$

where $a_i = 0$ $(i > M)$, $b_i = 0$ $(i > L)$. Although the noise process u_n is locally stationary, its stochastic structure may be changed gradually in long time. By detecting the nonstationarity in accordance with the procedure shown below, let it be considered that the models are to be renewed.

Now, suppose that a data set of n observations is observed and the residues $u_1, \ldots, u_n$ are calculated for the data set using (4.15). Also suppose that an autoregressive model AR_0 is fitted to the noise process u_n and m residues $u_{n+1}, \ldots, u_{n+m}$ are, newly

Kohei Ohtsu

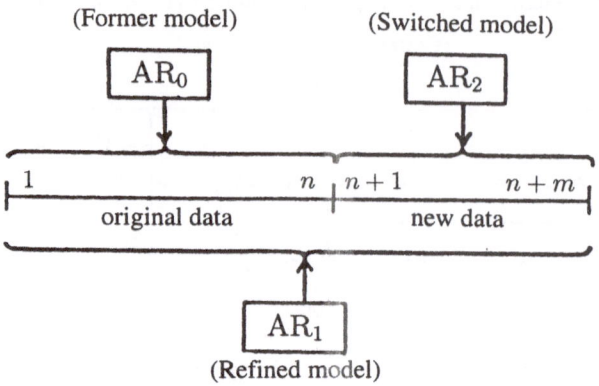

Figure 4.10 Algorithm of the adaptation model choice

obtained. Here, by the MAICE method, we decide whether the 2 types of the data sets are homogeneous or not.

First if the 2 types of the data sets are homogeneous, then an autoregressive model AR_1 is obtained for $u_1, \ldots, u_{n+m}$ which is obtained by pooling the 2 data sets (Figure 4.10). The goodness of the model at that time is evaluated by

$$AIC_1 = (n + m) \log \sigma_1^2 + 2(K_1 + 1), \tag{4.19}$$

where σ_1^2 is the variance of the residual of AR_1, K_1 is the order of the model. Contrarily to the above, AIC in case that the data sets are assumed to be non homogeneous is given as the sum of AIC_0 and the AIC of the autoregressive model AR_2 for the noise process $u_{n+1}, \ldots, u_{n+m}$ of the second half. Namely, it is expressed as

$$AIC_2 = n \log \sigma_0^2 + m \log \sigma_2^2 + 2(K_0 + K_2 + 2), \tag{4.20}$$

where σ_0^2 and σ_2^2 are respectively the variances of the residual process of AR_0 and AR_2. Meanwhile, K_0 and K_2 are the orders of the individual models.

Comparison is hereby made between AIC_1 and AIC_2. The judgement rule to select the next model are as follows; (1) if $AIC_2 \leq AIC_1$, then the model AR_2 is preferred by rejecting the model AR_0. (2) If $AIC_2 > AIC_1$, then the model AR_1 is to be used. By this method, it is possible to locally correspond to the state change of the external disturbances by selecting the models one by one in this system. Immediately after a new model is obtained, updated gain is calculated and steering is made in the next term using the updated gain. In the state space expression, the following state vector

$$Z_n = \left[x_n, \; x_{n-1}, \; \ldots, \; x_{n-M+1}, \; y_{n-1}, \; \ldots, \; y_{n-L+1} \right]^T \tag{4.21}$$

is used.

To realizes the method, it is necessary to run two programs simultaneously. One of them implements the ordered angle to be sent to the rudder real time. The other runs at every settled term on the background to do construction of the model and the gain. After that, the model and the gain that are updated are sent to the former program.

Figure 4.11 gives an example of the actual sea test by a noise-adaptive autopilot composed in a manner shown above. The ship used for the experiment is of almost

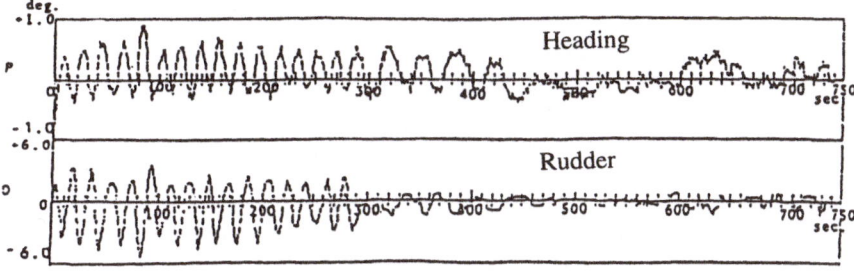

Figure 4.11 Actual sea test by the noise-adaptive autopilot system (upper half: heading deviation ±1° lower half: rudder angle amount ±6°)

the same type as Shioji-Maru with $n = m = 200$sec and the sampling rate $\Delta t = 1$sec. Illustrated in the figure is a part of the result, which shows that the control result is improved from the vicinity of 300sec that the heading deviation change having a considerable degree of course deviation is switched to a new model, and the deviation from the desired course of the heading angle becomes little with the small steering angle.

4.6   Rudder-Roll Control System

With the control system that has been referred to up to now, its design target has been placed on the high course-keeping ability by small rudder angle amount. However, actually the yaw motion caused by taking the rudder, also induce roll motion. The hydrodynamic force is applied to the rudder surface, when it is taken, and pressure is produced on the rudder surface. This results in the production of the roll moment between the point, onto which pressure is applied, and the center of the gravity of the hull, onto which the centrifugal force acts. Contrarily, it is expected that skillful use of this moment will reduce the roll. In the simulation, it is confirmed that this roll-reduction effect will appear by the proper selection of the gain. When rudder angle amount is small, roll is also small. From this reason, it can be said that making the rudder amount smaller might also be one of the methods to reduce the roll. However the system considered here is an active roll-control which also takes into account the effect of the rudder to the roll motion. Development of the rudder-roll control system (RRCS) having the roll effect caused by the rudder is now underway with Oda *et al.* (Oda, Ohtsu, Sasaki, Seki and Hotta 1991).

Difficulty residing with this type of the control system is that there is a possibility of the rudder being taken frequently and largely when rolling is severe because roll is controlled by the rudder motion. As a result, the course-keeping ability as an intrinsic function is decreased. It is necessary for the trade-off points of the both to be determined by the repetition of the many simulations. Figure 4.12 illustrates the result of the actual sea test of the rudder-roll control autopilot executed by using Shioji-Maru III. As the controlled variables in this experiment, the heading deviation and roll angular velocity are chosen. Furthermore as the control variable, the ordered rudder angle is chosen. The roll angular velocity was preferred rather than the roll angle itself, which is due to the fact that the greater rudder-roll control effect is gained

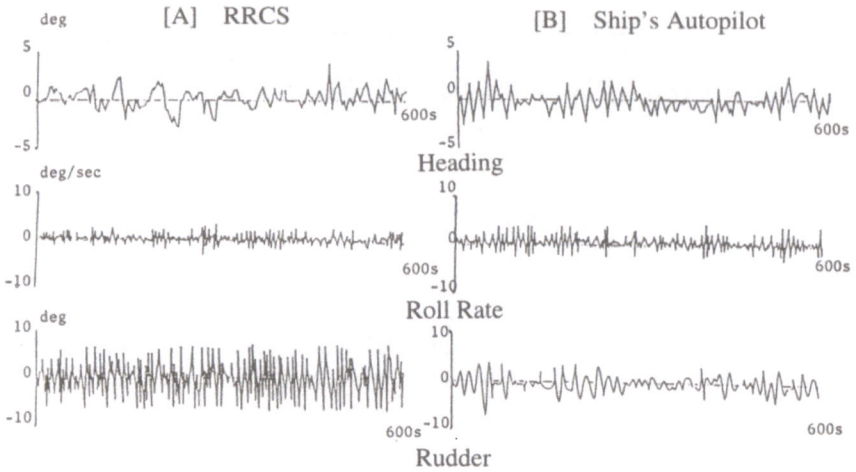

Figure 4.12 Comparison between the actual sea test by the rudder-roll control system [A] and the conventional system [B] (from the upper side, yawing (heading deviation), roll angular velocity, rudder angle)

as a result owing to the simulation in case that the roll angular velocity is used more than in case of the roll angle. Comparison is made between the rudder-roll actual sea test in the left half and the actual sea test by the conventional autopilot system executed immediately after the right half.

It is construed from this that the variation of the rudder-roll angular velocity shown in the center of this figure is decreased much more in case of RRCS. In addition to this, no deterioration of the course-keeping ability is seen with the heading deviation in comparison with the conventional autopilot system of the ship. Thus it is understood that no problem occurs when the system is put to practical use. However it is comprehended that the motion of the rudder has become more frequent and greater than the conventional one. It is apprehended that too frequent use of the rudder with too rapid velocity will destroy the actuator. To avoid such trouble, a countermeasure is taken by introducing the weight T of the performance index (4.9).

The system is at present put to practical use and is now in operation. And it has been explained that (1) the longer the distance between the center of the pressure directly applied to the rudder and the center of the gravity of the hull and (2) the rapider the rudder angle speed is, the stronger the rudder roll effect is.

4.7   Application to the Marine Main Engine Governor System

4.7.1   Experimental System

In this section, we consider the control of the engine governor for the regulating propeller revolution speed of the marine main engine which is as important as the autopilot system in the field of the ships-control system. The engine governor is a device to suppress the variation of the propeller revolution speed by regulating the fuel rate to the piston. The regulation is, at a final stage, done by adjusting the rack

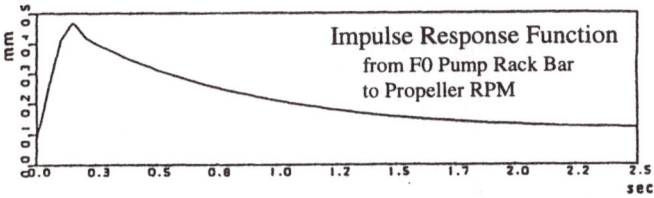

Figure 4.13   Impulse response function of the governor-propeller revolution system

bar connected to the throttle valve. While the propeller revolution speed is converted into electric signals to a computer, the ordered signal from the computer is sent to the rack bar in the actuator. To allow such computer-aided control to be possible, we developed an experimental system with which the controller of the existing governor (this shall hereafter be called the ship's governor) of Shioji-Maru III in Figure 4.3 is replaced with the computer. To implement the procedure to carry out experiments, first of all the system is transferred, for the sake of safety, from the control by the ship governor to the computer-aided governor using a reliable discrete-type PID control rule. Secondly, a method to convert to this experiment was adopted.

Following this procedure, fundamental data to fit a control-type autoregressive model in which signals to the rack bar of the governor as an input variable and the propeller revolution speed as an output one are adopted. In this experiment, the input signal is, as was already referred to also in the design of the autopilot system, required to be as random as possible. However inputting the random signals without control causes instability with the engine, and therefore random signals are placed on the command signals from the discrete-type of PID controller. In this preliminary experiment, the data of 1000 points were obtained at the sampling rate 0.1sec.

4.7.2   Dynamic Characteristics of the Main Engine Governor System

Today, desire to obtain larger-sized marine engines has successfully attained and reliability of such engines has also been increased, but almost none of examples of the analysis of the dynamic characteristics of the engines of the ships in navigation have been available up to now. Such being the case, a main engine governor system is analyzed by using the autoregressive model

$$X_n = \sum_{m=1}^{M} A_m X_{n-m} + U_n \,, \tag{4.22}$$

fitted to the data discussed in the previous section using MAICE method (Ishizuka, Ohtsu, Horigome 1991). Here, X_n is a 2-dimensional vector process and is composed of the propeller revolution speed and the governor command (hereafter the propeller rotation speed shall be expressed by rotation per minute, rpm). By using this model, the frequency response function of the governor-propeller rpm and the impulse response function are calculated. They can be easily calculated from the autoregressive model (Akaike and Nakagawa 1988).

Figures 4.13 and 4.14 illustrate the impulse response and frequency response function of the propeller rpm to the governor's desired signal (rack-bar position). From the above-mentioned figures, as a shape of the transfer function of this system, a form

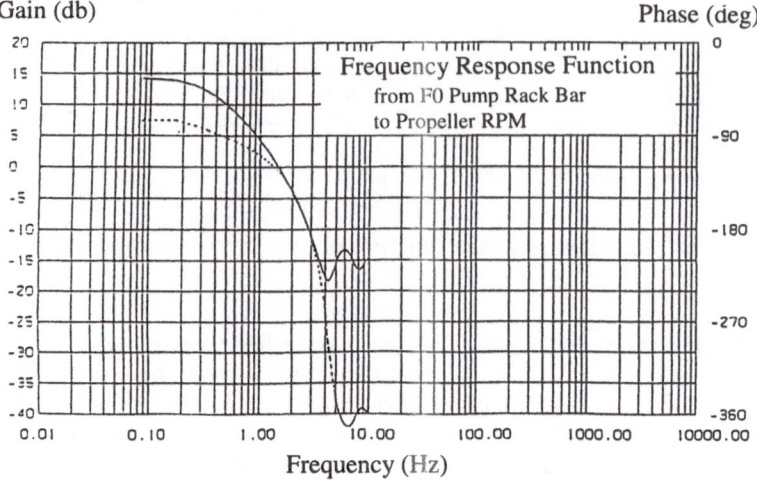

Figure 4.14 Frequency response function of the governor-propeller revolution system (solid line: amplitude, dashed line: phase)

Table 4.3 Actual sea test result by the optimal governor (variance)

Gain's name	Q/R	Variance of RPM	Variance of Rack Bar
A0001	0.001	7.195	0.0196
A0005	0.005	2.251	0.0254
A001	0.010	1.273	0.0756
Ship's Governor	–	4.012	0.1710

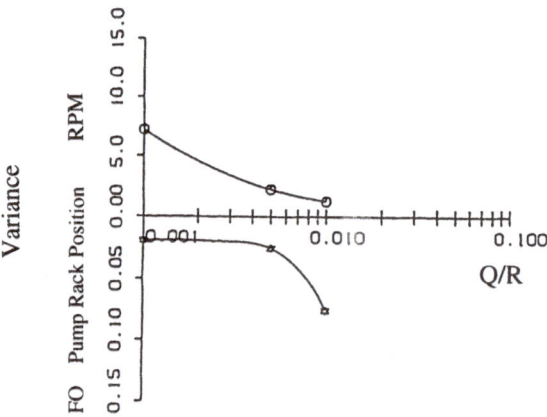

Figure 4.15 Relation between the gain ratio Q/R and the control result (variance value, upper half: propeller frequency variation, lower half: governor amount)

as

$$D(s) = e^{-Ls} \frac{K}{s(T_1 s + 1)(T_2 s + 1)} \qquad (4.23)$$

is expected. These kinds of knowledge will be available with the projects including the computer-aided control system design by the PID control rule described in the previous section.

4.7.3   Propeller Revolution Control Experiment by a New Governor

At the next stage, the control-type autoregressive model was applied by selecting the control variable as the input signal to the governor actuator and by selecting the propeller rpm as a state variable in the model (4.1). After that, the optimal control gain was calculated using a performance index with $T = 0$. With Shioji-Maru III in navigation, an actual sea test was implemented (Ishizuka, Ohtsu, Horigome 1991).

Table 4.3 shows the result of the whole experiments with the variance of the deviation from the mean values. Q/R gives a ratio of the weight coefficient to the governor command to the one to the variation of propeller rpm in the performance index. The experiment by the ship's installed governor shown in the lowest row was implemented immediately after the experiment of the optimal-type governor. In case of $Q/R = 0.001/1.0$ of the experiment No. A0001, it is noticed that the gain is too weak and the variation of the propeller rpm is great. However in other occasions, it is observed that decrease of the variation both in the propeller rpm and in the governor input are seen for the optimal controller compared with the case of the ship governor. In Figure 4.15, the variance of the propeller rpm versus the gain ratio is plotted in the upper half. The variance of the governor input is plotted in the lower half in the same figure. Figure 4.16 shows the experiment by the optimal governor in case of the experiment No. A001 followed by the experiment of the ship's governor immediately after that. It is seen that with the optimal governor the variation of the low frequency is suppressed, but variation with high frequency occurred. The cause of this is believed to be a problem between a combustion timing in the cylinder related with the propeller rpm (700rpm) of the propeller in question and the sampling rate of the control.

4.8   Conclusions

In this chapter, we show a design method and actual sea tests of the ship's fixed-gain type and noise-adaptive type autopilot systems, rudder-roll control system, and marine-engine governor system developed by the authors by applying a statistical optimum control theory based on a control-type autoregressive model introduced by Akaike. In the development of these systems, it becomes clear that the design method based on the control-type autoregressive model described in this chapter is an excellent method for the optimum control system design of a large and complicated system as a ship moving under strong external disturbance, particularly thanks to facility in designing it, the effect expected for the designed control system, assured attainment of the improvement, etc. Since now on, the authors are planning to proceed their study aiming at designing a robust and symmetrically unified marine optimum control system allowing, for example, the attitude control system and engine system to be unified as a ship's controlling system under strong disturbances.

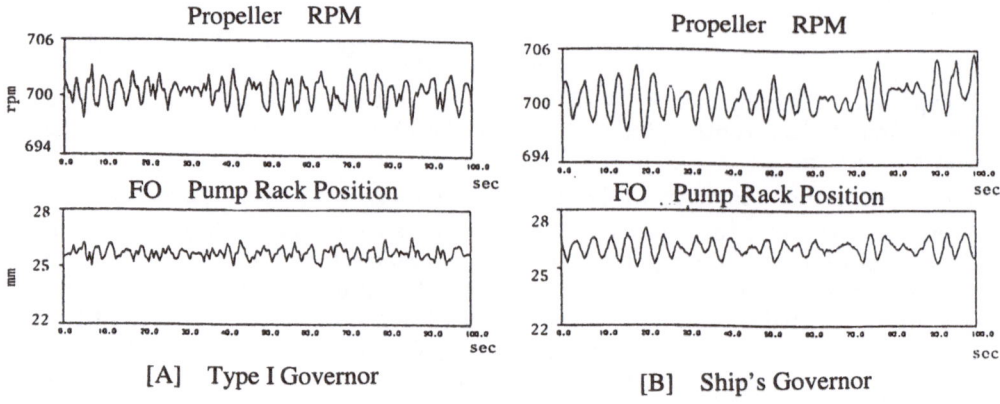

Figure 4.16 Actual sea test of the optimal governor [A] and conventional governor [B] (upper side: propeller rotation frequency, lower side: governor amount)

References

Akaike, H. (1974), "A New Look at the Statistical Model Identification," *IEEE, Transactions on Automatic Control*, Vol. AC-19, 716–723.

Akaike, H. (1971), "Autoregressive Model Fitting for Control," *Annals of the Institute of Statistical Mathematics*, Vol. 23, 163–180.

Akaike, H. and Nakagawa, T. (1988), *Statistical Analysis and control of Dynamic System*, KTK scientific Publishers, Tokyo.

Ishizuka, M. Ohtsu, K., Hotta, T. and Horigome, M., (1991, 1992), "Statistical Identification and Optimal Control of Marine Engine System, Part 1, Part 2," *Journal of the Society of Naval Architects of Japan*, Vol. 170, 211–220, Vol. 171, 425–433.

Isobe, T, (ed.) (1960), *Cavariance Function and Spectrum*, Tokyo University Press.

Oda, H., Ohtsu, K., Sasaki, M., Seki, Y. and Hotta, T., (1991), "Roll Stabilization by Rudder Control Through Multi-variate Auto-Regressive Model," *Journal of Kansai Society of Naval Architects*, Vol. 216, 165–173.

Otsu, K., Kitagawa, G. and Horigome, M. (1976a, 1976b), "Statistical Identification of Ship's Course Keeping Motion and Optimal Control," *Journal of the Society of Naval Architects of Japan*, Vol. 139, 31–44, Vol. 143, 216–224.

Ohtsu, K., M. Horigome and K. Kitagawa (1979), "A New Ship's Auto Pilot through a Stochastic Model," *Automatica*, Vol. 15, No. 3, 255–268.

The author is grateful to cooperative researches of this project, especially to Dr. Genshiro Kitagawa of the Institute of Statistical Mathematics. He also thanks the crews of Shioji-Maru II and III of Tokyo University of Marine who helped him in actual sea tests for many years and Mrs. Michiko Oda and Mr. Masatoshi Kurokawa who helped him in drawing figures.

Ohtsu, K., M. Horigome and K. Kitagawa (1979), "A Robust Autopilot System against the Various Sea Conditions," *Proceedings of ISSOA Symposium*, Tokyo, 118–123.

Ohtsu, K. (1983), "The Study on a Statistical Optimal Control of Ship's Motion – Part 1", *Journal of the Society of Naval Architects of Japan*, Vol. 152, 216–228.

Ozaki, T. and H. Tong (1978), "On the Fitting of Non Stationary Autoregressive Models in Time Series Analysis," *Proceedings of 8th Hawaii International Conference on System Science*, Western Periodical Company, 224–226.

Yamanouchi, Y. (1961a, 1961b), "On the Analysis of Ship Oscillations Among Waves – Part 1 & Part 2", *Journal of the Society of Naval Architects of Japan*, Vol. 109, 169–183, Vol. 110, 19–29.

Chapter 5

High Precision Estimation of Seismic Wave Arrival Times

Tetsuo Takanami
Institute of Seismological and Volcanology,
Graduate School of Science, Hokkaido University
ttaka@eos.hokudai.ac.jp

5.1 Introduction

When an earthquake occurs, vibrations propagate from the source in all directions. For large earthquakes, vibrations are soon observed even at sites far from the source. Generally, such a vibration is called a seismic wave. The seismic waves which propagate through the interior of the earth are comprised of two types: longitudinal P waves and transverse S waves. Additionally, surface waves (Rayleigh waves, Love waves, etc.) which propagate close to the surface of the earth also exist. These body waves and surface waves propagate with velocities which depend on the physical properties of the earth interior. Generally the velocity of the seismic wave increases with depth in the earth. Furthermore, the velocity of P waves is approximately 1.73 times than that of S waves, while the velocity of surface waves is 0.92 times that of the S waves. Therefore, depending on the distance of the observation point from the hypocenter and the depth of the hypocenter, there are differences in the arrival times of the individual waves. At present a table (travel time table) showing the relation between the distance, depth, and arrival times is available, and may be used to estimate the hypocenter. Observation of the arrival times at many observation points together with the use of the table allows the time and location of the earthquake to be estimated. Also, in contrast, a detailed velocity structures of the earth can be estimated by using many of the observation data of accurate arrival times. This results in the generation of a more accurate travel time table, thus more accurate hypocenter locations.

Precise determination of arrival times is difficult in practice because the ground on which seismometers are placed is incessantly in irregular vibration, produced by automobiles and other forms of transportation, and by various human activities such as noise-emitting factories, as well as by natural phenomena such as wind and wave.

Seismometers constantly record such kinds of ground vibrations. Arrival time identification is therefore very difficult for small and weak seismic waves. But one objective of the universities' high sensitivity seismic observation networks is the detection of changes in the activities of small earthquakes of magnitudes less than 3 since this may be important for earthquake prediction. It is necessary to determine any such change in seismic activities from very large data sets and to perform the determinations very rapidly.

The number of small earthquakes increases exponentially as the magnitude decreases. Therefore, to obtain earthquake information accurately and rapidly from the huge number of earthquake data recorded by micro-earthquake observation networks, real-time processing with the aid of computers is now being performed. From the viewpoint of earthquake prediction, it is necessary to have an efficient seismic wave identification algorithm for the purpose of obtaining information as rapidly as possible after the generation of the earthquake. Thus there has been continuous development of such an algorithm (e.g. Yokota et al. 1981, Takanami and Kitagawa 1988, Takanami and Kitagawa 1991).

The earthquake observation data are comprise a time series and an autoregressive model (AR model) is used to provide high-accuracy estimate of seismic wave arrival times in real time. The aftershock of the Hokkaido-Nansei-Oki Earthquake, which occurred on July 12, 1993 (Figure 5.1), is used to illustrate the method. Moreover, some description is given on the significance of the arrival times of P waves and S waves in the field of earth science: the apparent velocities of the P and S waves in the upper part of the crust in the vicinity of the hypocenter are hereby first obtained, and the Poisson's ratio indicating the elastic properties of the crust is estimated: it is then explained why these physical constants are fundamental data to be used with the other important physical constants.

Most of the time series of the seismic event recorded by a seismometer are nonstationary. The simplest method of modeling nonstationary time series is to divide a time interval into small enough sub-intervals such that the signal may be regarded stationary in the individual sub-intervals. Application of the AR model allows development of a model to express approximately the nonstationary time series. In this chapter, first an automatic method to determine the approximate arrival time of a seismic wave is given, and then a method to determine precisely the arrival times of P waves and S waves.

5.2 Locally Stationary AR Model

The time series $\{y_1, \ldots, y_N\}$ of the vibration recorded by a seismometer, even though it is not stationary, can usually be regarded as stationary provided that it is divided into adequately small sub-intervals. As described above, the time series with interval-wise stationary properties is called a locally stationary time series. Ozaki and Tong (1975), Kitagawa and Akaike (1978), and other researchers have established a concept of a locally stationary AR model, and have proposed a procedure to divide a time series of the recorded seismic event into several adequate AR models. Let hereunder the number of sub-intervals be k, and the number of data in the individual intervals be N_i ($N_1 + \cdots + N_k = N$). Where, the values of these of k or N_i are unknown. To estimate the locally stationary AR model, it is necessary to estimate the positions of the dividing points and the models for the individual intervals.

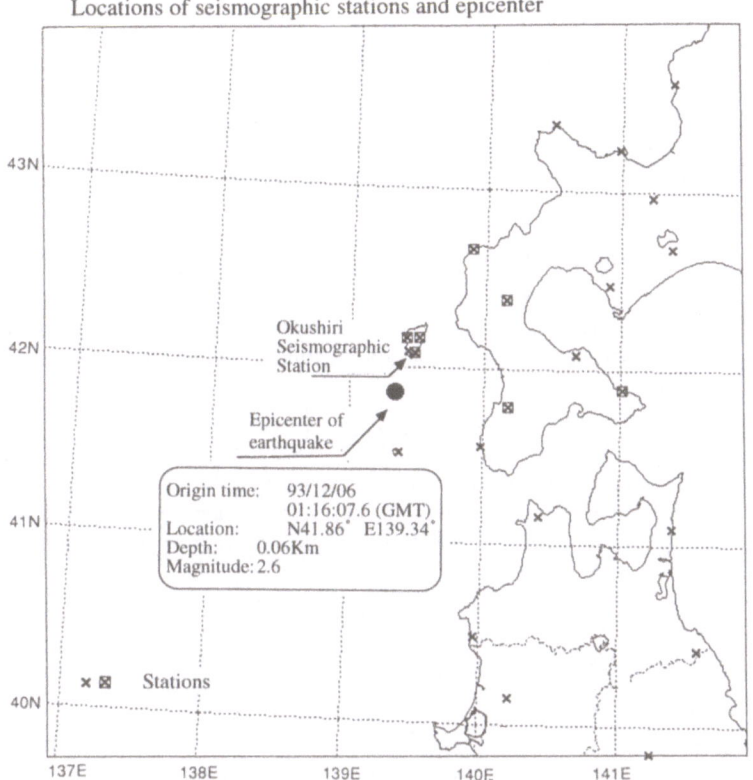

Locations of seismographic stations and epicenter

Figure 5.1 The epicenter of an earthquake and the seismic stations used for explanation. Solid circle and crosses are the locations of epicenter and stations, respectively. The source parameters are: origin time (12/6/1993 10h 16m 07.6s in JST), location (41.86N, 139.34E, depth=0.06km), and M=2.6. The mark of cross within a square means that the stations were used in the calculation for these hypocenter parameters.

Denote the i-th sub-interval as $[n_{i0}, n_{i1}]$, with starting point n_{i0} and ending point n_{i1} given by $n_{i0} = \sum_{j=1}^{i-1} N_j + 1$, and $n_{i1} = \sum_{j=1}^{i} N_j$. Suppose that the time series y_n is given by the AR model for in the j-th sub-interval

$$y_n = \sum_{i=1}^{m_j} a_{ji} y_{n-i} + v_{nj}, \tag{5.1}$$

where v_{nj} in the individual sub-intervals is white noise with the assumption that the expectations $E[v_{nj}] = 0$, $E[v_{nj}^2] = \sigma_j^2$, and $E[v_{nj} y_{n-m}] = 0$ are satisfied.

Supposing that $y_{n_{j0}-m_j}, \dots, y_{n_{j0}-1}$ are known in the interval j, the likelihood concerning $y_{n_{j0}}, \ldots, y_{n_{j1}}$ is given by

$$L_j(\theta) = p(y_{n0}, \ldots, y_{n1}) = \prod_{n=n_{j0}}^{n_{j1}} p_j(y_n | y_{n-m_j}, \ldots, y_{n-1}), \tag{5.2}$$

where $p_j(y_n | y_{n-m_j}, \ldots, y_{n-1})$ is the conditional distribution of y_n given $y_{n-m_j}, \ldots, y_{n-1}$, and is a density function of the normal distribution with the variance σ_j^2. Accordingly, the likelihood of the locally stationary AR model consisting of the number k of the intervals is given by

$$L(\theta) = \prod_{j=1}^{k} L_j(\theta) = \prod_{j=1}^{k} \prod_{n=n_{j0}}^{n_{j1}} p_j(y_n | y_1, \ldots, y_{n-1}). \tag{5.3}$$

Furthermore, ignoring the distribution of the first m_1 data, replacing N_1 with $N_1 - m_1$, and replacing n_{10} with $m_1 + 1$ gives the approximate likelihood of this model as

$$\prod_{j=1}^{k} \left(\frac{1}{2\pi\sigma_j^2} \right)^{N_j/2} \exp\left\{ -\frac{1}{2\sigma_j^2} \sum_{n=n_{j0}}^{n_{j1}} \left(y_n - \sum_{i=1}^{m_j} a_{ji} y_{n-i} \right)^2 \right\}. \tag{5.4}$$

If the above mentioned likelihood is considered to be a function of the number k of the sub-intervals, the length N_j of the sub-interval, the degree m_j, the autoregressive coefficient $a_j = (a_{j1}, \ldots, a_{jm_j})^t$, and the variance σ_j^2 of the white noise, then the log-likelihood is expressed as

$$\ell\left(k, N_j, m_j, a_j, \sigma_j^2; (j=1,\ldots,k)\right)$$

$$= -\frac{1}{2} \sum_{j=1}^{k} \left\{ N_j \log 2\pi\sigma_j^2 + \frac{1}{\sigma_j^2} \sum_{n=n_{j0}}^{n_{j1}} \left(y_n - \sum_{i=1}^{m_j} a_{ji} y_{n-i} \right)^2 \right\}. \tag{5.5}$$

The maximum likelihood $\hat{\sigma}_j^2$ of σ_j^2 for arbitrary a_j, is obtained as shown below by partially differentiating Equation(5.5) with respect to σ_j^2 set equal to 0.

$$\hat{\sigma}_j^2 = \frac{1}{N_j} \sum_{n=n_{j0}}^{n_{j1}} \left(y_n - \sum_{i=1}^{m_j} a_{ji} y_{n-i} \right)^2 \tag{5.6}$$

Furthermore, the substitution of (5.6) into (5.5) yields the log-likelihood of $a_{j1}, \ldots, a_{jm_j}$

$$\ell\left(k, N_j, m_j, a_j, \hat{\sigma}_j^2; (j=1,\ldots,k)\right)$$

$$= -\frac{1}{2} \sum_{j=1}^{k} \left(N_j \log 2\pi\hat{\sigma}_j^2 + N_j \right)$$

$$= -\frac{N - m_1}{2} (\log 2\pi + 1) - \frac{1}{2} \sum_{j=1}^{k} N_j \log \hat{\sigma}_j^2. \tag{5.7}$$

Hence the maximum likelihood estimates of $a_{j1}, \ldots, a_{jm_j}$ are obtained by minimizing σ_j^2 using of the method of least squares.

As the j-th AR model has m_j AR coefficients and the variance as the parameters, the AIC (Akaike Information Criterion) of the locally stationary AR model is given by

$$\text{AIC} = (N - m_1)(\log 2\pi + 1) + \sum_{j=1}^{k} N_j \log \hat{\sigma}_j^2 + 2 \sum_{j=1}^{k} (m_j + 1). \tag{5.8}$$

Thus the parameters, the number of the sub-intervals, k, length of the sub-intervals, N_j, and the order of the AR model m_j, are chosen so that AIC is minimum.

5.3 Automatic Division of a Locally Stationary Interval

In principle, the optimal locally stationary AR model can be estimated by employing both the method of least squares and AIC. However it is not realistic, as Kitagawa (1993) explains, to try the number k of intervals and length of the sub-interval $N_1, \ldots, N_k$ of the locally stationary model for all possible combinations in order to look for the factors which would allow AIC to be minimized, because an enormous amount of calculation would be required. In such a situation, it is practical to allow the dividing points of the locally stationary AR model to be automatically determined by designating in advance the minimum unit L of the division with $n_i = iL$ to be the dividing point. For example, it is considered how to determine efficiently and automatically, the dividing point between a locally stationary model prior to the arrival time of a seismic wave and a locally stationary model including a seismic wave. The procedure for this automatic determination is:

1) Let the length L of a sub-interval on which the stationarity of the series is assumed and the highest order m of the AR model to be fitted in the individual intervals be determined. Let L be of a length long enough to allow the AR model with order m to be fitted.

2) Let innovation variances $\hat{\sigma}_0^2(0), \ldots, \hat{\sigma}_0^2(m)$ be obtained by fitting the AR models of orders $0, \ldots, m$ to $y_1, \ldots, y_L$, and let $\text{AIC}_0(0), \ldots, \text{AIC}_0(m)$ be calculated by $\text{AIC}_0(j) = (L - m) \log \hat{\sigma}_0^2(j) + 2(j + 1)$. Moreover, define $\text{AIC}_0 = \min_j \text{AIC}_0(j)$, put $k = 1$, $n_{10} = m + 1$, $n_{11} = L$, $N_1 = L - m$.

3) By fitting AR models of orders $0, \ldots, m$ to $y_{n_{k1}+1}, \ldots, y_{n_{k1}+L}$, obtain the innovation variances $\hat{\sigma}_1^2(0), \ldots, \hat{\sigma}_1^2(m)$, and calculate $\text{AIC}_1(0), \ldots, \text{AIC}_1(m)$ by $\text{AIC}_1(j) = L \log \hat{\sigma}_1^2(j) + 2(j + 1)$. Define $\text{AIC}_1 = \min_j \text{AIC}_1(j)$. AIC_1 is the AIC of the new model under the assumption that the model is changed at the time $n_{k1} + 1$. AIC of the locally stationary AR model obtained by dividing the interval $[n_{k0}, n_{k1} + L]$ into 2 sub-intervals, $[n_{k0}, n_{k1}]$ and $[n_{k1} + 1, n_{k1} + L]$ is given by $\text{AIC}_D = \text{AIC}_0 + \text{AIC}_1$. This model is called a divided model.

4) By regarding $y_{n_{k0}}, \ldots, y_{n_{k1}+L}$ as a single interval and by fitting AR models of orders $0, \ldots, m$ and compute the innovation variances $\hat{\sigma}_p^2(0), \ldots, \hat{\sigma}_p^2(m)$. Let $\text{AIC}_P(0), \ldots, \text{AIC}_P(m)$ be calculated by $\text{AIC}_p(j) = (N_k + L) \log \hat{\sigma}_p^2(j) + 2(j + 1)$. Here, $\text{AIC}_P = \min_j \text{AIC}_P(j)$ is the AIC of the pooled model on the assumption that the interval $[n_{k0}, n_{k1}+L]$ is a single sub-interval without dividing this interval at the time $n_{k1} + 1$.

5) By comparing the values of AIC obtained in procedures 3) and 4), it is judged which model is better.

   (a) In the case $\text{AIC}_D < \text{AIC}_P$, the divided model is judged to be better. Thus $n_{k1} + 1$ becomes a new dividing point. Therefore, first of all, substitute it for $k \equiv k + 1$. Secondly, put $n_{k0} \equiv n_{k-1,1} + 1$, $n_{k1} = n_{k-1,1} + L$, $N_k = L$, $\text{AIC}_0 = \text{AIC}_D$. This is equivalent to the case in which seismic wave arrives in the interval $[n_{k0}, n_{k1} + L]$.

   (b) In the case $\text{AIC}_D \geq \text{AIC}_P$, a pooled model is preferred. Thus let a new sub-interval $[n_{k1} + 1, n_{k1} + L]$ be pooled with the original sub-interval to make $[n_{k0}, n_{k1} + L]$. Accordingly, put $n_{k1} \equiv n_{k1} + L$, $N_k = N_k + L$, $\text{AIC}_0 = \text{AIC}_P$. This is equivalent to the case of an interval in which no seismic wave has

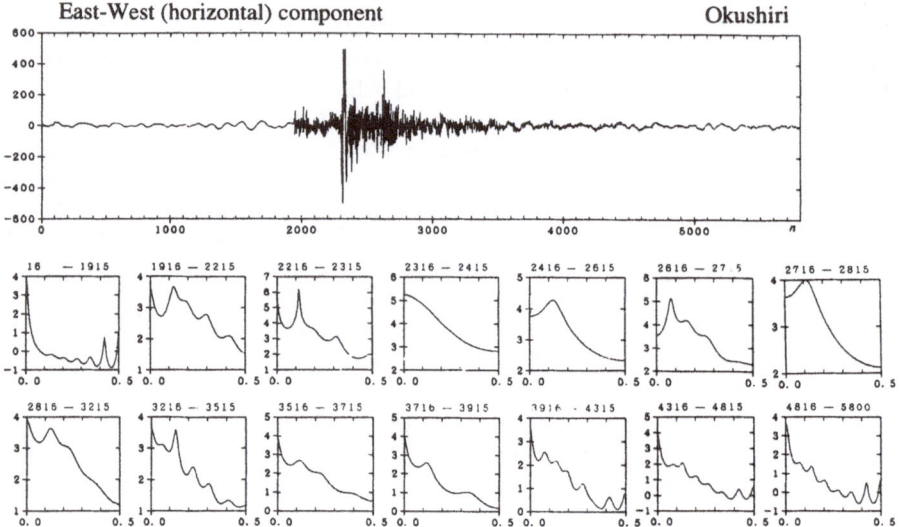

Figure 5.2 Record (east-west) of an aftershock of the 1993 Hokkaido-Nansei-Oki Earthquake (top), and power spectra for the locally stationary autoregressive models at sub-intervals (in the boxes). Numbers shown in above the spectra are the first and end points of these sub-intervals.

arrived or of an interval where a seismic wave has already arrived and a re-defined interval is established.

6) If more than L pieces of observations are available, then return to step 3), else, k is the dividing number, and $[1, n_{11}], [n_{20}, n_{21}], \ldots, [n_{k0}, N]$ give k pieces of the division.

• **Example** Figure 5.2 shows the aftershock record (the east-west motion component, $N = 5800$, at the Okushiri observation site) of the Hokkaido-Nansei-Oki Earthquake which occurred on July 12, 1993, and the results obtained by fitting the locally stationary AR model under the conditions that $L = 100$, $m = 15$. The record includes a noise part called microtremor and 3 types of waves, i.e. the P wave, S wave and converted wave. In the individual plots below the aftershock record, the place of the sub-intervals regarded as stationary by the locally stationary AR model and the power spectra estimated for those intervals are shown.

The change is detected at 13 locations, $n = 1915$, 2215, 2315, 2415, 2615, 2715, 2815, 3215, 3515, 3715, 3915, 4315, and 4815. The change before and after $n = 1900$ corresponds to the change of the variance and spectrum caused by the arrival of the P wave. While the interval of $n = 1900, \ldots, 2300$ is composed of the P wave and its coda part, the interval of $n = 2000, \ldots, 2200$ in the vicinity of the P wave initial motion forms a relatively smooth spectrum distribution. Here it is apparent that a single periodic component dominates in $n = 2200, \ldots, 2300$ in the 2nd half. Although a portion of the S wave is noted in $n = 2200, \ldots, 2300$, it is revealed that not only is the power spectrum lowered as the amplitude of seismic signal is decreased but also the position of the principal peak of the spectrum is shifted from low frequency to high frequency side. A converted wave, suggesting a velocity discontinuity in the crustal structure, is seen in the vicinity of $n = 2600, \ldots, 2700$, and gives a change of spectrum. This is believed to be because the earthquake occurred in an exceedingly shallow place, and therefore the surface wave was superposed on the coda part of the S wave. From that time, the spectrum undergoes complicated change with lower power. After $n = 4800$, there is almost no change.

5.4 Precision Estimation of Seismic Wave Arrival Times

In the previous section, a method was given which allows the nonstationary time series divided automatically into a number of intervals, each of which can be regarded as stationary. Now, on the assumption that the arrival of a seismic wave is known to be in an interval $[n_0, n_1]$, a method to estimate precisely the arrival time is considered in this section. On the assumption, that a seismic wave has arrived at time n, satisfying $n_0 \leq n \leq n_1$, AR models are fitted to two intervals $[1, n-1]$ and $[n, N]$. By calculating the sums of the two AIC's, the goodness of the model (an arrival time n) can be evaluated. Hence to make a precision determination of the arrival time by making use of a locally stationary AR model, the minimum AIC is determined by calculating for all n of $n_0 \leq n \leq n_1$.

Suppose, first, that the residual variances obtained by fitting the AR models of orders $0, \ldots, m$ to $y_1, \ldots, y_{n_0}$ by the method of least squares are $\hat{\sigma}_0^2(0), \ldots, \hat{\sigma}_0^2(m)$. AIC's of these models can be calculated by

$$\mathrm{AIC}_0(j) = (n_0 - m) \log \sigma_0^2(j) + 2(j + 1). \tag{5.9}$$

Therefore, the AIC of the best AR model with order j in the first half under the assumption that a new wave arrives at the time $n_0 + 1$,

$$\mathrm{AIC}_0^1 \equiv \min_j \mathrm{AIC}_0(j). \tag{5.10}$$

Next if the AR models are fitted to $n_0 + 1$ observations $y_1, \ldots, y_{n_0+1}$ and likewise the residual variances $\hat{\sigma}_1^2(0), \ldots, \hat{\sigma}_1^2(m)$ are calculated, then AIC of the AR model with order j fitted to $y_1, \ldots, y_{n_0+1}$ can be calculated by

$$\mathrm{AIC}_1(j) = (n_0 - m + 1) \log \sigma_1^2(j) + 2(j + 1). \tag{5.11}$$

Therefore, the AIC of the best AR model with order j in the first half on the assumption that a new wave arrives at the time $n_0 + 2$ is given by

$$\mathrm{AIC}_1^1 \equiv \min_j \mathrm{AIC}_1(j). \tag{5.12}$$

By repeating likewise, AIC's of the AR models fitted to $\{y_1, \ldots, y_{n_0}\}$, $\{y_1, \ldots, y_{n_0+1}\}$, $\ldots, \{y_1, \ldots, y_{n_1}\}$, i.e., $\mathrm{AIC}_0^1, \mathrm{AIC}_1^1, \ldots, \mathrm{AIC}_\ell^1$, can be calculated. Here $\ell = n_1 - n_0$.

Likewise, let the AR models be fitted to $y_{n_1+1}, \ldots, y_N$. Then by adding successively the data $y_{n_1}, y_{n_1-1}, \ldots, y_{n_0+1}$. The AIC's of the model in the second half, i.e., $\mathrm{AIC}_\ell^2, \mathrm{AIC}_{\ell-1}^2, \ldots, \mathrm{AIC}_0^2$, can be obtained.

Then

$$\mathrm{AIC}_j = \mathrm{AIC}_j^1 + \mathrm{AIC}_j^2 \tag{5.13}$$

expresses the AIC of the locally stationary AR model on the assumption that a seismic wave arrived at the time $n_0 + j + 1$. Therefore, by looking for the minimal value among $AIC_0, \ldots, AIC_\ell$, the optimal time interval is obtained. This optimal dividing point is equivalent to the arrival time of the seismic wave.

In the above, a large amount of computation is required if the least squares is applied every time the number of the data and the order of the models are changed. However by using the Householder method algorithm (Kitagawa, 1993) for least squares, the calculation can be made successively with extreme efficiency.

• **Example**   Figure 5.3(a)–(h) gives the results obtained by examining precisely the changing AIC value in the vicinity of $n = 1950$ and $n = 2300$ where large changes are seen in Figure 5.2. (b) shows the magnified part of $n = 1900, \ldots, 2000$ in the vertical component record, which contains the P wave arrival, (a) shows the values of AIC obtained by (5.13). The minimum value is achieved at $n = 1941$, indicating that the P wave arrives at that time.

(d) shows the magnified part of $n = 2250, \ldots, 2350$ in the same vertical component seismograph record. The first half contains the coda part of the P wave, whereas the second half contains the S wave. From the values of AIC of (c), it can be deduced that the S wave arrived by $n = 2290, \ldots, 2305$. However, while it is shown in the case of (a) that the change of AIC is quite sharp and estimation accuracy is excellent, it is clear in (c) that the change is relatively slow and detection of the S wave onset is more difficult than for the P wave. (f) and (h) show the magnified diagram for the east-west and north-south horizontal components in the same interval as (d), and the coda part (succeeding waves such as scattered waves produced by the inhomogeneity crust, multi-reflecting waves, etc.) of the P wave together with the S wave is depicted. From the values of AIC's of (e) and (g), it can be inferred that the S wave has arrived at $n = 2299$ or $n = 2300$. (e) and (g) of the horizontal motion estimate the arrival time of the S wave with better accuracy than with that of (c) of the vertical motion.

As can also be estimated from the wave theory, the above description implies that it is preferable to determine the arrival time of the P wave from the vertical-motion component and that of the S wave from the horizontal-motion component. In the vicinity of the arrival time of the real S wave, which is superposed on the coda part of the P wave, there exist in a mixed style several conversion waves due to the complicated crustal structure. In such a case, use of multi-variate time series of the three components (one vertical and two horizontal directions) enables the determination of accurate arrival times. For a complicated S wave, Takanami and Kitagawa (1991) developed an estimation method of the arrival time by a 3-component autoregressive model.

5.5   Application: Earthquake Location and Velocity Structure Determination from Precise Arrival Time Estimates

The precise determination method of the arrival time of a seismic wave, described above, was applied to a real seismic wave to estimate the hypocenter position from the arrival times of the P and S waves determined for the individual earthquake observation points. The result of the estimation is shown in a small black circle in Figure 5.1. This earthquake was one of the aftershocks occurred at the southernmost

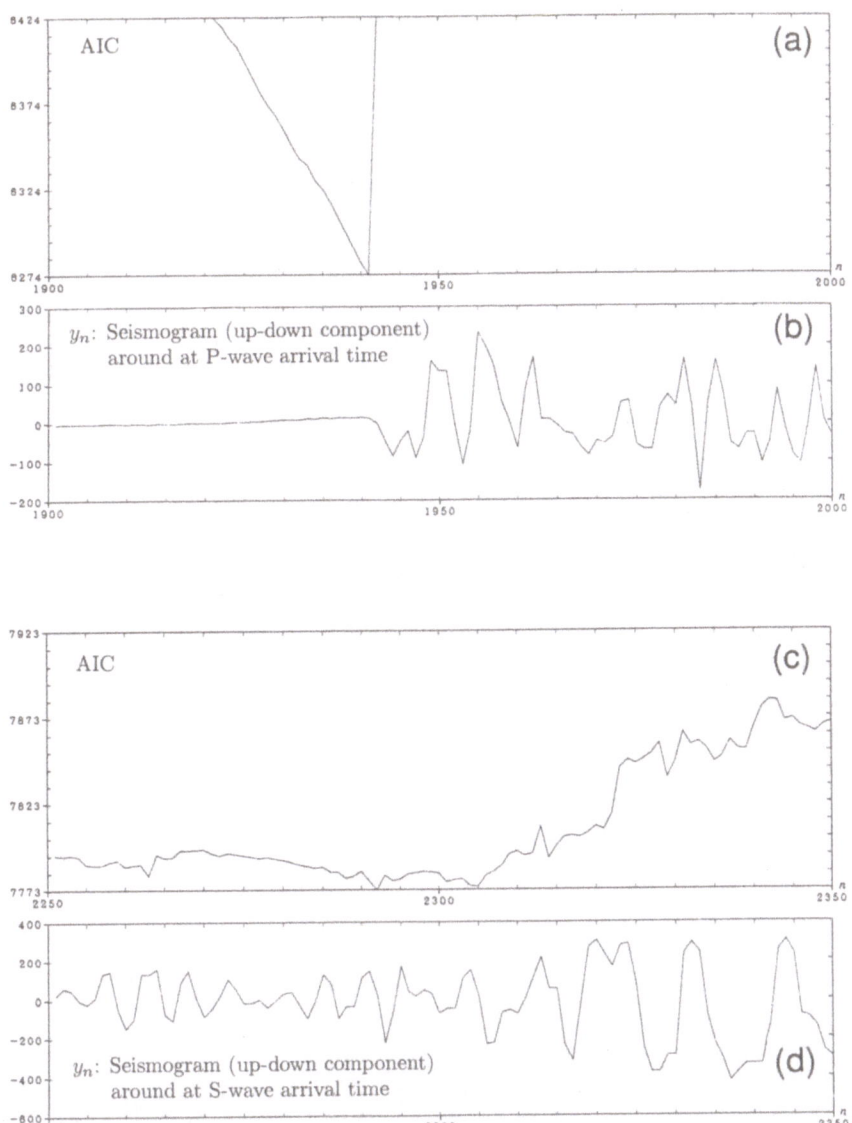

Figure 5.3 Determination of onset times for P and S waves by the AIC curve. (a) and (b) are the AIC curve and the seismogram of vertical component near the onset time of P wave, respectively. (c) and (d) are the AIC curve and seismogram of vertical component near the onset of S wave.

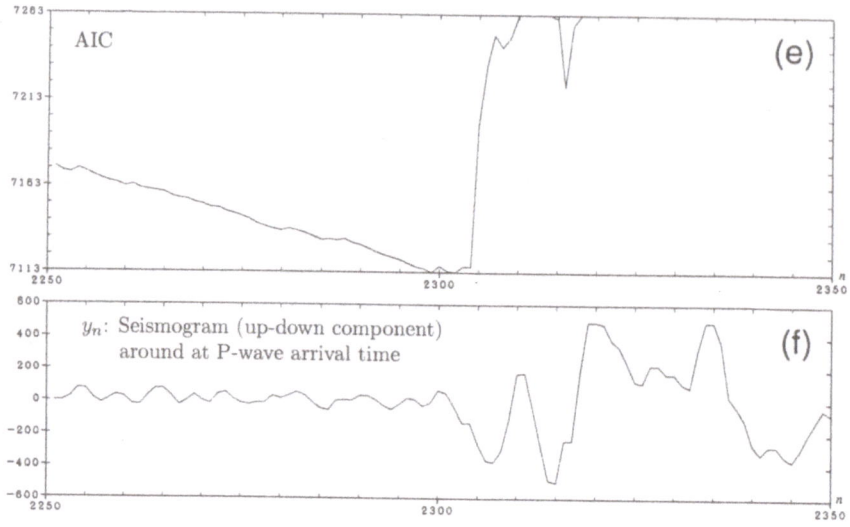

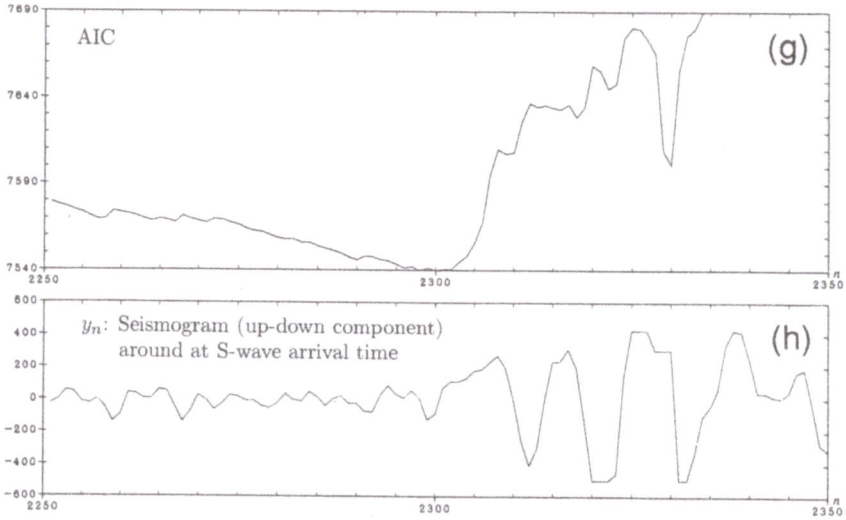

Figure 5.3 Determination of onset times for P and S phases by the AIC curve. (e) and (f), (g) and (h) are the AIC curves and seismograms for east-west and north-south components, respectively **(continued)**.

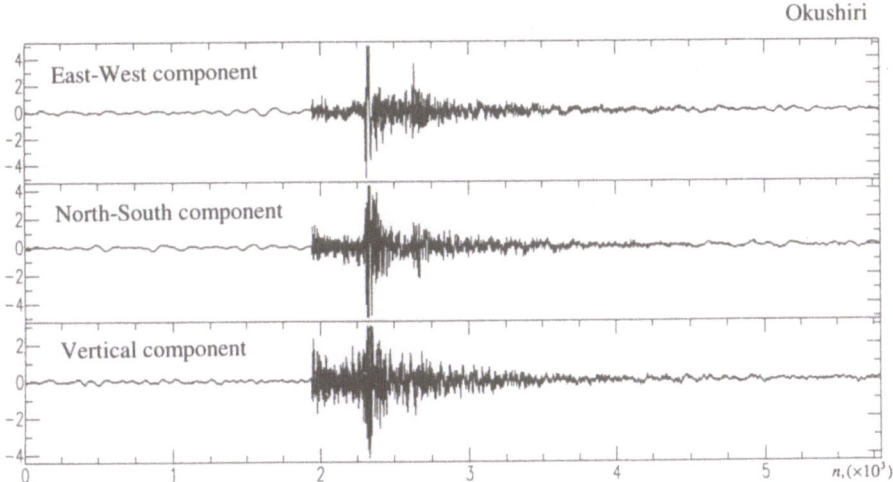

Figure 5.4 An example of 3-component seismograms recorded at Okushiri observation site.

end of the aftershock area in the Hokkaido-Nansei-Oki Earthquake of July 12, 1993; its magnitude and depth were found to be 2.6 and 0.06 km, respectively. This data reveal that the earthquake in question was a small earthquake that occurred almost at the surface of the earth. In the case of such a small earthquake, the seismic energy radiated by the source is very small, and the waveform becomes, in general, very complicated at the observation site being influenced by the shallow crustal structure.

In the figure, the positions of the observation sites (the earthquake observation sites of Hokkaido University, Tohoku University, and Hirosaki University) are indicated by × marks; those for which the arrival time of the P wave was used for the hypocenter calculation are located inside a square. That subset is restricted to Okushiri Island and to within the Oshima Peninsula, since the S/N is very low at the other observation sites, preventing them from being used for the hypocenter calculation. The epicentral distance for the Okushiri observation site (Aonae, Okushiri-cho), the nearest observation site to the hypocenter, was found to be 27.5km.

• **3-component records**   Figure 5.4 shows the 3-component records of the east-west, north-south, and vertical motions obtained at the Okushiri observation site. This record was used in the preceding pages to illustrate the method of high precision estimation of seismic wave arrival times for an actual event. The 3-component records also make clear that the arrival times can be obtained more accurately by analyzing the vertical component for the P wave and the horizontal component for the S wave. Also, another clear wave group is seen in the coda of the S wave, suggesting discontinuities of the velocity structure of the crust. Consequently, the precision arrival time method can also be utilized to determine the arrival times of succeeding waves, and additionally provide greater detail of the structure of the earth by revealing velocity anomalies.

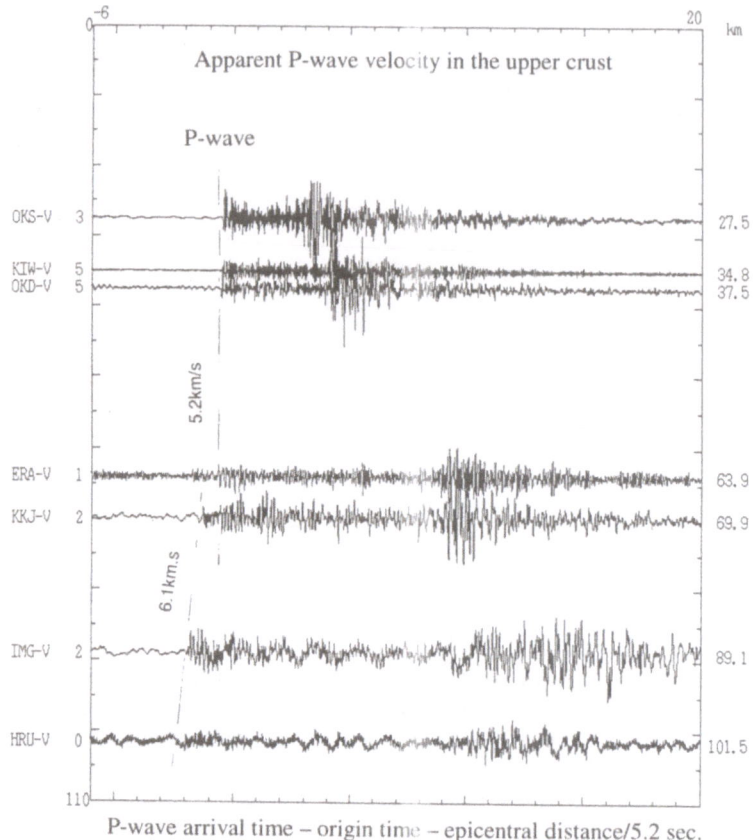

Figure 5.5 Determination of apparent P-velocity by using the onset times of P-waves. Seismograms (vertical component) at the stations within the epicentral distances from 0 to 110 km. The axis of abscissa is that of the reduced travel times by the To−O.T−(Δ/5.2), where To is the observed onset time for the first arrival of P wave, O.T is the origin time, Δ is the epicentral distance, and 5.2 is the apparent velocity for the first arrival P-waves. The time ranges from −6 to 20 sec.

● **Travel-time analysis of the P wave**   Figure 5.5 shows the waveforms of the vertical components for the sites aligned from top to bottom by epicentral distance. Going on the assumption that, in general, the velocity distribution of the medium through which the P wave propagates is homogeneous, the arrival time of the P wave as a function of epicentral distance is obtained. The time axis for each trace is shifted by the epicentral distance divided by a propagation velocity, so that the wave arrives at almost the same time on all traces. In reality, the velocity is determined by trial and error, viewing the figure on the CRT in the hope that the arrival time will be as close to the same time at many of the observation sites. Furthermore, by subtracting the origin time of the earthquake obtained by the hypocenter calculations from those arrival times, the first arrivals of the P wave can be aligned at almost the same point on the time axis.

Thanks to this simple expression method, it is possible to magnify a restricted time range so that attention may be paid to the vicinity of the first arrival of the seismic wave. Furthermore, from the pattern of the alignment of the first arrival, it can be determined whether the velocity distribution in the region of the individual observation sites is homogeneous or not. By the operation of this waveform expression method, the existence of 2 layers is apparent; one layer comprised of the apparent velocity $V_{P1} = 5.2$km/sec for the P wave in the upper crust and the second layer, located beneath the first layer, having a velocity $V_{P2} = 6.1$km/sec. V_{P1} of the first layer coincides with a $V_{P1} = 5.2$km/sec obtained from the seismic earthquake observation of quarry blasting conducted in the northern part of the Oshima Peninsula (Takanami et al, 1991).

• **Travel-time analysis of the S wave**   Figure 5.6 shows the waveforms of the S waves with a similar time-axis shift (Figure 5.5). The figure also indicates the existence of two layers, comprised of the apparent velocities $V_{S1} = 3.0$km/sec and $V_{S2} = 3.6$km/sec. In both cases observation sites with larger epicentral distances are used to determine the velocity of the second layer. Larger earthquakes give better signals over larger distance ranges and thus are more useful for determining the deeper velocity structure. Nevertheless this small earthquake allowed the velocity of the second layer to be estimated. For earthquake records with very low S/N, it is preferable to use a waveform separation method based on a state space model developed by Kitagawa and Takanami (1985). Here, the Poisson's ratio σ_i of each layer, calculated from the apparent velocity of the two pairs of the P waves and S waves is 0.25 and 0.23, respectively. Poisson's ratio is determined from the velocity ratio of the P wave and S wave by

$$\frac{V_{Pi}}{V_{Si}} = \sqrt{\frac{2(1 - \sigma_i)}{1 - 2\sigma_i}}, \qquad (i = 1, 2). \tag{5.14}$$

Furthermore, from these physical constants, the following have been obtained: density $\rho_1 = 2.5$g/cm^3, $\rho_2 = 2.7$g/cm^3 (The Society of Exploration Geographics of Japan 1989), bulk modulus $K_1 = 3.76 \times 10^{11}$dyne · cm^{-2}, $K_2 = 5.4 \times 10^{11}$dyne · cm^{-2}. However, there is a possibility that the second layer velocities, which are obtained under the assumption of flat-layered structure for the sake of convenience, might be different from the true values. Despite the present uncertainly, more accurate values can be estimated also for deeper layers by the tomography method using earthquake arrival times at many observation sites.

From the physical constants of the first layer, existence of rocks equivalent to limestone or granite is suggested. This is consistent with evidence that such rocks occur in the Paleozoic/Mesozoic era Strata, (Committee for Regional Geology of Japan, [Nihon'no Chishitsu, 1990]) found in the region.

Under investigation, significant effort has been devoted to develop various systems capable of rapidly obtaining accurate information about hypocenters for the field of earthquake prediction. Today, automated systems using automatic detection method for seismic wave arrival times by means of autoregressive models are in widespread use.

As demonstrated above, from the specific examples, it is shown that the arrival times of seismic waves provide very important information for obtaining the physical constants for the interior of the earth, in addition to maintaining a vigilant and

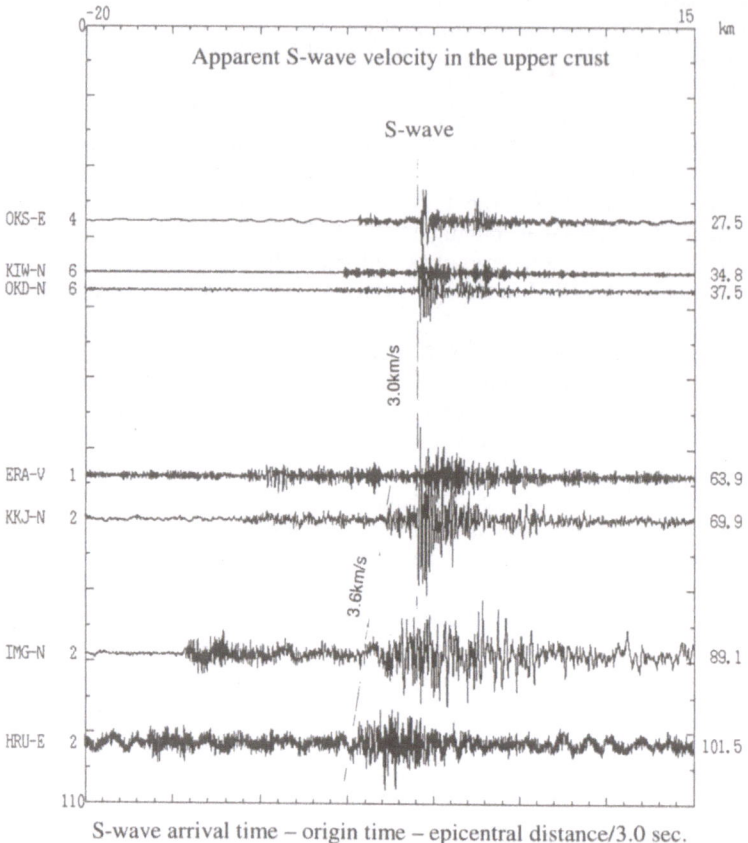

Figure 5.6 Determination of apparent S-velocity by using the onset times of S-waves. Seismograms (horizontal component) at the stations within the epicentral distances from 0 to 110 km. The axis of abscissa is that of the reduced travel times by the To–O.T–($\Delta/3.0$), where To is the observed onset time for the first arrival S wave, O.T. is the origin time, Δ is the epicentral distance, and 3.0 is the apparent velocity for the first arrival S-waves. The time ranges from -20 to 15 sec.

increasingly sharp eye on the detection of changes in earthquake activities. The key to success for a practical application lies in the use of Akaike Information Criterion, as a rational statistical expression of seismic waves, a very important pioneering role in introducing an autoregressive model coupled with a locally stationary modeling.

5.6   Conclusions

This chapter demonstrates a locally stationary autoregressive model that can be applied to the high precise determination of the arrival times of P and S waves from an earthquake. Details of the algorithm that can be applied to real-time processing with great accuracy are explained. Accurate arrival times are important both for determining the wave velocity structure of the earth and for accurate earthquake locations, changes in which may be significant for the earthquake prediction problems.

References

Akaike, H. and T. Nakagawa (1972), Statistical analysis and control of dynamic system, Saiensu-sha, Tokyo (in Japanese, with the list of TIMSAC in a FORTRAN IV type language).

Akaike, H. (1973), "Information theory and an extension of the maximum likelihood principle," *2nd International Symposium on Information Theory* (eds. B. N. Petrov and F. Csaki), 267–281.

Akaike, H. (1976), "What is Akaike's information criterion AIC?", Mathematical Sciences, No. 153, 51–57, Saiensu-sha, Tokyo (in Japanese).

Akaike, H. (1979), "A new look at the statistical model indentification," *IEEE Transactions on Automatic Control*, Vol. AC-19, No. 6, 716–723.

Akaike, H. *et al.* (1979), TIMSAC-78, *Computer Science Monographs*, No. 11, The Institute of Statistical Mathematics.

Committee for Regional Geology of Japan (1990), Regional Geology of Japan, Part 1 Hokkaido, 337pp, Kyoritsu Shuppan, Tokyo, (in Japanese).

Kitagawa, G. and H. Akaike (1978), "Procedure for the modeling of non-stationary time series," *Annals of the Institute of Statistical Mathematics*, Vol. 30, 351–363.

Kitagawa, G. (1983), "Changing spectrum estimation," *Journal of Sound Vibration*, Vol. 89, 433–445.

Kitagawa, G. and T. Takanami (1985), "Extraction of signal by a time series model and screening out micro earthquakes," *Signal Processing*, Vol. 8, 303–314.

Kitagawa, G. (1993), FORTRAN 77 time series analysis programming, The Iwanami Computer Series, 390 pp, Iwanami Shoten in Japan, (in Japanese).

Ozaki, T. and H. Tong (1975), "On the fitting of non-stationary autoregressive models in the time series analysis," *Proceedings of the 8th Hawaii International Conference on System Science*, Western Periodical Hawaii, 224–226.

Sakamoto, Y., M. Ishiguro and G. Kitagawa (1983), Information statistics, Joho-Kagaku-Kouza A.5.4. 337pp, Kyoritsu Shuppan, (in Japanese).

Society of Exploration Geophysicist of Japan (1989), An explanatory diagram: Geophysical exploration, 240pp, (in Japanese).

Takanami, T. and G. Kitagawa (1988), "A new efficient procedure for the estimation of onset times of seismic waves," *Journal of Physics of the Earth*, Vol. 36, 267–290.

Takanami, T. (1991), "A study of detection and extraction methods for microearthquake waves by autoregressive models", *Journal of the Faculty of Science*, Hokkaido University, Series VII (Geophysics), Vol. 9, No. 1, 67–196.

Takanami, T. and G. Kitagawa (1991), "Estimation of the arrival times of seismic waves by multivariate time series model," *Annals of the Institute of Statistical Mathematics*, Vol. 43, 403–433.

Takanami, T. and M. Yamauchi (1996), "Shallow crustal structure derived from explosion observations by using quarry blasts in the southernmost part of the Oshima peninsula, Hokkaido, Japan," *Geophysical Bulletin of Hokkaido University*, No. 59, 189–209, (in Japanese with English abstract).

Yokota, T., S. Zhou, M. Mizoue, and I. Nakamura (1981), "An automatic measurement of arrival time of seismic waves and its application to an on-line processing system", *Bulletin of the Earthquake Research Institute*, University of Tokyo, Vol. 55, 449–484, (in Japanese with English abstract).

Chapter 6

Analysis of Dynamic Characteristics of a Driver-Vehicle System

Hitoshi Soma
Japan Automobile Research Institute
2530 Karima, Tsukuba, Ibaraki 305-0822, Japan
hsoma@jari.or.jp

6.1 Introduction

An automobile can never be driven into motion until the driver manipulates an accelerator pedal, brake pedal or steering wheel. Therefore when we analyze the motion characteristics of an automobile, it becomes very important not only just to pay attention to the vehicle itself in its operation characteristics but also to identify it as a driver-vehicle system including the driver's operation characteristics.

A variety of methods for analyzing the driver-vehicle system are available, but one of the classical methods is to obtain the transfer function of the driver and/or the automobile. The transfer function gives a large amount of useful information such as transient response, stationary response, frequency response, etc. of the driver or the automobile. From these information, the optimum design of automobiles that are easy for drivers to drive and are safer than the conventional ones becomes possible. Besides, such information is quite effective in examining the properties of the advanced control systems such as 4-wheel steering, traction control, anti-lock brake, active suspension, etc. that are vigorously being developed in recent years.

However it is not easy to obtain a transfer function of a driver or automobile from the data observed at an automobile under actual motion. While the dynamics of the machines can be analyzed by an impulse response test by adding arbitrarily designed impulsive noise, it is in general impossible to take such a method for the driver-vehicle system. This is because that man as a major element of the system can determine the desired path in accordance with his own will and therefore he does not necessarily respond to the input force from the outside. On the other hand since he changes his own characteristics as freely as possible responding to the environment or affairs, the affairs in a limited time play a role of stimulus for the driver to cause a certain steering action.

Figure 6.1 Lateral-wind generator

Therefore in the analysis of driver-vehicle system, it is necessary to find out the signal that acts as the input under certain affairs in a short time, and a transfer function should be estimated from the input signal and the measurable output signal using some time series analysis method.

An example of analyzing the characteristics of the driver-vehicle system by means of a transfer function is shown in this chapter. The dynamics of a driver's steering behavior and the dynamics concerning lateral and yaw motion of an automobile under lateral-wind disturbance are analyzed by the time series analysis method based on a multivariate AR model (hereafter called the AR method). An automobile is often subjected to a lateral wind while it is running especially on a highway or in an elevated road. This is an important problem that drivers actually encounter.

6.2 Dynamics of Automobile under Lateral-Wind Disturbance

Research on the application of an AR model to vehicle motion especially to the active safety, has hardly been available. Complying with such a situation, dynamics identification of the (open-loop) vehicle system using the AR method is described, prior to the identification of the (closed-loop) driver-vehicle system. An automobile affected by lateral-wind disturbance is specifically discussed.

6.2.1 Lateral-Wind Test with Fixed Steering

Test Method   The lateral-wind test was conducted as shown below following JASO-Z108 (Test Procedure of Crosswind Stability for Passenger Car 1976). First of all, let lateral wind be generated in a direction perpendicular to the running course by the lateral-wind generator (Figure 6.1). Secondly, let an automobile proceed along the running course perpendicularly in the direction of the wind velocity. The driver of the automobile let his car to pass through the blowing-zone at a specified velocity with fixed steering (Figure 6.2), under the condition as follows: Wind velocity; 15m/s

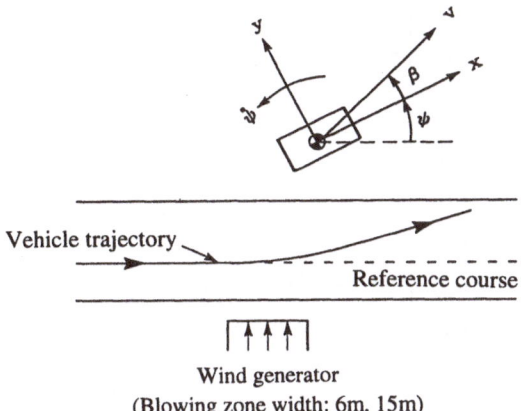

Vehicle trajectory

Reference course

Wind generator
(Blowing zone width: 6m, 15m)

Figure 6.2   Test course and coordinate system in lateral wind test, x: longitudinal axis of vehicle, y: lateral axis of vehicle, v: velocity, ψ: direction of vehicle (yaw angle), $\dot{\psi}$: directional angular velocity (yaw velocity), β: side slip angle

and 22.5m/s. Blowing zone width: 6m and 15m. Vehicle velocity: 40, 60, 80, and 100km/h. Driving were repeated 4 to 6 times under each condition. The automobile used for the test was a one box car relatively vulnerable to the influence of lateral wind.

Wind velocity, lateral acceleration, directional angular velocity (yaw velocity) and vehicle velocity are measured. The wind velocity was measured using a 2-dimensional ultrasonic anemometer. In order not only to suppress the measuring errors caused by reflection of the ultrasonic wave from the car body but also to minimize the influence of the flow in the vicinity of the car body, the ultrasonic anemometer was installed on the car in a style protruded approximately 1.5m from the front of the car. In this connection, it was confirmed that attachment of the anemometer exercises does not have any influence on the lateral-wind characteristics of the vehicle, because lateral deviations transfer distance of the center of gravity do not depend on whether the ultrasonic anemometer was equipped or not. The lateral acceleration is the acceleration component in the lateral direction of the car body and was measured by accelerometer. The directional angular velocity is the rotating velocity around the vertical axis of the car body and was measured by rate-gyro.

Features of the Measured Signals   Figure 6.3 shows the signals measured in case of the wind velocity 15m/s and blowing-zone width 15m, and automobile velocity 80km/h. The time span where the vehicle passes through the blowing-zone, i.e. the time from the rise of the lateral-wind velocity to the time when the wind velocity returns to almost 0, is approximately 0.7s. Besides, the time span between the generation of the response of the lateral acceleration and the settling point where the response almost converged is 2s. Throughout the test, the sampling time is about 10ms, the numbers of the effective data points for the lateral-wind velocity and the lateral acceleration are about 70 and 200, respectively.

When identification is made by the conventional method using the data with short

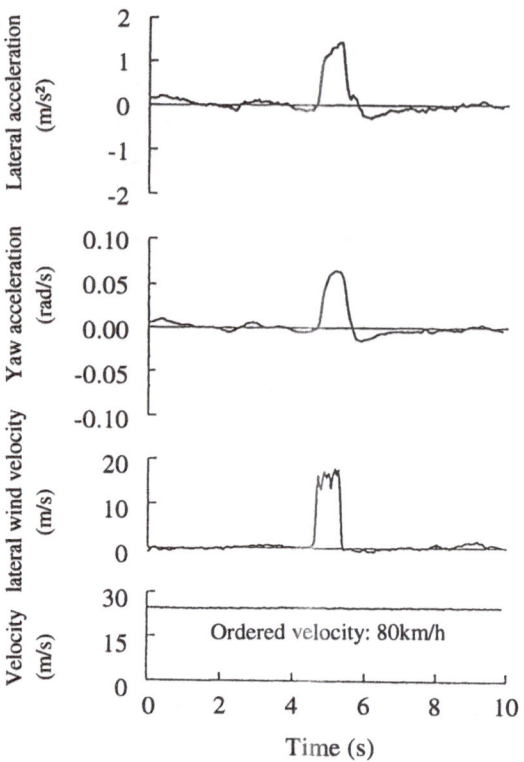

Figure 6.3 Time series data in lateral wind test (fixed steering, wind velocity: 15m/s, blowing zone width: 15m)

effective measurement time the frequency resolution often becomes large. On the contrary, it is reported that sufficient resolution is obtained (Suzuki *et al.* 1982) even with very few data points when the AR method is used, and the method is effective for the analysis of the vehicle dynamics under the lateral-wind disturbance. Furthermore although the AR method is developed exclusively for a stationary time series, it is known that the method gives effective analytical results even when the observed data are nonstationary provided that a relation between the variables is stationary. This study is expected to give a typical example of such an analysis.

6.2.2 Analytical Procedure of the AR Method

We shall present here a method to identify the vehicle dynamics under lateral wind by applying a multivariable AR model,

$$x_n = \sum_{i=1}^{m} A_i x_{n-i} + \varepsilon_n,$$

where x_n is a vector time series and A_i is the AR coefficient matrix. The component of the time series vector is defined by

$$x = [\, \theta v_r^2, \, \ddot{y}, \, \dot{\psi} \,]^T,$$

where

θ :   angle formed by the resultant wind velocity and the longitudinal direction of the car body,

v_r: resultant wind velocity $(= \sqrt{w_v^2 + v^2},\ w_v$: wind velocity, v : vehicle velocity),

$\ddot{y}$:   lateral acceleration,

$\dot{\psi}$:   directional angular velocity.

The input signal is not just wind velocity but θv_r^2. This can be explained from the fact that θv_r^2 can be regarded as pseudo-aerodynamic force based on the following reasons: (1) The aerodynamic force applied onto the automobile is increased in almost proportion to the increase of v_r^2. (2) The value of the aerodynamic coefficient (an index indicating how much the force or moment is exerted from the flow of air, and the larger the value is, the greater the force or moment is) is increased as the value of θ is increased. (Strictly speaking, θ is increased almost between 0 and 40°.)

Hereafter a multivariate AR model is identified, and then the spectrum density function and the transfer function are obtained from the estimated AR model by the well-known procedure (Akaike and Nakagawa 1988). The transfer function obtained here is $\ddot{y}(s)/\theta v_r^2(s)$ and $\dot{\psi}(s)/\theta v_r^2(s)$, where the order m of the AR model is selected by minimizing the FPE (the final prediction error) criterion or equivalently the AIC criterion.

The data span for the analysis is ranging from 3s to 5s including the straightly running part before and after the blowing-zone. The sampling interval of the data is 10ms, and the data are pre-processed by passing through a low-pass filter with cut-off frequency 20Hz.

6.2.3   Identification Result of Dynamics

Comparison between the AR and the FFT Methods   In Figure 6.4, transfer functions respectively identified by the AR method and the FFT method are shown. The result of the calculation (dotted line) obtained from the transfer function based on a single track 2-degrees-of-freedom model (Abe 1992) almost coincided with the result obtained by the AR method. Here, the FFT method obtains the transfer function as the ratio of the cross-spectrum density function between the input/output signals and the power spectrum density function of the input signals obtained by the fast Fourier transform. The result of Figure 6.4 is obtained using 256 data points.

Vehicle Velocity and Dynamics   In Figure 6.5, transfer functions of the lateral acceleration and directional angular velocity for θv_r^2 in case of the wind velocity 15m/s and blowing width 15m are shown. The difference of the characteristics owing to the variation of the vehicle velocity is distinctly shown. From this fact, it is comprehended that the gain curve is as a whole increased as the vehicle velocity is expedited, and the automobile is sensitive to the lateral wind. According to Wallentowitz (1980) or Tsuchiya (1973), lateral-wind sensitivity increases with the increase of the vehicle velocity for an under-steer vehicle such as the one used in this test, and the increase of the gain curve shown in Figure 6.5 corresponds to this matter. (Under-steer means the characteristics of the radius of the turning circle of a car being increased compared with the radius of the original circle when the car velocity is increased during the circular rotation with the steering angle fixed. Almost all the commercial cars exhibit this characteristics.)

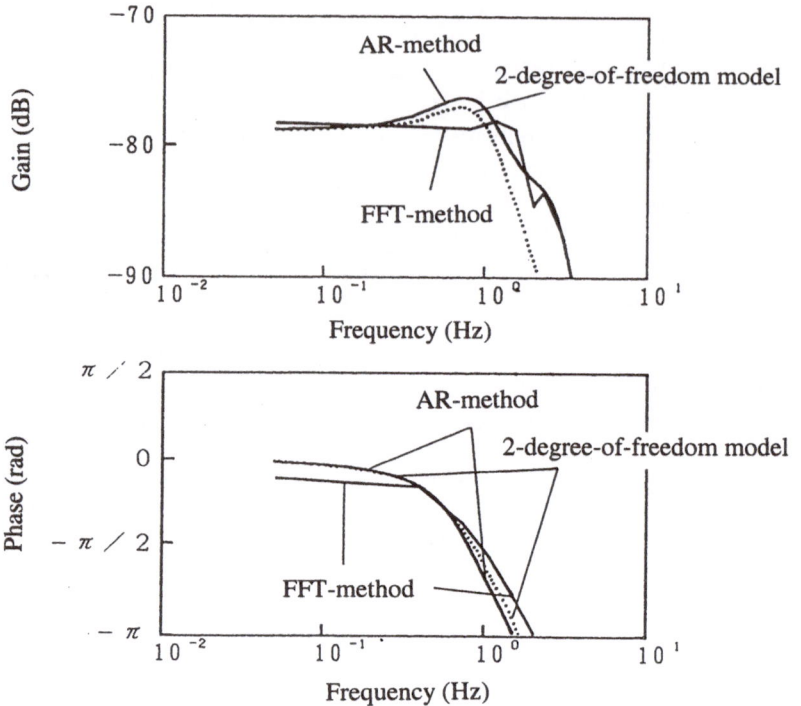

Figure 6.4 Comparison between AR method and FFT method (wind velocity: 15m/s, blowing-zone width: 15m, velocity: 100km/h)

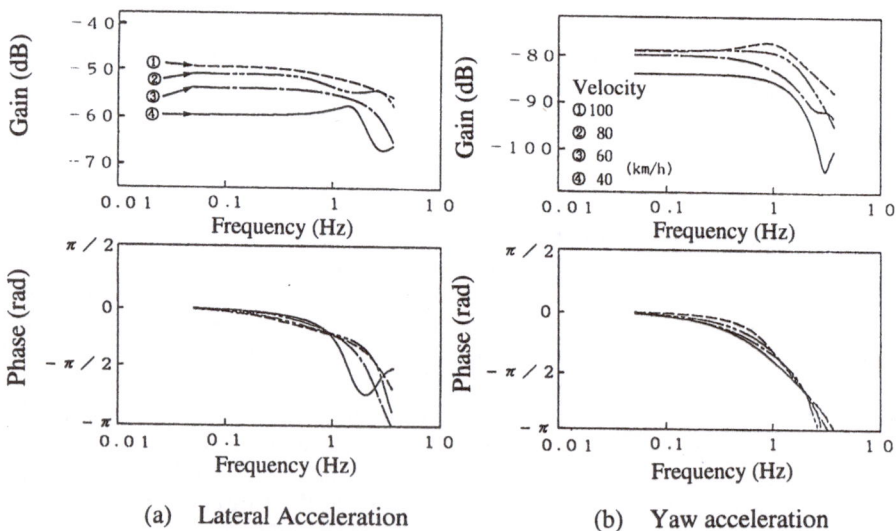

(a)   Lateral Acceleration   (b)   Yaw acceleration

Figure 6.5 Transfer function of vehicle affected by lateral wind: result identified by AR method (wind velocity: 15m/s, blowing-zone width: 15m)

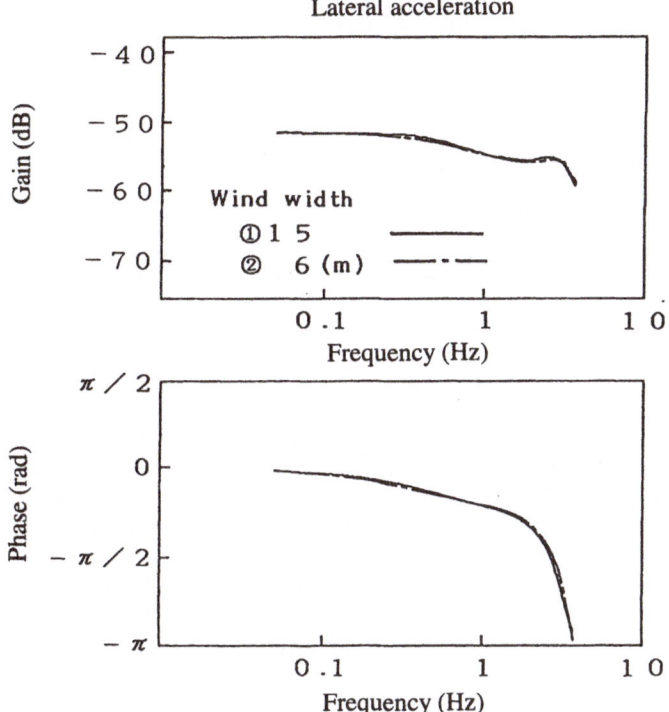

Figure 6.6 Variation of transfer function for blowing-zone width (wind veloc-ity: 15m/s, velocity: 80km/h)

Blowing-zone Width and Dynamics   Figure 6.6 shows the effect of the dif-ferent blowing-zone width (6m and 15m) when the car velocity is 80km/h. The lateral-wind input exhibits a shape of pulse when the blowing-zone width is as narrow as 6m, whereas the input waveform becomes rectangular as the blowing-zone width is widened. The result in Figure 6.6 shows, based upon the fact that the results are the same regardless of the blowing-zone width, that the AR method is very robust to the influence in a style of the lateral-wind input giving good results even in a narrow blowing-zone.

FPE   Figure 6.7 shows the change of FPE when the AR models with various orders are fitted to the signals in Figure 6.3. Since FPE took the minimum value at the order 9, 9 was adopted as the order of the AR model. The selected orders of the AR model, which was slightly subject to change depending on the data used for the identification, ranges from 7 to 15 in this experiment.

6.2.4   Comparison with Time Response

To examine whether the multivariate AR model is properly identified or not, the time series data reproduced by the AR model is compared with the experimental re-sult. By defining a state space model (Akaike and Nakagawa 1988) from the identified autoregressive coefficient matrix and by substituting the measured value at the time 0s for the initial value of the state variables, a time series data was generated.

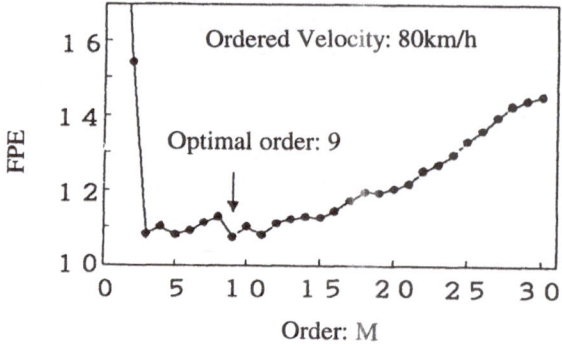

Figure 6.7   Order of AR model and variation of FPE

Figure 6.8 shows results of the comparison in case of the wind velocity of 15m/s and the blowing-zone width of 6m at the car velocity of 100km/h. The solid line indicates the time series data reproduced from the identified AR model, whereas the dots denote the experimental result. Both coincide with each other, indicating that the identified multivariable AR model is a proper one.

When the AR method is applied to the analysis of the dynamic characteristics of an automobile affected by the lateral-wind disturbance, the change of the characteristics is clearly detected. Besides, the analysis is possible in the event of the blowing-zone width being as short as 6m.

6.3   Application of Multivariate AR Model to Driver-Vehicle System

In this section, we show a method to obtain the dynamics of a driver's steering behavior and the dynamics concerning lateral and yaw motion of an automobile by utilizing the multivariate AR model.

6.3.1   Model of Driver-Vehicle System and Analytical Principle

A block diagram of a driver-vehicle system affected by lateral-wind disturbance is shown in Figure 6.9. Drivers control the automobile, when the lateral-wind disturbance displaces the car by manipulating the steering wheel (regulated steering). Thus the automobile is forcibly subjected to 2 types of inputs, i.e. the lateral-wind disturbance and steering-wheel angle. On the other hand, the driver perceives the lateral displacement and the directional angle (yaw angle) to regulate the steering wheel.

Referring to Figure 6.9, the 3 types of the dynamics shown below are obtained from w_v (lateral-wind velocity), δ (steering-wheel angle), $\ddot{y}$ (lateral acceleration), and $\dot{\psi}$ (direction-angle velocity).

1) Dynamics of the automobile lateral and yaw motion $(\ddot{y}, \dot{\psi})$ to the lateral-wind disturbance input (θv_r^2).

2) Dynamics of the automobile lateral and yaw motion $(\ddot{y}, \dot{\psi})$ to the steering-wheel angle input (δ).

3) Dynamics $(\delta/y, \delta/\psi)$ of the driver.

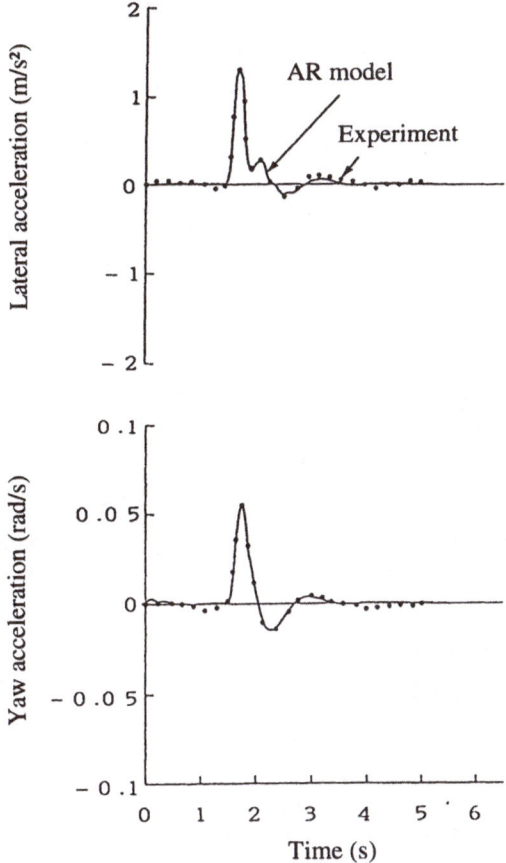

Figure 6.8   Comparison of time series data between AR model and experiment (wind velocity: 15m/s, blowing-zone width: 6m, automobile velocity: 100km/h)

Here the lateral displacement, y, and the directional angle, ψ, are calculated by

$$\psi = \int \dot{\psi} dt$$
$$\beta = \int (\ddot{y}/v - \dot{\psi}) dt \tag{6.1}$$
$$y = \int v \sin(\psi + \beta) dt,$$

where v is the car velocity (Hiramatsu *et al.* 1991).

We shall explain here the principle to identify the dynamics. From the description by the discrete time series among the individual signals based on the block diagram in Figure 6.9, we obtain

$$u_n = G_{nw}(z)\varepsilon_{wn} \tag{6.2}$$
$$\delta_n = -G_p(z)G_{Hy}(z)y_n - G_p(z)G_{H\psi}(z)\psi_n + \varepsilon_{\delta n} \tag{6.3}$$

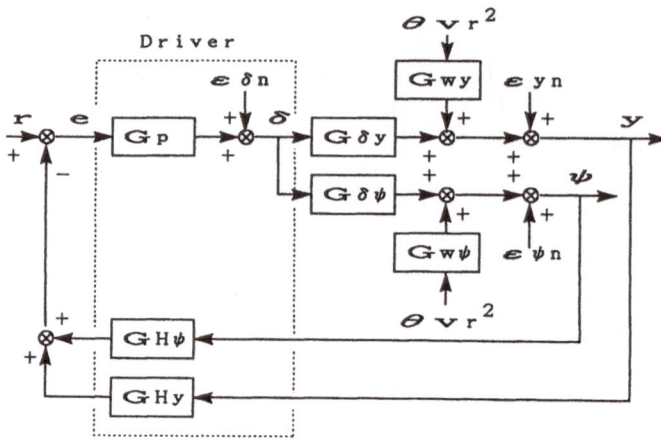

Figure 6.9 Block diagram of driver-vehicle system (θv_r^2: lateral wind input, δ: steering wheel angle, y: lateral displacement of vehicle, ψ: yaw angle of vehicle, r: desired path, e: course deviation, ε_{δ_n}: operation noise of driver, ε_{y_n}: noise of y, ε_{ψ_n}: noise of ψ)

$$y_n = G_{\delta y}(z)\delta_n + G_{wy}(z)u_n + \varepsilon_{yn} \tag{6.4}$$

$$\psi_n = G_{\delta\psi}(z)\delta_n + G_{w\psi}(z)u_n + \varepsilon_{\psi n} \tag{6.5}$$

$$u_n = \theta v_{rn}^2 = \theta(w_{vn}^2 + v^2) ; \qquad (v : \text{a constant}) \tag{6.6}$$

Here, the suffix n indicates the time, whereas z denotes the time-shift operator defined by $z^i x_n = x_{n-i}$. It is noticed that w_v, which is lateral-wind velocity generated by the wind generator, is not ideal white noise. Therefore, it is assumed that the resultant wind u_n is produced from the white noise ε_w by (6.2). In reality, u_n can be related with the white noise ε_w by using the AR model

$$u_n = \sum_{i=1}^{m} a_i u_{n-i} + \varepsilon_{wn}.$$

On the definition that

$$x_n = [\, \theta v_{rn}^2, \delta_n, y_n, \psi_n \,]^{\mathrm{T}}, \tag{6.7}$$

Equations (6.2) through (6.5) can be expressed as

$$x_n = G(z)x_n + E(z)\varepsilon_n, \tag{6.8}$$

where

$$G(z) = \begin{bmatrix} 0 & 0 & 0 & 0 \\ 0 & 0 & -G_y(z) & -G_\psi(z) \\ G_{wy}(z) & G_{\delta y}(z) & 0 & 0 \\ G_{w\psi}(z) & G_{\delta\psi}(z) & 0 & 0 \end{bmatrix} \tag{6.9}$$

$$E(z) = \begin{bmatrix} G_{nw}(z) & 0 & 0 & 0 \\ 0 & 1 & 0 & 0 \\ 0 & 0 & 1 & 0 \\ 0 & 0 & 0 & 1 \end{bmatrix} \tag{6.10}$$

$$\varepsilon_n = [\, \varepsilon_{wn}\, \varepsilon_{\delta n}\, \varepsilon_{yn}\, \varepsilon_{\psi n}\,]^{\mathrm{T}} \tag{6.11}$$
$$G_y(z) = G_p(z)G_{Hy}(z)$$
$$G_\psi(z) = G_p(z)G_{H\psi}(z).$$

By modifying (6.8) as

$$x_n = \{I - G(z)\}^{-1}E(z)\varepsilon_n\,, \tag{6.12}$$

and by putting

$$C(z) \equiv \{I - G(z)\}^{-1}E(z), \tag{6.13}$$

$\{I - G(z)\}^{-1}E(z)$ in the right-hand side is a 4×4 matrix and therefore $C(z)$ in the left-hand side is also 4×4 matrix. By substituting (6.9) and (6.10) to the above equation, the individual elements C_{ij} $(i, j = 1, \cdots, 4)$ of the transfer function $C(z)$ are obtained as

$$G_y = \frac{C_{24}C_{43} - C_{23}C_{44}}{C_{33}C_{44} - C_{34}C_{43}}, \qquad G_\psi = \frac{C_{23}C_{34} - C_{24}C_{33}}{C_{33}C_{44} - C_{34}C_{43}}$$

$$G_{\delta y} = \frac{C_{32}C_{11} - C_{31}C_{12}}{C_{11}C_{22} - C_{12}C_{21}}, \qquad G_{\delta\psi} = \frac{C_{42}C_{11} - C_{41}C_{12}}{C_{11}C_{22} - C_{12}C_{21}} \tag{6.14}$$

$$G_{wy} = \frac{C_{31}C_{22} - C_{32}C_{21}}{C_{11}C_{22} - C_{12}C_{21}}, \qquad G_{w\psi} = \frac{C_{41}C_{22} - C_{42}C_{21}}{C_{11}C_{22} - C_{12}C_{21}}.$$

Note that in the above expressions the operator z is omitted.

On the other hand, the multivariate AR model is generally expressed as

$$x_n = \sum_{i=1}^{m} A_i x_{n-i} + \varepsilon_n\,. \tag{6.15}$$

Here, since

$$A(z) = \sum_{i=1}^{m} A_i z^i \tag{6.16}$$

we obtain

$$x_n = \{I - A(z)\}^{-1}\varepsilon_n. \tag{6.17}$$

By comparing (6.12) with (6.17), we obtain

$$\{I - A(z)\}^{-1} = C(z) \equiv \{I - G(z)\}^{-1}E(z). \tag{6.18}$$

Therefore, the elements of $C(z)$ are obtained from the relation of (6.18), if the multivariate AR model of (6.15) or (6.17) can be identified, and then the transfer functions (the dynamics) can be calculated from (6.14).

6.3.2 Analysis Procedure

A concrete procedure to identify the individual transfer function is hereunder described, using the method shown above. First of all, let the multivariate AR model be identified after the individual elements of the time series signal x_n are given as (6.7). The innovation series at that time becomes (6.11). Secondly, calculate the (6.16) from the identified AR coefficient matrix by putting $z^i = \exp(-j2\pi f i)$. Furthermore calculate the multivariate spectrum density function matrix using the (6.16) and covariance matrix of the innovation. Thus the elements of the spectrum density

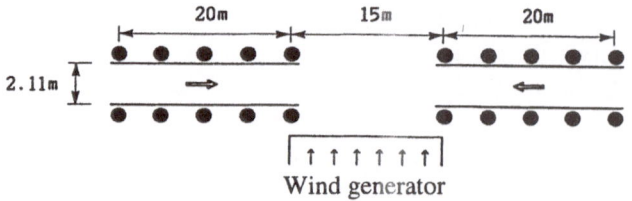

Figure 6.10 Lateral wind test course with free steering

function matrix become the elements in the complex domain of $C(z)$ in (6.18). Finally if (6.14) is calculated in the complex domain, the individual transfer functions shown in the left-hand side of these are obtained.

However to confirm that the identification is correctly made in the process shown above, it should be examined:

1) That the innovation series can be considered as a white noise or can be approximated by a white noise within the important frequency range.

2) That the correlation coefficients between the individual elements of the innovation are small enough.

6.4 Dynamics of Driver-Vehicle System under Lateral-Wind

In this section, the result of the identification of the dynamics concerning the automobile's lateral and yaw motion and the ones regarding the driver's steering behavior is shown using the method described in the previous section.

6.4.1 Running Test Under Lateral-Wind with Free Control

Test Method   In Figure 6.10, a test course is illustrated. The drivers allow the automobiles to enter the straight lane in the figure alternately from the right and left sides, and make regulating steering so that the automobiles will not deviate from the course even after they are subjected to the lateral-wind disturbance. The course width was determined as 2.11m in the regulation (car width $\times$ 1.1 + 0.25m) of the lane change test method (TR-3888) of ISO. A recording staff who rode with the driver instructed the car velocity at each run, and the driver drove the automobile so that the change of the instructed car velocity will be less than $\pm$3km/h complying with the velocity. Each one time of the left entrance and right entrance was made at the same car velocity. 40, 60, 80, 90, and 100km/h were the ordered car velocities.

Common male drivers who are not professional drivers were engaged in the runs for the experiment. These males were comprised of 10 youths with the age ranging from 20 to 30 and 10 middle-aged men between 40 and 50, totaling to 20 people. The test was conducted under the condition that the blowing-zone width 15m and 2 types of the wind-velocity 15m/s and 22.5m/s are available. The steering-wheel angle was measured to examine the drivers' steering state in addition to the ones referred to the items for the measurement in the section 6.1.

Features of the Measured Signals   An example of the individual signals measured in this experiment is shown in Figure 6.11 with respect to the cases of the wind velocity 15m/s, blowing-zone width 15m, and instructed car velocity 80km/h. The time while the lateral-wind velocity is rising and the time while the response of

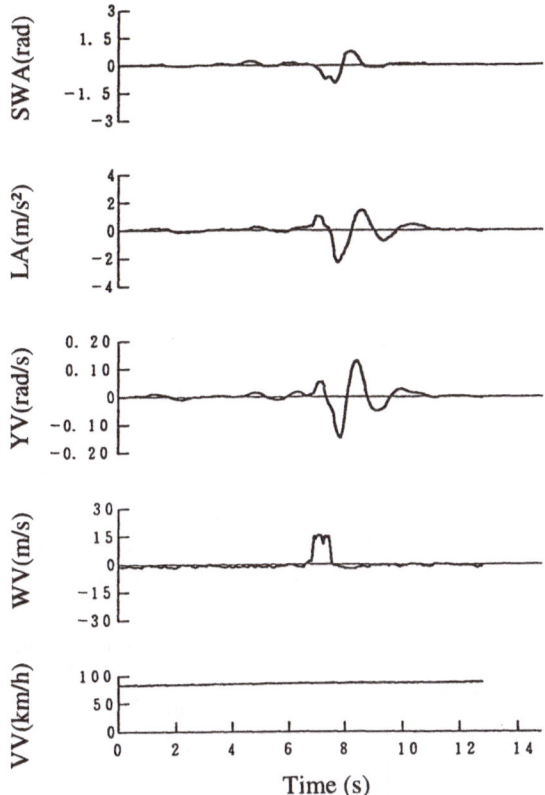

Figure 6.11   Time series data in lateral wind test with free steering (wind velocity: 15m/s, blowing zone width: 15m, velocity: 80km/h), SWA: steering wheel angle, WV: wind velocity, LA: lateral acceleration, VV: vehicle velocity, YV: directional angular velocity (yaw velocity)

the direction angle velocity is appearing are approximately 0.7s and approximately 4.5s, respectively. Although the response time is elongated due to drivers' regulating steering in comparison with the case of the automobile in section 6.1, the frequency range is restricted under 10Hz and it can hardly be said that the response time is long enough in dynamics analysis.

6.4.2   Automobile's Dynamics Obtained from Driver-Vehicle System

Dynamics for Steering-Wheel Angle Input   Firstly, Figure 6.12 shows the dynamics of the test car for the steering-wheel angle input. This is the result obtained when the wind velocity is 15m/s, blowing-zone width is 15m, and car velocity is 100km/h. The same as in the case of section 6.1, it is noticed that the result gives a gain curve and a phase curve almost equal to those of the transfer function (dotted line) obtained from the 2-degrees-of-freedom single track model.

Furthermore it is comprehended that the dynamics of the automobile can be ob-

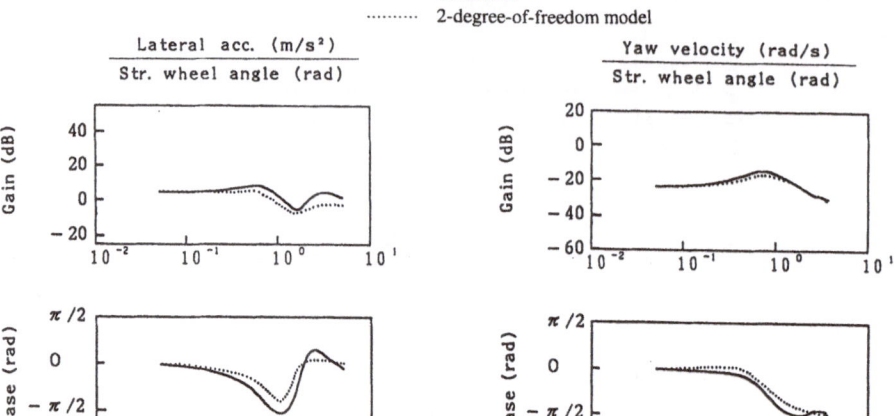

Figure 6.12 Transfer function of vehicle for steering wheel angle input (wind velosity: 15m/s, blowing zone width: 15m, velocity: 100km/h)

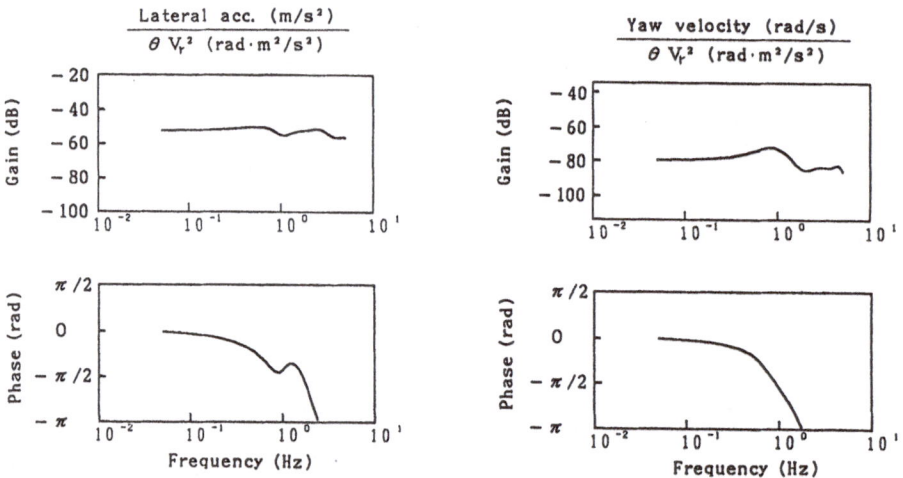

Figure 6.13 Transfer function of vehicle for lateral wind input (wind velocity: 15m/s, blowing zone width: 15m, velocity: 100km/h)

tained from either of the test on the automobile itself and the test in the driver-vehicle system the same as above.

Dynamics for Lateral-Wind Disturbance Input   An analytical result is shown in Figure 6.13, in relation to the lateral acceleration and direction angle velocity for the lateral-wind disturbance input under the test condition the same as above. This dynamics has been brought about simultaneously with the ones mentioned above.

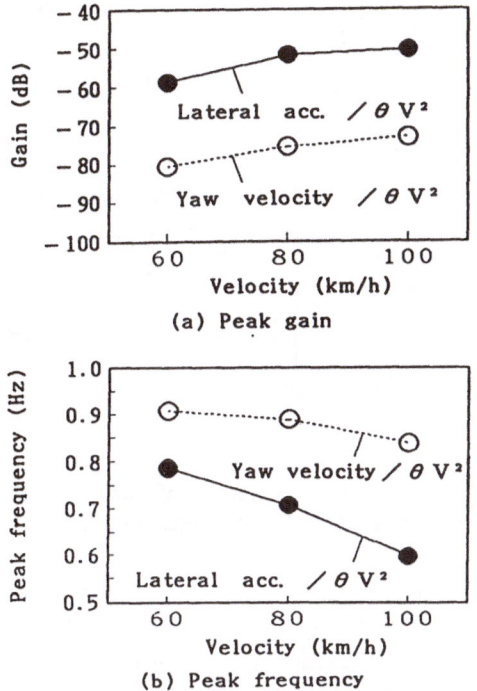

Figure 6.14 Dynamic characteristics of vehicle affected by lateral wind

As evaluation of the vehicle motion characteristics for the lateral-wind disturbance, a lateral-wind sensitivity coefficient has been pointed out (Tsuchiya *et al.* 1973) as the index that has conventionally been used. The coefficient is defined by the factors including the stationary values of the lateral acceleration or direction angle velocity per unit lateral force. As is seen the same as this, it is revealed that the value of the maximum gain (the peak gain) becomes most remarkable in the vehicle response when it is subjected to the pressure of the lateral wind of the frequency component, and the value can be regarded as the dynamic lateral-wind sensitivity.

A relation between the value of the peak gain and its frequency is summarized in Figure 6.14. As the car velocity is increased, the value of the peak gain becomes greater, and it is noticed that damping characteristics for the lateral-wind disturbance is being degraded. Besides, the peak frequency is gradually lowered as the car velocity is increased. In general, drivers' steering frequency is believed to be in a domain lower than 1Hz. Therefore there is a possibility that lowering of the peak frequency induces resonance of the drivers' steering and vehicle response for lateral-wind disturbance. This is considered to be unfavorable from a viewpoint of the driver-vehicle system.

6.4.3 Dynamics of Drivers' Steering Behavior

Drivers' Dynamics for Vehicle Velocity   Figures 6.15 through 6.17 show drivers' dynamics when the car runs at the speed of 60, 80, and 100km/h under the condition of the blowing-zone width 15m and wind velocity 15m/s. The drivers are all middle-aged persons. The characteristics in the left side in the individual figures

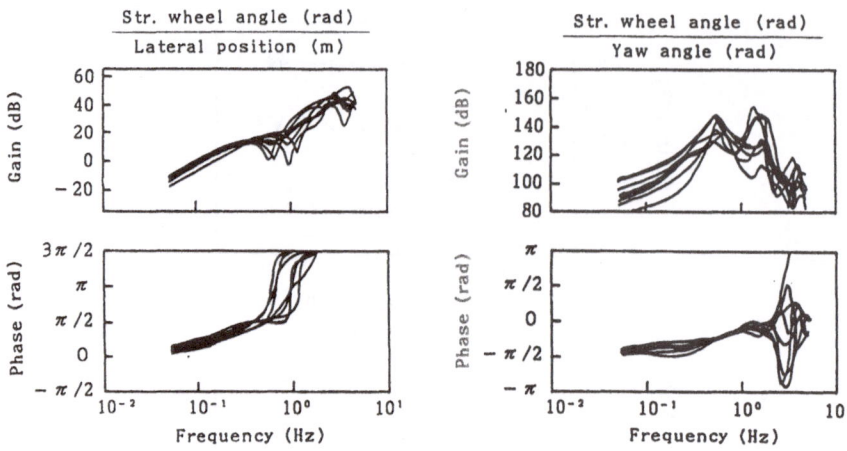

Figure 6.15 Dynamics of steering behavior of middle aged drivers (wind velocity: 15m/s, blowing zone width: 15m, velocity: 60km/h)

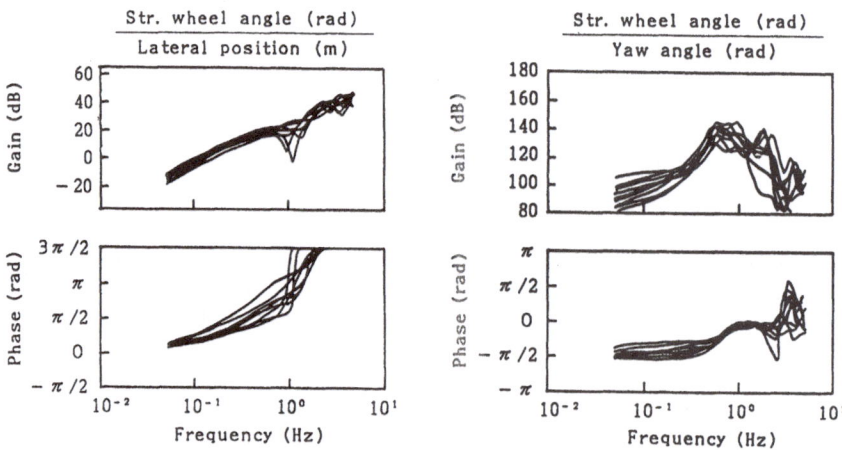

Figure 6.16 Dynamics of steering behavior of middle aged drivers (wind velocity: 15m/s, blowing zone width: 15m, velocity: 80km/h)

indicates G_y with the vehicle lateral displacement as feedback, whereas the right side shows G_ψ with the directional angle of the automobile as feedback.

The gain of G_y shows an increasing curve and the phase also goes ahead, from which it is construed that differential characteristics of drivers are exhibited. The fact means the drivers' prediction characteristics to compensate the tracking performance onto the course and the stability of the vehicle behavior. On the contrary, remarkable peaks of G_ψ gain are seen at around 0.7Hz and 1.4Hz, and the feedback effect of the automobile's direction is strengthened with the frequency. As is seen above, the feedback of the vehicle lateral displacement and the feedback of automobile's direction

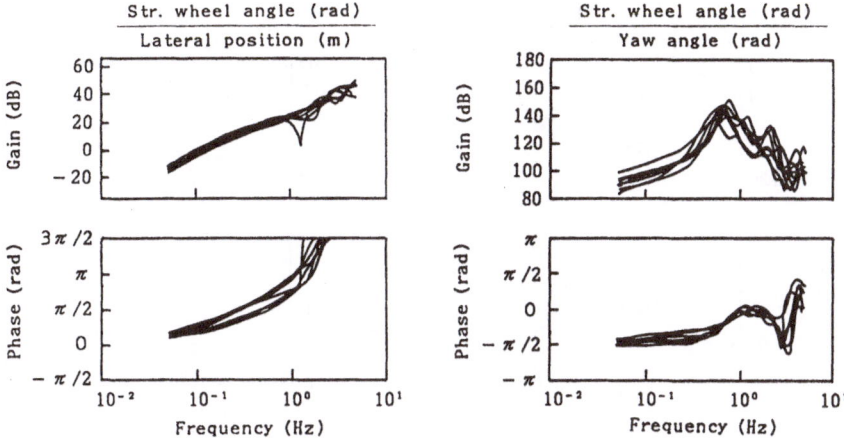

Figure 6.17 Dynamics of steering behavior of middle aged drivers (wind velocity: 15m/s, blowing zone width: 15m, velocity: 100km/h)

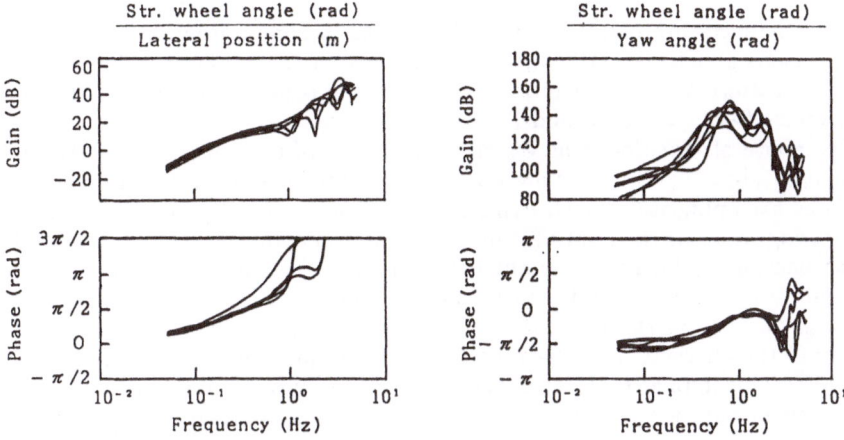

Figure 6.18 Dynamics of steering behavior of middle aged drivers (wind velocity: 22.5m/s, blowing zone width: 15m, velocity: 80km/h)

reveal different effects in drivers' steering behavior when the car is affected by the lateral wind.

Drastic change is hardly seen with G_y when the difference owing to the car velocity of the individual characteristics is scrutinized, but the peak is decreased in the vicinity of approximately 1.4Hz with G_ψ as the car velocity is increased. The increase of the car velocity is considered to be influential to the feedback of the direction of automobile.

Difference Among Subject Drivers   Transfer functions of the middle-aged people and youths under the condition of the blowing-zone width 15m, wind velocity 22.5m, and car velocity 80km/h are shown in Figures 6.18 and 6.19, respectively.

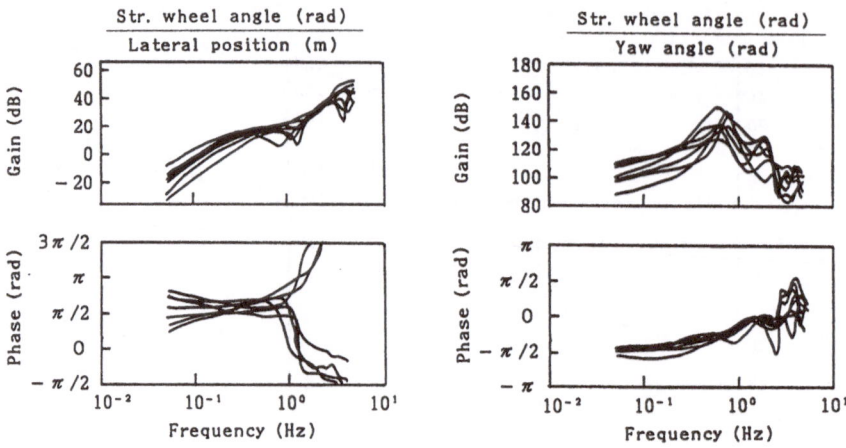

Figure 6.19 Dynamics of steering behavior of young drivers (wind velocity: 22.5m/s, blowing zone width: 15m, velocity: 80km/h)

Major differences are pointed out in the youths in comparison with in the middle-aged persons as follows: [1] Individual difference of the response is large. [2] Some phase characteristics of G_y exhibit phase lag in the region of more than 1Hz. [3] The peak value of G_ψ gain in around 1.4Hz is small.

The above observation is an example of the use of the multivariate AR model to the driver-vehicle system. Taking up an automobile affected by lateral-wind disturbance as exemplification, such dynamics of [1] automobile lateral and yaw motion for steering-wheel angle input, [2] automobile lateral and yaw motion for lateral-wind disturbance, and [3] drivers' steering behavior have been obtained. The feature of the exemplification resides with the fact that the dynamics in [1]–[3] have been identified simultaneously from the test result of the closed-loop system, which means that a variety of the characteristics have been grasped from only a single type of the running test. This implies that use of such analysis enables us to obtain lots of precious information from the test of the driver-vehicle system more than by an open-loop test.

6.5 Conclusions

In the analysis of a complicated closed-loop system such as a driver-vehicle system, it is very important to obtain the dynamics of the individual elements composing the system by means of isolation. The examples shown in this chapter suggest that the AR method is quite effective for the analysis of the dynamics. An example of the investigation of the dynamics concerning the lateral and yaw motion of an automobile itself also suggests that the AR method which is adequate for the analysis of the short duration data is quite effective since long-time measurement of the response of an automobile is usually difficult in this kind of the tests. In this chapter only the case when the lateral-wind disturbance is applied as input, but the method is believed to be applicable to the analysis of vehicle motion or driver-vehicle system under the other circumstances.

References

Abe, M. (1992), Vehicle dynamics and control, Sankaido Pub.

Akaike, H. and Nakagawa, T. (1988), Statistical Analysis and Control of Dynamic Systems, Kluwer Academic Publishers, Dordrecht.

Hiramatsu, K. and H. Soma (1991), "Measurement of vehicle lateral deviation caused by side wind generators," *Proceedings of JSAE*, No. 912156.

Soma, H. and K. Hiramatsu (1993), "Identification of vehicle dynamics under lateral wind disturbance using autoregressive model," *SAE paper*, No. 931894.

Soma, H. and K. Hiramatsu (1995), "Dynamic identification of driver-vehicle system using AR-methods," *Vehicle System Dynamics*, Vol. 24, No. 4/5, 263–282.

Soma, H. and K. Hiramatsu (1996), "Dynamics of driver's steering behavior under lateral wind disturbance," *Transaction of JSAE*, Vol. 27, No. 2, 107–112.

Suzuki, K. and A. Nakajima (1983), "Damping estimation method using the spectral analysis technique based on the auto-regressive model fitting," *Bulletin of the JSME*, Vol. 26, No. 215, 832–838.

Tsuchiya, S. and H. Iwase (1973), "On the cross wind sensitivity of the automobile," *Transactions of the JSME*, Vol. 39, No. 324, 2372–2380.

Wallentowitz, H. (1980), "Zusammenwirken von fahrer und fahrzeug bei normaler straßen-fahrt unter natürlichem seitenwind," *Proceedings of 18th FISITA Congress*, 237–247.

JSAE (1976, 1989 revision), "Test procedure of crosswind stability for passenger car," *JASO-Z108-89*.

Chapter 7

Estimation of Directional Wave Spectra Using Ship Motion Data

Toshio Iseki
Tokyo University of Mercantile Marine
2-1-6 Etchujima, Koto-ku, Tokyo 135-8533, Japan
iseki@ipc.tosho-u.ac.jp

7.1   Introduction

In rough seas, it is very important for mariners to grasp the sea state around the ship and to secure the safety of the hull and cargo by suitable maneuvers, such as changing the course, slowing down the speed and so on. The difficulty of decision making for these maneuvers arises from the relation between the safety and the cost of the voyage. Therefore the mariners are required to have profound knowledge concerning the sea-keeping characteristics of the ship.

Systems to support safe navigation of ships have intensively been studied in recent years, and some of them have been put to practical use. These systems are based on the researches concerning ship motions or prediction of the wave load in a field of naval architecture, and the information considered to be effective in operating the ship is successively offered for the ship's navigators by analyzing the data obtained by various types of sensors. However, no systems on a commercial basis is still available and many problems are left to be overcome before spreading such a kind of system to general mercantile ships. Among the information required by these systems, the wave direction is one of the most difficult types of the information to be had. Furthermore, most of the systems referred to above require the help of visual observation by crewmen.

The directional wave spectrum expressing the distribution of wave energy according to frequency and direction simultaneously, draws much interest in many research fields such as oceanography, coastal/harbor engineering, naval architecture, etc., and remarkable development of analyzing techniques are now underway based on the development of the measurement technology. Among these analyzing techniques, the extended maximum likelihood method (EMLM) proposed by Isobe *et al.* (1984) for a wave probe array generalizes the maximum likelihood method developed by Capon (1969) so that it can be applied to an arbitrary wave motion data. Because of the simple calculation procedure, this method is widely used for the directional analysis of actual sea waves. On the other hand, the estimation method based on Bayesian models proposed by Hashimoto (1987) has very high estimation accuracy compared with the other method. This method attracted much attention, because it is not only applicable to an arbitrary wave motion data as the EMLM but also is robust to the influence of the observation error.

Meanwhile a ship that is navigating in waves can also be regarded as a kind of a wave probe. If a linear input/output relation is supposed between the waves and ship motions, then a transfer function of the ship motions to the wave input can be obtained with a considerable accuracy by theoretical calculation. Thus the theoretical method referred to above becomes applicable. Since a variety of inexpensive and maintenance-free sensors including vibrating element gyroscopes have been developed in recent years, it becomes relatively easy to measure ship motions such as pitching, rolling, etc. Therefore if a system which can estimate the directional wave spectrum only from the ship motion data is developed, it will become a ship-borne wave meter to be widely usable for general mercantile ships. In Japan, it is under obligation by law for navigating ships with more than a level of tonnage to release meteorological reports on the sea, and the ship-borne wave meter is believed to be not only useful for the purpose but also useful as a sub-system to compose a part of the ship-operation-guidance system referred to above.

An attempt to estimate a directional wave spectrum only from ship motion data has already been made by Hirayama (1987) and Iseki *et al.* (1992), and a variety of results of actual sea tests and tank tests have been reported. However several problems to be solved are left untouched before the system is in practical use. In this chapter, we describe a method to apply a Bayesian modeling procedure to estimate a directional wave spectrum only from the ship motion data. Then we show the results of an estimation for a tank test using a model ship to reveal the usefulness and problems encountered on applying.

7.2   Cross-spectrum Analysis by a Multivariate AR Model

In this section, we show a method to fit a multivariate AR model to a stationary vector time series, and numerical method to compute a cross spectrum and power contribution from the fitted multivariate AR model. Moreover an example of the computation for actual ship-motion data is given to exemplify the method.

Suppose that the ship motions of k variables ($k \leq 6$) including pitching, rolling, etc. are observed on an actual ship. This vector time series $\boldsymbol{y}(s)$ ($s = 1, \ldots, N$) can be expressed by a k-dimensional AR model as

$$\boldsymbol{y}(s) = \sum_{m=1}^{M} \boldsymbol{A}(m)\boldsymbol{y}(s-m) + \boldsymbol{w}(s), \qquad (7.1)$$

where N denotes the data length

$$\boldsymbol{y}(s) = \begin{bmatrix} y_1(s) \\ y_2(s) \\ \vdots \\ y_k(s) \end{bmatrix}, \quad \boldsymbol{w}(s) = \begin{bmatrix} w_1(s) \\ w_2(s) \\ \vdots \\ w_k(s) \end{bmatrix}$$

$$A(m) = \begin{bmatrix} a_{11}(m) & a_{12}(m) & \cdots & a_{1k}(m) \\ a_{21}(m) & a_{22}(m) & \cdots & a_{2k}(m) \\ \vdots & \vdots & \ddots & \vdots \\ a_{k1}(m) & a_{k2}(m) & \cdots & a_{kk}(m) \end{bmatrix}$$

$A(m)$ is an $M \times M$ AR coefficient matrix, and $w(s)$ is assumed to be a k-variate white noise. The AR coefficient matrix can be calculated effectively for a variety of orders M successively by Levinson-Durbin algorithm. On the other hand, the optimal order of the model can be determined by the MAICE method. That is to say, it is obtained by minimizing the AIC of the k-variate AR model,

$$\text{AIC}(M) = N \log |\Sigma_M| + 2k^2 M + k(k+1). \tag{7.2}$$

Here, $|\Sigma_M|$ is a determinant of the prediction error variance covariance matrix. A concrete calculation procedure is detailed in Akaike and Nakagawa (1988).

At the next stage, consider the relation

$$y(f) = B(f) \cdot w(f) \tag{7.3}$$

where $B(f) = (I - A(f))^{-1}$, and $y(f)$, $w(f)$ and $A(f)$ are defined by Fourier transformation as

$$y(f) = \sum_{s=1}^{N} y(s)e^{-2\pi isf}, \quad w(f) = \sum_{s=1}^{N} w(s)e^{-2\pi isf}, \quad A(f) = \sum_{s=1}^{M} A(s)e^{-2\pi isf}.$$

The power spectrum $\Phi_{YY}(f)$ is then given by

$$\Phi_{YY}(f) = \text{E}\left[y(f) \cdot y^*(f)^T\right], \tag{7.4}$$

where * represents a conjugate complex. Paying attention to the diagonal element $\phi_{ii}(f)$ of $\Phi_{YY}(f)$ gives an equation

$$\phi_{ii}(f) = \text{E}\left[\sum_{j=1}^{k} |b_{ij}(f)|^2 w_i^2(f)\right] + \text{E}\left[\sum_{j_1=1}^{k} \sum_{j_2=1}^{k} b_{ij_1}(f) b_{ij_2}(f)^* w_{j_1}(f) w_{j_2}^*(f)\right], \tag{7.5}$$

where $b_{ij}(f)$ represents the (i,j) component of $B(f)$. Therefore if $E[w_{j_1}(f)w_{j_2}^*(f)] = 0$, that is to say, the noise of $y_{j_1}(s)$ and $y_{j_2}(s)$ is not correlated, then the second term in the right hand side of the above equation becomes zero, and it is comprehended that the power spectrum $\phi_{ii}(f)$ of $y_i(s)$ is expressed as the sum of spectra of $w_j(s)$ as

$$\phi_{ii}(f) = \sum_{j=1}^{k} |b_{ij}(f)|^2 \phi_{w_j}(f). \tag{7.6}$$

Using this expression, the relative power contribution

$$\gamma_{ij}(f) = \frac{|b_{ij}(f)|^2 \phi_{w_j}(f)}{\phi_{ii}(f)} \tag{7.7}$$

can be defined.

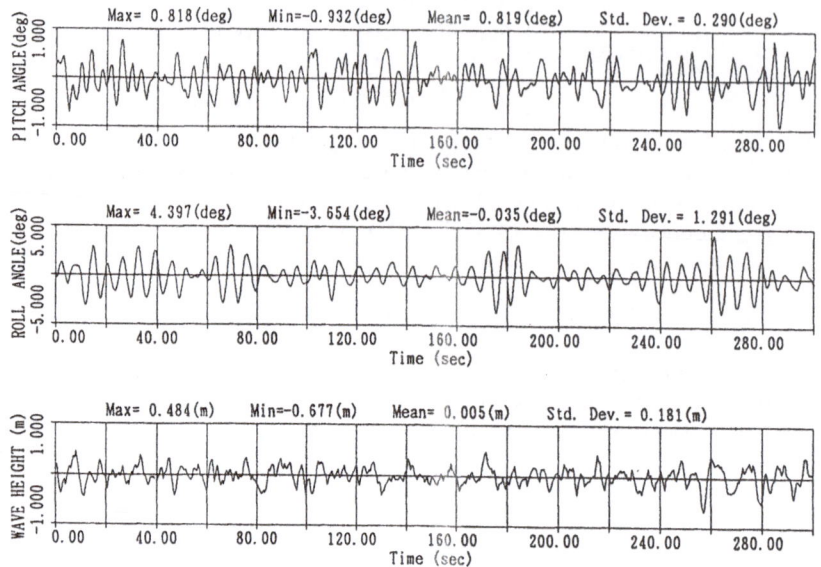

Figure 7.1   Time series data obtained by onboard tests

Shown in Figure 7.1 is a part of the time series obtained in the experimental voyage of the training ship, Shioji-Maru of Tokyo University of Mercantile Marine. The figure shows the time series of the pitch, roll, and wave height in order from top to bottom. The wave height is measured by a micro-wave type ship-borne wave height recorder (Kuwashima 1988) attached at the bow. It is known that the pitch frequency of a ship is susceptible to the influence of the surrounding waves, whereas the roll frequency is almost free from the influence of the waves. Figure 7.1 clearly reveals such a tendency, and it is clearly seen that the roll has a natural frequency.

Figure 7.2 shows the cross-spectrum of the data obtained from a multi-variate AR model. The sampling period in the analysis is 0.5sec, and the data length is 600 points. On the other hand, the order of the model determined by MAICE method was 16. The boldfaced solid line and lightfaced solid line in the figure represent the real part and imaginary part of the cross spectrum, respectively. From the figure, it is perceived that the peak frequency of the power spectrum of the pitch almost coincides with that of the wave height. Meanwhile it is also seen from the figure that peaks appear exclusively at the natural frequency for the power spectrum of the roll. This implies that the tendency described above is endorsed.

Figure 7.3 shows the power contribution of the wave height to the power spectra of the pitch and roll. In the figure, the contribution of the power from the wave height is exhibited by rough hatching, whereas the contribution of the motions themselves is shown by dense hatching. On the other hand, the proportion of the contribution to the whole power spectrum is shown in percentage. From these plots, it is explained how severely the motions of the hull is influenced by the surrounding waves. The angle of encounter of the hull and the waves at the time when the data are recorded is about

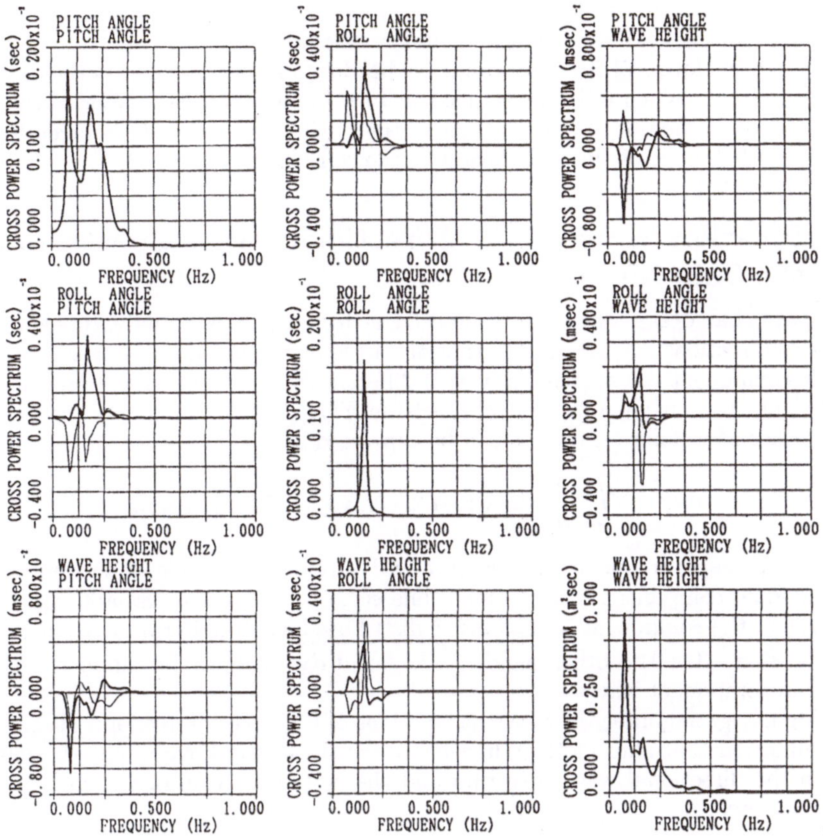

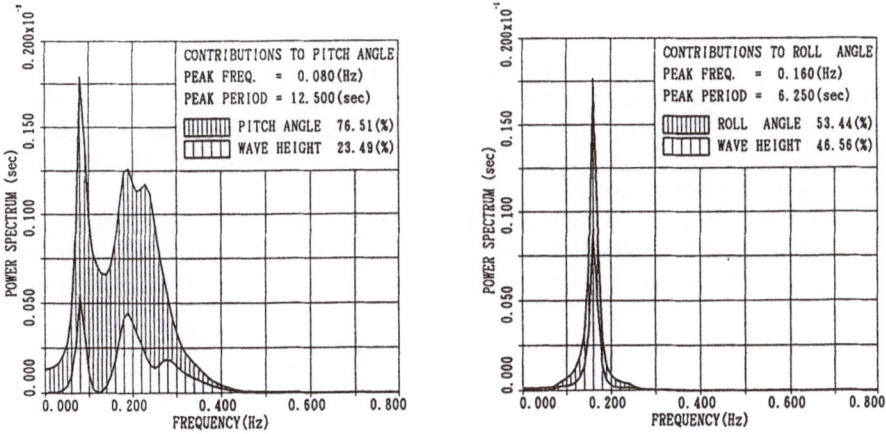

Figure 7.2 Cross spectra of the actual ship time series

Figure 7.3 Contribution of wave height to ship motions

$\chi = 106°$ according to visual observation (χ takes the values ranging from $-180°$ to $180°$, the following seas is $0°$ and the counterclockwise state is plus), and the hull is subjected to the forces of the waves from the heading slightly deviated in the direction of the starboard. When the encounter angle with the waves comes closer to the head seas, the contribution of the wave height to the pitch is increased. Concurrently with this, the contribution of the wave height to the roll becomes lowered.

7.3 Relation Between the Directional Wave Spectrum and the Ship Motions

Suppose that ocean waves are composed of irregular waves that propagate in a variety of directions. Also suppose that the irregular waves are expressed by innumerable component waves (regular waves) different from each other in frequency and wave height. Then using a directional spectrum $E(f, \chi)$, elevation $\eta(t)$ of the irregular wave surface at a point on the sea can be expressed as

$$\eta(t) = \int_{-\pi}^{\pi} \int_{0}^{\infty} \cos\{2\pi f t + \epsilon(f, \chi)\}\sqrt{2E(f, \chi)df d\chi}, \tag{7.8}$$

where f represents frequency of the wave and χ represents angle of encounter with the individual component waves. Meanwhile $\epsilon(f, \chi)$ is a random variable expressing the phase of the component waves, and is distributed as an uniform distribution over $-\pi \leq \epsilon \leq \pi$. On the other hand, $\sqrt{2E(f, \chi)df d\chi}$ denotes the amplitude of the individual component waves.

A ship navigating in ocean waves has motions of 6 degrees of freedom. However on the assumption that ship motions linearly responses to the wave input, the relation between the directional wave spectrum and the cross spectrum of the ship motions can be expressed as

$$\phi_{ij}(f) = \int_{-\pi}^{\pi} H_i(f, \chi)H_j^*(f, \chi)E(f, \chi)d\chi, \tag{7.9}$$

where $\phi_{ij}(f)$ is the cross spectrum of the ship motions and i and j represent roll and pitch, respectively. In the meantime, $H_i(f, \chi)$ expresses a response function (transfer function) of the ship motion i and $*$ represents the complex conjugate. The response function can be obtained theoretically such as the strip method, provided that hull shape is given. The strip method is a simple and convenient quasi-3-dimensional calculation method widely used in a field of naval architecture. For the details of the method, refer to Price (1974).

By dividing the integral range of χ into N pieces of infinitesimal intervals and by regarding the response function and directional wave spectrum in the infinitesimal integral element as constant in (7.9), the expression can be re-written in a discrete form as

$$\phi_{ij}(f) = \sum_{n=1}^{N} H_{in}(f)H_{jn}^*(f)E_n(f), \tag{7.10}$$

where

$$H_{in}(f) = \sqrt{\Delta\chi}H_i(f, \chi_n), \quad H_{jn}^*(f) = \sqrt{\Delta\chi}H_j^*(f, \chi_n)$$
$$\Delta\chi = 2\pi/N, \quad E_n(f) = E(f, \chi_n), \quad \chi_n = -\pi + (n-1)\Delta\chi.$$

Now that, supposing that the 3 factors as arbitrary ship motions, i.e. ζ as heave, θ as pitch, and ψ as roll are measured, the cross spectrum of the ship motions is given as 3×3 matrix $\boldsymbol{\Phi}_{YY}$. Accordingly the relation between the directional wave spectrum and the cross spectrum of the ship motions is given by the matrix expression as (hereafter, for simplicity of notation the frequency f will be omitted.)

$$
\begin{bmatrix} \phi_{\zeta\zeta} & \phi_{\zeta\theta} & \phi_{\zeta\psi} \\ \phi_{\theta\zeta} & \phi_{\theta\theta} & \phi_{\theta\psi} \\ \phi_{\psi\zeta} & \phi_{\psi\theta} & \phi_{\psi\psi} \end{bmatrix} = \begin{bmatrix} H_{\zeta1} & H_{\zeta2} & \cdots & H_{\zeta N} \\ H_{\theta1} & H_{\theta2} & \cdots & H_{\theta N} \\ H_{\psi1} & H_{\psi2} & \cdots & H_{\psi N} \end{bmatrix}
$$

$$
\times \begin{bmatrix} E_1 & 0 & \cdots & 0 \\ 0 & E_2 & \cdots & 0 \\ \vdots & \vdots & \ddots & \vdots \\ 0 & 0 & \cdots & E_N \end{bmatrix} \begin{bmatrix} H_{\zeta1}^* & H_{\theta1}^* & H_{\psi1}^* \\ H_{\zeta2}^* & H_{\theta2}^* & H_{\psi2}^* \\ \vdots & \vdots & \vdots \\ H_{\zeta N}^* & H_{\theta N}^* & H_{\psi N}^* \end{bmatrix},
$$

$$(7.11)$$

where $\boldsymbol{\Phi}_{YY}$ is Hermite matrix, and therefore only the components on and the above the diagonal are considered and the real part and imaginary part are expressed separately. Furthermore the error term (7.11) can be re-written as a linear regressive model as

$$\boldsymbol{\varphi} = \boldsymbol{A}\boldsymbol{\delta}(\boldsymbol{x}) + \boldsymbol{w}, \tag{7.12}$$

where

$$
\boldsymbol{\varphi} = \begin{bmatrix} \phi_{\zeta\zeta} \\ \phi_{\theta\theta} \\ \phi_{\psi\psi} \\ R_e(\phi_{\zeta\theta}) \\ R_e(\phi_{\zeta\psi}) \\ R_e(\phi_{\theta\psi}) \\ I_m(\phi_{\zeta\theta}) \\ I_m(\phi_{\zeta\psi}) \\ I_m(\phi_{\theta\psi}) \end{bmatrix}, \quad \boldsymbol{\delta}(\boldsymbol{x}) = \begin{bmatrix} \exp(x_1) \\ \exp(x_2) \\ \vdots \\ \\ \\ \vdots \\ \\ \exp(x_N) \end{bmatrix}, \quad \boldsymbol{w} = \begin{bmatrix} w_1 \\ w_2 \\ w_3 \\ w_4 \\ w_5 \\ w_6 \\ w_7 \\ w_8 \\ w_9 \end{bmatrix}
$$

This expression will be used in the next section.

$$
\boldsymbol{A} = \begin{bmatrix} |H_{\zeta1}|^2 & |H_{\zeta2}|^2 & \cdots & |H_{\zeta N}|^2 \\ |H_{\theta1}|^2 & |H_{\theta2}|^2 & \cdots & |H_{\theta N}|^2 \\ |H_{\psi1}|^2 & |H_{\psi2}|^2 & \cdots & |H_{\psi N}|^2 \\ R_e(H_{\zeta1}H_{\theta1}^*) & R_e(H_{\zeta2}H_{\theta2}^*) & \cdots & R_e(H_{\zeta N}H_{\theta N}^*) \\ R_e(H_{\zeta1}H_{\psi1}^*) & R_e(H_{\zeta2}H_{\psi2}^*) & \cdots & R_e(H_{\zeta N}H_{\psi N}^*) \\ R_e(H_{\theta1}H_{\psi1}^*) & R_e(H_{\theta2}H_{\psi2}^*) & \cdots & R_e(H_{\theta N}H_{\psi N}^*) \\ I_m(H_{\zeta1}H_{\theta1}^*) & I_m(H_{\zeta2}H_{\theta2}^*) & \cdots & I_m(H_{\zeta N}H_{\theta N}^*) \\ I_m(H_{\zeta1}H_{\psi1}^*) & I_m(H_{\zeta2}H_{\psi2}^*) & \cdots & I_m(H_{\zeta N}H_{\psi N}^*) \\ I_m(H_{\theta1}H_{\psi1}^*) & I_m(H_{\theta2}H_{\psi2}^*) & \cdots & I_m(H_{\theta N}H_{\psi N}^*) \end{bmatrix}.
$$

Here, considering the fact that the directional wave spectrum is non-negative, we put $E_n = \exp(x_n)$.

7.4 Estimation of the Directional Wave Spectrum Using a Bayesian Model

It turns out that the estimating problem of a directional wave spectrum from the cross spectrum of ship motions is reduced to a fitting problem of the linear regressive model. In solving this problem, according to the Bayesian estimation method formulated by Akaike (1980), the estimator of the directional wave spectrum is obtained by minimizing the product of the likelihood of the model and the adequately defined prior distribution. The calculation method for the above description is hereunder explained following Hashimoto (1987).

Under the assumption of the normality of the observation error, the likelihood function of the model expressed by (7.12) is given as

$$L(x|\sigma^2) = \left(\frac{1}{2\pi\sigma^2}\right)^{\frac{9}{2}} \exp\left\{-\frac{1}{2\sigma^2}||A\delta(x) - \varphi||^2\right\}, \tag{7.13}$$

where $||a||$ represents the norm of the vector a.

On the other hand, as a prior distribution, assuming that the estimate of $\delta(x)$ changes smoothly, with the change of the angle of encounter χ. We consider the total sum of the quadratic difference, that is to say,

$$\sum_{n=1}^{N} \varepsilon_n^2 = \sum_{n=1}^{N} (x_n - 2x_{n-1} + x_{n-2})^2 \tag{7.14}$$

Now assuming that ε_n is distributed as the normal distribution with the mean 0 and the variance σ^2/u^2, the a prior distribution $p(x)$ is given as

$$p(x|u^2) = \left(\frac{u^2}{2\pi\sigma^2}\right)^{\frac{N}{2}} \exp\left\{-\frac{u^2}{2\sigma^2}\sum_{n=1}^{N}\varepsilon_n^2\right\}$$

$$= \left(\frac{u^2}{2\pi\sigma^2}\right)^{\frac{N}{2}} \exp\left\{-\frac{u^2}{2\sigma^2}||Dx||^2\right\}, \tag{7.15}$$

where

$$D = \begin{bmatrix} 1 & 0 & 0 & \cdots & 0 & 1 & -2 \\ -2 & 1 & 0 & \cdots & 0 & 0 & 1 \\ 1 & -2 & 1 & \cdots & 0 & 0 & 0 \\ \vdots & \vdots & \vdots & \ddots & \vdots & \vdots & \vdots \\ 0 & 0 & 0 & \cdots & 1 & -2 & 1 \end{bmatrix}, \quad x = \begin{bmatrix} x_1 \\ x_2 \\ \vdots \\ x_N \end{bmatrix}.$$

Here, u^2 that plays a role of a weighting coefficient determining the balance of the fit of the model and the smoothness of the estimate is called a hyperparameter. In this connection, the hyperparameter u^2 is determined by minimizing the ABIC (Akaike's Bayesian information criterion) given by

$$\text{ABIC} = -2\log\int L(x|\sigma^2)p(x|u^2)dx. \tag{7.16}$$

Therefore, in order to determine the unknown parameter $\delta(x)$ in Bayesian method, the procedure shown below is recommended. (1) For a fixed u^2, find x that maximizes

$$L(x|\sigma^2)p(x|u^2) = \left(\frac{1}{2\pi\sigma^2}\right)^{\frac{9}{2}}\left(\frac{u^2}{2\pi\sigma^2}\right)^{\frac{N}{2}}$$

$$\times \exp\left[-\frac{1}{2\sigma^2}\left\{||A\delta(x) - \varphi||^2 + u^2||Dx||^2\right\}\right]. \quad (7.17)$$

(2) Among the several u^2, the one allowing ABIC to be minimum is selected as the optimal hyperparameter, and it is selected as the optimal estimate. Meanwhile when attention is paid to the exponent part in (7.17), the maximization of (7.17) is nothing but the minimization of

$$J(x) = ||A\delta(x) - \varphi||^2 + u^2||Dx||^2. \quad (7.18)$$

The first term of the right-hand member of (7.18) is nonlinear with respect to x. Therefore, by the Taylor expansion of $\delta(x)$ around x_0 assuming that the initial value x_0 is sufficiently close to the estimate $\hat{x}$, we obtain the approximation,

$$\delta(x) \simeq \delta(x_0) + \delta'(x_0)(x - x_0), \quad (7.19)$$

where

$$\delta'(x) = \frac{\partial \delta(x)}{\partial x} = \begin{bmatrix} \exp(x_1) & 0 & \cdots & 0 \\ 0 & \exp(x_2) & \cdots & 0 \\ \vdots & \vdots & \ddots & \vdots \\ 0 & 0 & \cdots & \exp(x_N) \end{bmatrix}. \quad (7.20)$$

Substituting (7.19) into (7.18) yields

$$J(x) \simeq ||\hat{A}x - \hat{\varphi}||^2 + u^2||Dx||^2$$
$$= \left|\left|\begin{pmatrix}\hat{A}\\uD\end{pmatrix}x - \begin{pmatrix}\hat{\varphi}\\0\end{pmatrix}\right|\right|^2, \quad (7.21)$$

where

$$\hat{A} = A\delta'(x_0), \qquad \hat{\varphi} = \varphi - A\delta(x_0) + \hat{A}x_0. \quad (7.22)$$

Thus in actual calculation, the procedure shown below is recommended. (1) Using an adequate initial value x_0, (7.21) is solved by the method of least squares. (2) Repeat the minimization of (7.21) by replacing the x_0 by the newly obtained x. (3) When x converges, it is considered as the estimate that maximizes (7.17).

7.5   Results of the Tank Test Using a Model Ship

To check the validity of the directional wave spectrum estimation method using the Bayesian model, a tank test using a model ship was conducted. Also the accuracy of the proposed method is checked, by comparing the result with that of the extended maximum likelihood method which is most widely used at present for the estimation of the directional wave spectrum.

In the experiment, the motions of the model ship floating at the center of the tank with 0 speed are measured by generating irregular waves (long crested irregular waves) that progress in a specific direction. Furthermore, by installing a wave probe array composed of 3 pieces of the capacity-type wave probes, the simultaneous data obtained by the probes are also measured. Since use of the wave probe makes it possible to measure accurately the directional wave spectrum in the tank, the estimate obtained from this data is considered to be the true directional wave spectrum of the waves

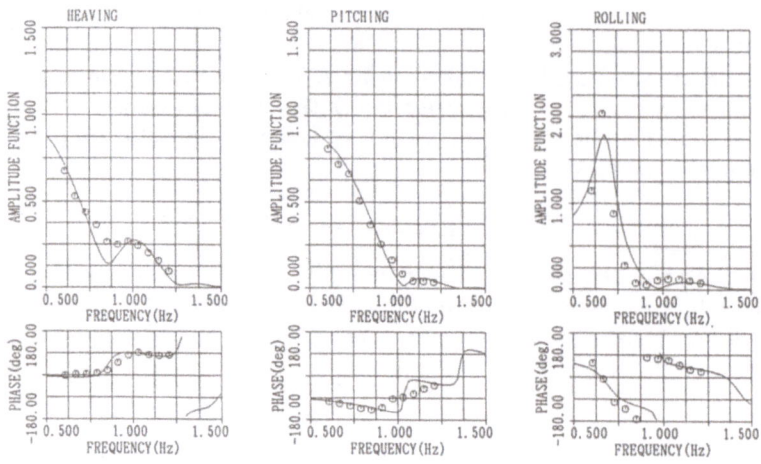

Figure 7.4   Response function of ship motions

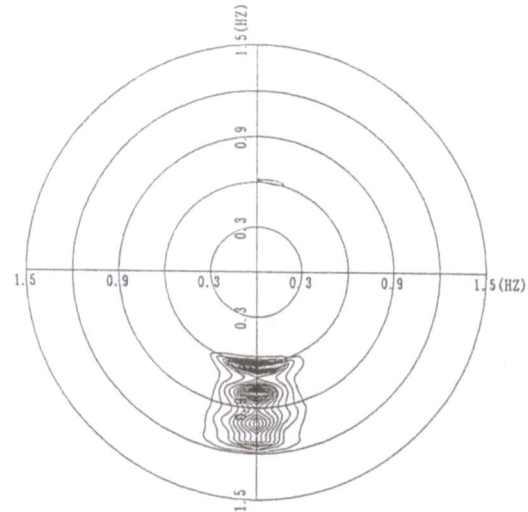

Figure 7.5   Directional spectrum in the tank estimated by wave-probe array

generated in the tank. In the experiment surging motion of the model ship was rigidly
fixed, and sway and yaw motions were lightly restricted by loose springs.

Figure 7.4 shows the comparison between the response function of the ship motions obtained by the theoretical calculation and the one obtained from the model experiment. The theoretical calculation method adopted here is the strip method. The figure shows the case of bow waves with the angle of encounter $\chi = 150°$, from which it is revealed that coincidence of the both is excellent. In the estimation of the

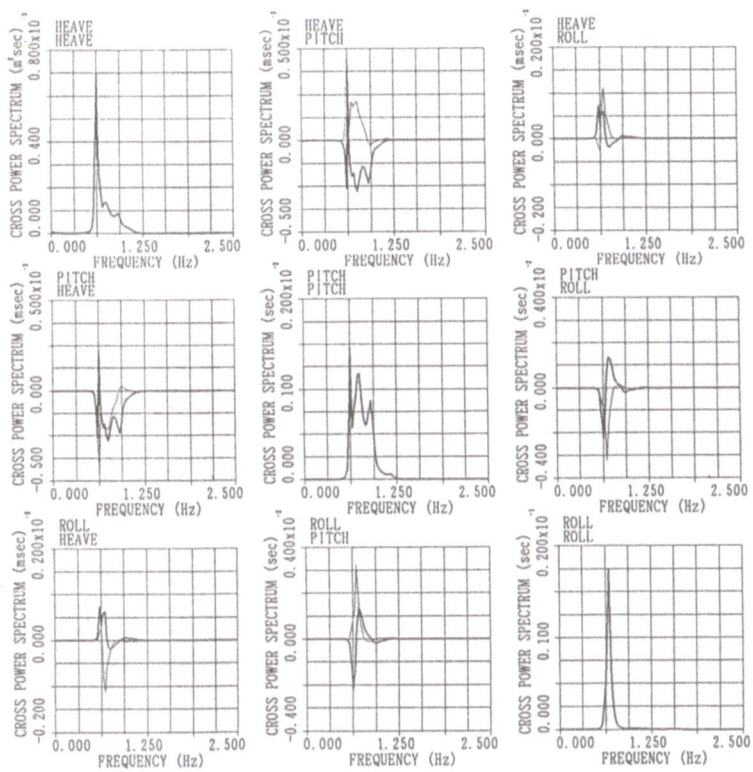

Figure 7.6   Ship motion cross spectrum of the model ship

directional wave spectrum, the response function based on this theoretical calculation
is used.

Figure 7.5 illustrates the directional wave spectrum in the water tank that is estimated from the data of the wave probe array. The figure illustrate the contours of
the directional wave spectrum, with the frequency ranging from 0.0Hz to 1.5Hz. In
this figure, the wave maker is placed downward, and it is perceived from the figure
as well that the direction of the incident wave is estimated with great accuracy. This
result reveals that the significant wave height (mean wave height of the highest 1/3
of the total number of maxima) in the tank was approximately 8cm and the average
frequency was approximately 1sec.

Figure 7.6 shows the cross-spectra of the ship motions that were obtained by using
a multivariate AR model. The ship motions used in the analysis are heave, pitch, and
roll, whereas the angle of encounter is 150°. In the analysis by the multivariate AR
model, the sampling period was 0.2sec and data length 600 points, and the order of
the model determined by MAICE method was 15.

Figures 7.7 and 7.8 illustrate the directional wave spectra in the tank that were
estimated from the cross spectrum shown in Figure 7.6. Figure 7.7 shows the result
estimated by using the Bayesian model, whereas Figure 7.8 reveals the estimate obtained by the extended maximum likelihood method. The arrow marks in both the

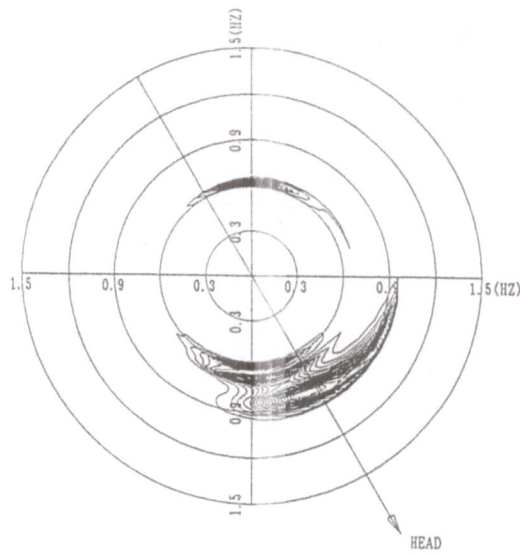

Figure 7.7 Estimation result of the directional wave spectrum by Bayesian model

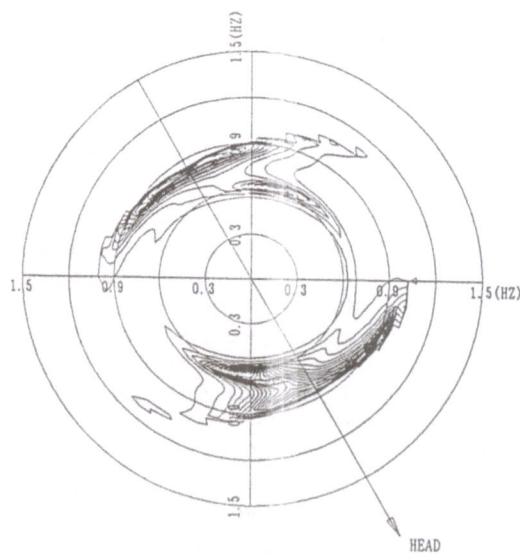

Figure 7.8 Estimation result of the directional wave spectrum by expanded maximum likelihood method

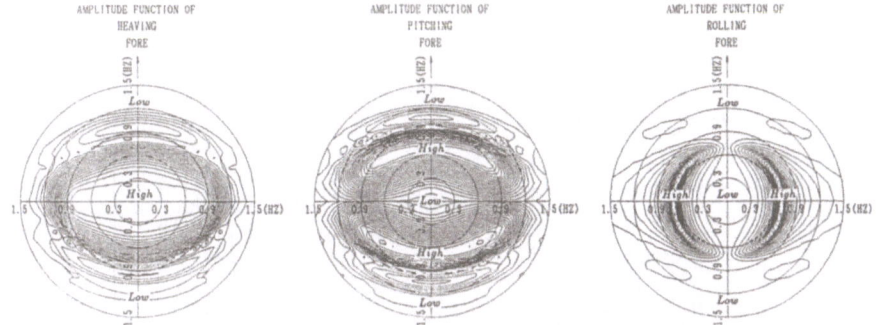

Figure 7.9   Response amplitude of the ship motions

figures indicate the heading direction of the model ship. Comparing these results with the result in Figure 7.5, it is conceived that the peak of the directional wave spectrum appears also in the directional wave spectrum estimated from the ship motion data. However estimation errors in an arc shape appear in the heading direction with Figure 7.7, and in both the directions of the bow and stern with Figure 7.8. It is comprehended that the appearance of the errors makes the conventional peak position of the directional wave spectrum not prominent.

To examine the cause of the estimation errors, the contour lines representing the response amplitudes by theoretical calculation are shown in Figure 7.9.

From the figure, it is explained that there exists in the vicinity of 0.9Hz in the bow-to-stern direction a domain where all the types of the response amplitude take very small values, which coincides with a region where large errors occur. Theoretically speaking, there might be almost none of possibility that ship motions will occur in this domain. However as seen in Figure 7.4, the actual data of the model ship has some power in this region. Therefore, the directional wave spectrum is over-estimated in this region. The estimate of the directional wave spectrum will be improved by increasing the accuracy of the response function.

Figures 7.10 and 7.11 show the comparison between the estimate of the directional wave spectrum by the ship motions and the one by the wave probe array. Shown in the figures is the 1-dimensional wave spectrum obtained by integrating the directional wave spectrum with respect to the angle of encounter. The boldface solid line represents the 1-dimensional wave spectrum estimated from the ship motions, whereas the light-faced solid line represents the one estimated from the wave probe array.

If the area of the part where the boldface and light-faced solid lines are inconsistent with each other is considered to be the error, the error accounts for 59.5% of the whole of the area enclosed by the 1-dimensional wave spectrum in Figure 10 and accounts for 123.3% in Figure 7.11. While deflection of the peak frequency is +0.02Hz in Figure 7.10, the value is 0.14Hz in Figure 7.11. That is to say, it is explained that the estimate by a Bayesian model is excellent in accuracy compared with that of the extended maximum likelihood method.

The accuracy of the method referred to above is shown with respect to all the angles of encounter in an arranged manner in Tables 7.1 and 7.2. The quantities shown in percent in the tables indicate the error ratio previously referred to, whereas the ○

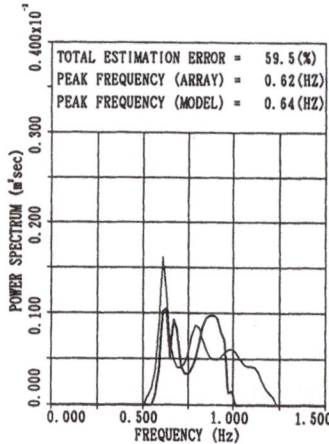

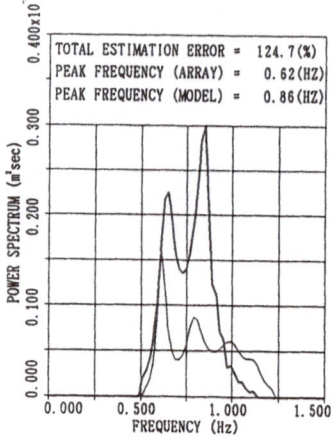

Figure 7.10 Estimation result of the 1-dimensional wave spectra by Bayesian model

Figure 7.11 Estimation result of the 1-dimensional wave spectra by the expanded maximum likelihood method

Table 7.1 Estimation accuracy of the directional wave spectra (Bayesian model)

Angle of encounter	Error	Peak frequency	Wave direction
180°	64.6%	+0.02Hz	○
150°	59.5%	+0.02Hz	+30°
120°	63.0%	+0.36Hz	+30°
90°	50.4%	+0.02Hz	−10°
60°	52.8%	○	−10°
30°	50.8%	+0.02Hz	−20°

marks mean that the estimations are perfectly in agreement with the true values. From the results shown in these tables, it is explained that the accuracy of the 1-dimensional wave spectrum by the Bayesian model ranges from approximately 50% to 60%. It clearly shows that the accuracy is improved for all angle of encounter with the wave compared with the results by the extended maximum likelihood method.

At the next stage, the peak frequency is almost in agreement except for the angle of encounter 120°. On the contrary, it is perceived that deflection is noted in the accuracy in Table 7.2. Finally, while the major direction of the wave (of the direction where there exists the maximum value of the directional wave spectrum) can be estimated with the error within ±30° in Table 7.1, deflection is noticed with the accuracy in Table7.2.

7.6   Conclusions

In this chapter we introduced a method to estimate the directional wave spectrum only from the ship motion data by regarding the ship as a wave probe by using a

Table 7.2 Estimation accuracy of the directional wave spectra (extended maximum likelihood method)

Angle of encounter	Error	Peak frequency	Wave direction
180°	78.6%	+0.04Hz	◯
150°	123.3%	+0.24Hz	+60°
120°	87.3%	+0.02Hz	+200°
90°	48.5%	+0.02Hz	◯
60°	111.6%	+0.24Hz	−20°
30°	86.1%	−0.04Hz	−10°

Bayesian model. To evaluate the accuracy of the estimate, it was compared with that of the directional wave spectrum obtained from the wave height array. Also by comparing with the estimate by the extended maximum likelihood method, superiority of the method based on the Bayesian model was demonstrated.

The results obtained are summarized as shown below.

1) It is shown that the estimate of the directional wave spectrum by a Bayesian model gives good result in accuracy compared with the one by the extended maximum likelihood method. Concurrently with this, it is explained that the estimation accuracy is stabilized for the change of the angle of encounter of the ship with the wave.

2) Use of the Bayesian model makes it possible to estimate the major direction of the wave.

3) Since estimation error becomes large on a part where the response amplitude of the ship motions by theoretical calculation is small, it is necessary to adopt a theoretical calculation method with sufficient accuracy on this part.

Although natural, a ship is, a wave probe whose accuracy is very poor. Especially on a frequency band where almost none of ship motions are induced, it is difficult to obtain good estimation accuracy. However in consideration of the fact that the factor exercising bad influence on ships' safe operation is mainly excessive ship motions caused by the waves in a long-frequency region, the above matter does not influence the utility of the ship-operation guidance system. Furthermore, by the development of a proper Bayesian model, the problem in 3) mentioned above may be solved.

References

Akaike, H. (1980), "Likelihood and Bayes procedure," *Bayesian Statistics*, Bernardo, J. M., De Groot, M. H., Lindley, D. U. and Smith, A. F. M. eds., University Press, Valencia, 143–166.

Akaike, H. and Nakagawa, T. (1988), Statistical analysis and control of dynamic systems, KTK Scientific Publishers.

Capon, J. (1969), "High-resolution frequency-wavenumber spectrum analysis," *Proceedings of IEEE*, Vol. 57, 1408–1418.

Hashimoto, N. (1987), "Estimation of directional spectra from a Bayesian approach", *Report of The Port and Harbour Research Institute*, Vol. 26, No, 2, 97–125.

Hirayama, T. (1987), "Real -time estimation of sea spectra based on motions of a running ship (2nd Report)–Directional wave estimation–", *Journal of The Kansai Society of Naval Architects*, No. 204, 21–27.

Iseki, T. Ohtsu, K. and Fujino, M. (1992), "A study on estimation of directional spectra based on ship motions", *The Journal of Japan Institute of Navigation*, Vol. 86, 179–188.

Iseki, T. Ohtsu, K. and Fujino, M. (1992), "A study on estimation of directional spectra based on ship motions-II–Experimental investigation for accuracy–", *The Journal of Japan Institute of Navigation*, Vol. 87, 197–203.

Iseki, T. Ohtsu, K. and Fujino, M. (1992), "Bayesian estimation of directional wave spectra based on ship motions", *Journal of The Society of Naval Architects of Japan*, Vol. 172, 17–25.

Isobe, M. , Kondo, H. and Horikawa, K. (1984), Expansion of MLM for estimation of directional spectra", *Proceedings of Coastal Engineering, JSCE*, Vol. 31, 173–177.

Kuwashima, S. and Yasuda, A. (1988), "Wave observation at oceangoing vessels", *Navigation* No. 96, 17–25, Japan Institute of Navigation.

Price, W.G. and Bishop, R.E.D. (1974), Probabilistic theory of ship dynamics, Chapman and Hall Ltd..

Chapter 8

Control of Filature Production Process

Akinori Shimazaki

691-2 Kokubu, Ueda-shi, Nagano-ken 386-0016, Japan

Although longer than 1,000m of a cocoon filament can be reeled from a cocoon obtained from a silkworm, the filament is thin enough just to weigh no more than 3g per 10,000m and textural raw threads called raw silk are produced by binding several filaments and by adding other filaments when the reeling filaments are broken. Since the property and style of the reeled cocoon filaments have greatly influence on the quality, production efficiency, yield, etc. of the raw silk, many of important control factors of the process depend on the reeling time. Therefore in settling a filature controlling criterion, a result obtained through time series analysis plays a very important role.

8.1 Dropping-end Control and Gap Process

A thread producing method called a multi-reeling method has long been available. The method always allows a thread to be produced from a fixed number of cocoon filaments in accordance with a measure to immediately add another cocoon filament when a cocoon filament is cut and its end falls, i.e., when dropping-end is made. The raw material in a factory is produced by combining similar ones selected from a lot of cocoons per farmhouse, generally the material is less than the amount of the filature for several days. When a measure to allow the dropping-end to be reduced is taken, the amount of the brushing waste is increased. Meanwhile when a measure to allow the flocks to be reduced, the quality of the raw silk is lowered due to multiplied fine irregularity. Based upon the occurrence of the dropping-end noticed in a filature process, engineers are engaged in the control of the process so that the raw silk as target quality will be economically produced.

Introduction of a statistical control method requiring the time for data collection for a settled term to determine a control criterion has been believed to be difficult in production of threads because the raw material is small. However in case of the filament reeling process, this problem has been solved by the analysis based on a viewpoint of the gap process (Akaike 1956, 1959) and dropping-end control system with good efficiency has been established.

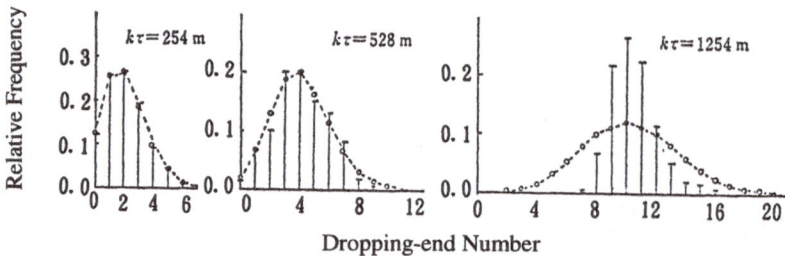

Figure 8.1 Frequency distribution of the dropping-end number created in the term length $k\tau$, bar chart: observed frequency, dotted line: Poisson distribution.

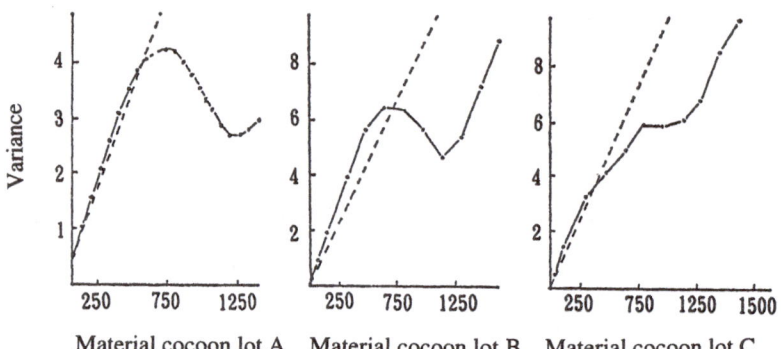

Figure 8.2 Variance-thread length curve.

8.1.1 Size of Research Interval and Distribution of Dropping-end Number

First of all, the determination of the length of the measurement of the dropping-end number was a problem. To comply with this, frequency distribution of the dropping-end number created in the test length $k\tau$, $(k = 1, 2, 3, \ldots)$ was made based upon the time series $\{x_n; n = 1, 2, \ldots\}$ of the dropping-end number produced while the raw silk is reeled with its length τ. An example of the distribution is shown in Figure 8.1. For the sake of comparison, Poisson distribution expected under the assumption that the dropping-end is caused at random is shown in a dotted line. From these it is shown that the dropping-end shows the distribution approximated to the Poisson distribution when the raw silk thread length is less than 500m. However it is also shown that the dropping-end starts to show the distribution concentrated onto the closer periphery of the mean value than the one in case of the Poisson distribution when the research length is elongated exceeding the above-mentioned distribution.

8.1.2 Variance of Dropping-end Number vs. Thread Length Curve

Figure 8.2 shows a variance vs. thread length curve (Akaike 1959) expressing how the dropping-end distribution accompanied with the research length is changed. Also since the mean value and the variance are equal to each other in the Poisson distribution, a mean value vs. thread length curve was shown as the variance of the Poisson distribution vs. thread length curve. From the figure, it is revealed that the variance with the research length to the extent of around 600m is, although affairs vary de-

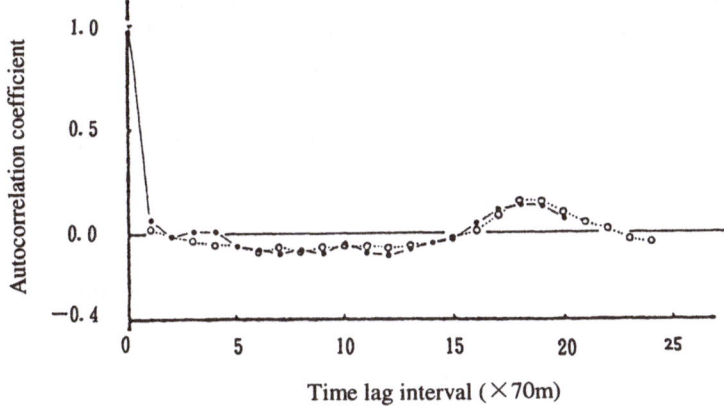

Figure 8.3   Correlogram of dropping-end number.

pending upon the material, equal to the mean value or takes a value higher than the mean value. However it is also revealed that the variance is started to be, after taking a minimum value, decreased when the value exceeds 750m and is converted again to heightening. When such changing structure of the variance is explained, dropping-end control with good efficiency can be done by using the research thread length where the distribution is minimized.

8.1.3   Correlogram of Dropping-end

To comprehend what mutual relationship exists between the dropping-end intervals seen in the filature process, a correlogram is estimated from the time series of the dropping-end and is shown in Figure 8.3. With respect to any of the raw materials, a small peak showing pseudo cycle on the point of more than 1,000m can be seen.

8.1.4   Single-cocoon Filature Process and Gap Process

When the structure of the correlogram of dropping-end is explained from the correlation between the property, style, etc. of the raw material and the filature condition, a controlling criterion of the dropping-end maintaining the optimum condition as per the raw material can be given prior to starting production by using the result of the test filature. The gap process (Akaike 1956, 1959) gives the structure of the dropping-end produced in the filature process of single-cocoon composing the raw silk.

Auto-correlation Coefficient of Dropping-end   Paying attention to the single-cocoon filature process, the cocoon filament is equally divided with an interval length τ short enough to allow the probability causing the dropping-end more than twice can be ignored. Thus a series of the number n $(n = 1, 2, \ldots)$ is given at those intervals. When the occurrence of the dropping-end is expressed with a random variable X_n taking a value $X_n = 1$ when the dropping-end is caused at an interval n and taking a value $X_n = 0$ when the dropping-end is not caused at the interval, the occurrence is approximated with Akaike's gap process. By putting

$$p_\nu \;=\; \Pr\{X_{n+1} = 0, \ldots, X_{n+\nu-1} = 0, X_{n+\nu} = 1 | X_n = 1\},$$

as the probability causing the first dropping-end at the interval $n + \mu$ after the dropping-end is caused at the interval n and furthermore by putting

$$P_\mu = \Pr\{X_{n+\mu} = 1 | X_n = 1\}, \quad P_0 \equiv 1$$

as the probability causing the dropping-end between the intervals n and $n + \nu$,

$$P_\mu = \sum_{\nu=1}^{\mu} p_\nu P_{\mu-\nu}$$

is established. On the other hand, the probability P causing the dropping-end at an interval distant enough from the starting point is given by

$$P = \lim_{n \to \infty} P_n = \frac{1}{\displaystyle\sum_{\nu=1}^{\infty} \nu p_\nu},$$

where

$$E[X_n] = \sum_{\nu=0}^{1} \nu \Pr\{X_n = \nu\} = P \tag{8.1}$$

$$V[X_n] = \sum_{\nu=0}^{1} (\nu - P)^2 \Pr\{X_n = \nu\} = P(1 - P) \tag{8.2}$$

$$\begin{aligned} \mathrm{Cov}(X_n, X_{n+\nu}) &= E[(X_n - P)(X_{n+\nu} - P)] \\ &= \sum_{k=0}^{1}\sum_{\ell=0}^{1} k\ell \Pr\{X_n = k, X_{n+\nu} = \ell\} - P^2 \\ &= P(P_\nu - P) \end{aligned} \tag{8.3}$$

with respect to the n distant enough from the starting point. The X_n is approximated with the stationary probability process called a gap process having auto-correlation coefficient.

$$R_\nu = \frac{\mathrm{Cov}(X_n, X_{n+\nu})}{V[X_n]} = \frac{P_\nu - P}{1 - P}. \tag{8.4}$$

The distribution p_ν is called gap distribution.

Gap Distribution   It has been explained that the dropping-end structure produced in the single-cocoon filature process continued to be reeled with new cocoon filament is perfectly restricted with the gap distribution, when a piece of the cocoon filament is dropped. The distribution is concurrently the distribution of the length of the cocoon filament reeled without being cut after the feeding-end. The distribution is the fundamental one obtained as distribution of the length of the filament reeled from cocoons, i.e. as the non-broken filament length distribution, at the time of the test reeling. The distribution of the length of the single-cocoon filament when the whole of the length of the filament is reeled without being cut is called a total length of cocoon filament distribution and is approximated with the normal distribution and the length in case of being cut in the process exhibits the change approximated to the

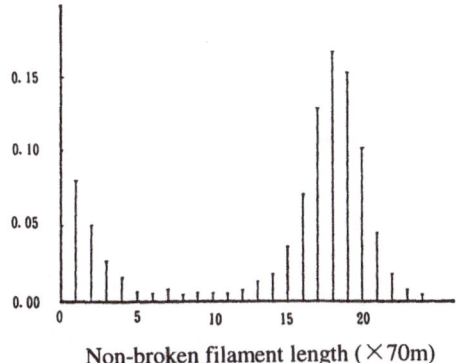

Non-broken filament length ($\times 70$m)

Figure 8.4   Distribution of non broken filament length.

exponential distribution. Thus the reelable thread length distribution, $f(x)$ is given as the mixture distribution of these two as shown in Figure 8.4 (Shimazaki 1956),

$$f(x) \; = \; \alpha_1 \frac{1}{\beta} e^{-\frac{x}{\beta}} + \alpha_2 \frac{1}{\sqrt{2\pi}\sigma} e^{-\frac{(x-\mu)^2}{2\sigma^2}} ,$$

where $\alpha_1 + \alpha_2 = 1$. The equation obtained by dividing the above equation into sections at the interval length τ gives the gap distribution referred to above. From these facts, it has been possible to estimate various characteristics of the dropping-end produced time-dependently thanks to the non-broken filament length distribution which has conventionally been regarded as the fundamental distribution of raw silk production.

Upon the correlogram of the dropping-end obtained from the filature process in Figure 8.3, the correlogram obtained from the gap-process was given by reelable thread length distribution was depict with the white dotted points.

8.1.5   Multi-reeling Process and Gap Process

Let attention be paid to the multi-reeling process allowing a string of raw silk to be obtained by binding k pieces of cocoon filament. At that time, the dropping-end produced upon the individual filaments can be regarded as mutually independent among the cocoon filaments (Shimazaki 1961). From this it is construed that the time-dependent occurrence of the dropping-end noticed in the multi-reeling process is also given from its gap distribution. That is to say, supposing that the dropping-end number produced on the n-th division of the i-th cocoon filament is X_{in} ($i = 1, 2, \ldots, k$) and the one of the raw silk thread with K cocoons is Z_n, we have

$$
\begin{aligned}
Z_n \;&=\; X_{1n} + X_{2n} + \cdots + X_{in} + \cdots + X_{Kn} \\
\mathrm{E}[Z_n] \;&=\; KP, \quad \mathrm{V}[Z_n] \;=\; KP(1-P) \\
\mathrm{Cov}(Z_n, Z_{n+s}) \;&=\; KP(P_s - P) \\
R_{ks} \;&=\; \frac{KP(P_s - P)}{KP(1 - P)} = \frac{P_s - P}{1 - P} = R_s .
\end{aligned}
\tag{8.5}
$$

8.1.6   Variance vs. Thread Length Curve

Let it be considered that a variance vs. thread length curve required for settling a control criterion. Now supposing that the dropping-end number produced on an

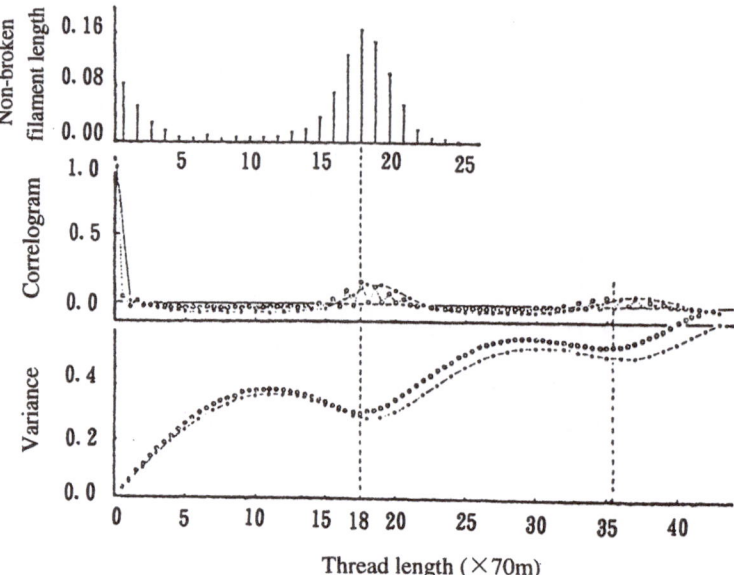

Figure 8.5 Non broken length distribution, correlogram and variance-thread length curve.

arbitrary division n of the multi-reeled raw silk obtained by the filature production is put as Z_n ($n = 1, 2, \ldots, k$) and the one produced on the k pieces of the terms following the precedent divisions, i.e. in the $k\tau$ test thread length is W_k,

$$
\begin{aligned}
W_k &= Z_1 + Z_2 + \cdots + Z_k \\
\mathrm{E}[W_k] &= kKP \\
\mathrm{V}[W_k] &= \mathrm{E}\left[(W_k - kKP)^2\right] \\
&= \mathrm{E}\left[\sum_{n=1}^{k}(Z_n - KP)^2 + 2\sum_{s=1}^{k-1}\sum_{n=1}^{k-s}(Z_n - KP)(Z_{n+s} - KP)\right] \\
&= KP(1 - P)\left\{k + 2\sum_{s=1}^{k-1}(k - s)\frac{P_s - P}{1 - P}\right\}
\end{aligned}
$$

(8.6)

(8.7)

(8.7) are obtained from (8.5). That is to say, it shows that the complicated change of the dropping-end variance accompanied with the test thread length seen in Figure 8.2 is caused by the reelability thread length distribution. An example of the dropping-end variance vs. the thread length curve of the raw silk estimated by obtaining an auto-correlation coefficient from the variance of the dropping-end vs. the thread length curve produced in a raw silk filature process of 8-reeling cocoon filament is shown in Figure 8.5. From the figure, it is revealed that the position of the minimized variance coincides with the mode of the non-broken filament length distribution, i.e. the average of total cocoon filament length.

8.1.7 Dropping-end Control

The droppings-end individually generated on the raw silk that is being produced can be in a sense regarded as mutually independent. Therefore also with the dropping-end number of the k_τ term generated on the raw silk of L pieces of raw silk thread that are being produced in a factory, both its average and its variance can also be obtained if the non-broken filament length distribution is given. Based on this, a control method as shown below allowing dropping-end control to be done simultaneously with the production has been settled by giving a control criterion of the raw material before starting production of raw silk.

1) Using approximately 1,000 pieces of test cocoon filament, the non-broken filament length distribution under the treatment condition appropriate enough for the material filature is built up.

2) The average dropping-end number $E[W_k] = kKP$ and the variance $V[W_k] = \sigma^2$ produced while single-string raw silk is reeled at the average of total cocoon filament length are calculated.

3) By investigating the dropping-end number while raw silk is reeled at the average cocoon filament length $(k\tau)$ with the raw silk of 100 strings, the average dropping-end number $\overline{W}$ and the variance, variance V are obtained.

4) $\Delta W = \overline{W} - kKP$ and $\Delta V = V - \sigma^2$ are calculated. That is to say, turbulence components of the process are extracted by excluding the raw cocoon property and style components composing the dropping-end variation components.

By using these results, it has become possible to detect the matters as shown below.

1) Change of ΔW: Deviation from the standard condition, influencing causes such as the cocoon boiling condition, reeling hot-water temperature, etc.

2) ΔV: Safety degree of working techniques such as turbulence of cocoon supplying, workers' idiosyncrasy, etc.

3) Time series of variation of ΔW and ΔV: Turbulence of homogeneity on the occasion of combining material cocoons.

As is observed in the above, influence of the material characteristics was greatly exercised upon the occurrence of the dropping-end while filature was in progress. At the same time, the raw material was so small that introduction of a statistical control method was difficult. However by excluding the influence of the dropping-end components depending on the material cocoon filament, the information concerning the control turbulence of the process was efficiently extracted with introduction of an idea of the gap process. Thus a dropping-end control method that can start accurate control as soon as the production was started was built up.

8.2 Size Control of Raw Silk

With long texture such as raw silk, nylon thread, etc., a value "denier" representing thickness has been established taking up as the unit the thread with the weight 0.05g per the length of 450m. The cocoon filament is, as shown in Figure 8.6, of the size thin enough with the average size of no more than 3 deniers taking the thickest diameter around the length ranging from 100m to 200m. Trend of the filature technology is steered into a direction of manufacturing raw silk with the required thickness almost free from size irregularity by combining several pieces of the filament with a varying cocoon filament size curve in accordance with the individual cocoons or the affairs inside such cocoons. Based on the control criterion called a fixed size control method,

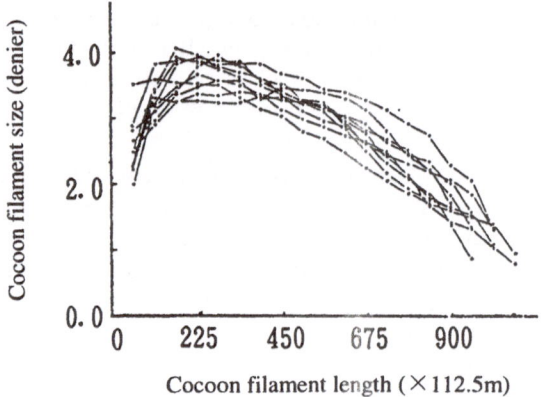

Figure 8.6   Cocoon filament size curve.

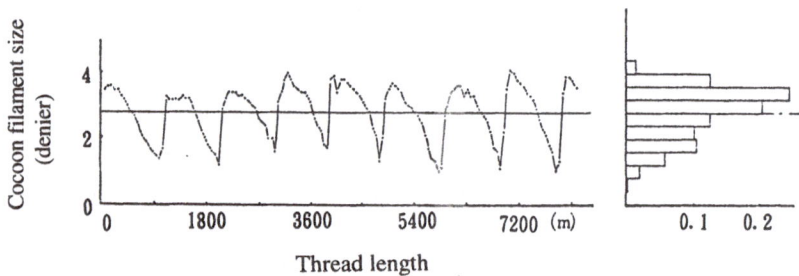

Figure 8.7   Size time series of cocoon filament formed by adding a single piece of filament.

raw silk size control is at present designed to allow the raw silk with desired size almost free from filament irregularity to be produced freely in any of raw materials.

8.2.1   Size Control of Multi-reeling Raw Silk

First of all, let us consider the size characteristics of the conventional multi-reeling raw silk creating a thread of raw silk by gathering and binding a fixed number of cocoon filament.

Size of Single-cocoon Filature   The size time series of the cocoon filament obtained by continuously collecting the filament of 56.25m from the long enough cocoon filament formed by adding a single cocoon filament is shown in Figure 8.7. In the meantime, the correlogram and power spectrum of the time series are shown in Figures 8.8 and 8.9. In the size time series of the cocoon filament, the cocoons are sometimes replaced with other ones because of cutting in the middle of the series. However it is understood from the correlogram or power spectrum that the single-cocoon size curve composes directly the main body of the change as the fundamentals.

Size Control of Multi-reeling Raw Silk   The correlogram of the size time series data obtained by continuously collecting the thread of 112.5m from the multi-reeling raw silk gained by collecting and by binding k pieces of the cocoon filament is shown in Figure 8.10. At the same time in Figure 8.9, the power spectrum of the

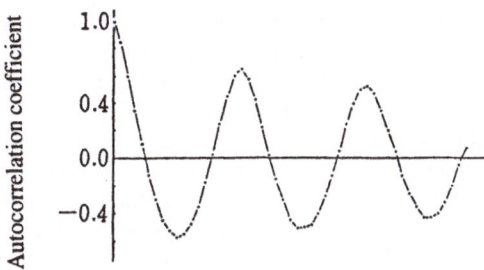

Figure 8.8   Size correlogram of cocoon filament.

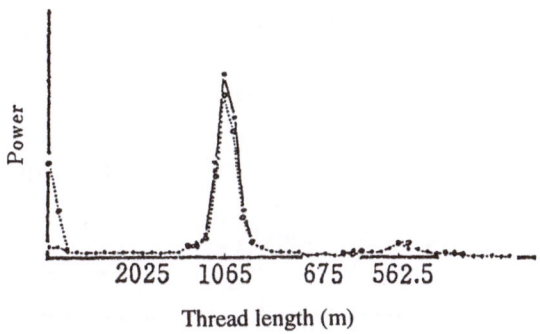

Figure 8.9   Power spectrum of cocoon filament size and raw silk size

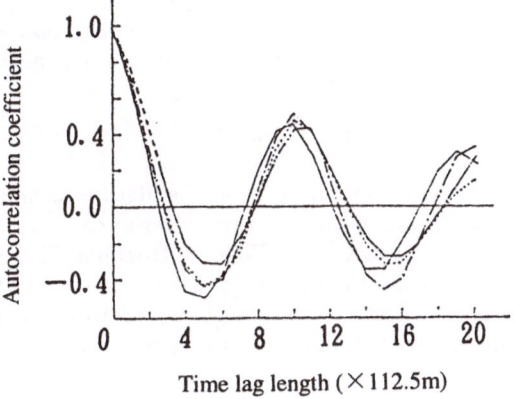

Figure 8.10   Size correlogram of fixed number of cocoon raw silk

multi-reeling raw silk obtained at the inspection thread length of 56.25m is shown. By combination of pieces of the cocoon filament, some irregularity great undulation of the wave never noticed with the cocoon thread size on the size wave of the multi-reeling raw silk can be seen. However from these figures, it is comprehended that the main body of the size wave is the cocoon filament size curve.

Senior engineers of filature technology make selection of the cocoons when the filament is bound and make judgment, based on whether the cocoon layer of the reeling cocoon is thick or thin, whether the cocoon filament that is being reeled is thick or thin. Thus the engineers can conduct the filature paying attention so that there will exist thickness and thinness of the cocoons composing a piece of filament in a mixed style.

Suppose that the average size of cocoon filament is μ and the standard deviation of the filament is σ. Also suppose that the i-th cocoon filament series of K pieces composing the raw silk, the size of the n-th position is X_{in} $(i = 1, 2, \ldots, K)$, the raw silk size on the position is Z_n. Then we obtain

$$
\begin{aligned}
Z_n &= X_{1n} + X_{2n} + \cdots + X_{in} + \cdots + X_{kn} \\
E[Z_n] &= K\mu \\
V[Z_n] &= E\left[(Z_n - K\mu)^2\right]
\end{aligned}
\tag{8.8}
$$

$$
\begin{aligned}
&= E\left[\sum_{i=1}^{K}(X_{in} - \mu)^2 + 2\sum_{i<\ell}(X_{in} - \mu)(X_{\ell n} - \mu)\right] \\
&= \sigma^2\left(K + 2\sum_{i<\ell}\rho_{i\ell}\right),
\end{aligned}
\tag{8.9}
$$

where $\rho_{i\ell}$ is the correlation coefficient of the cocoon size between the i-th and ℓ-th columns. Meanwhile since $2\sum_{i<\ell}\rho_{i\ell}$ takes the values from $-K$ to $K(K-1)$ when ρ takes the value 0 or ± 1, the size standard deviation of the multi-reeling raw silk is given by

$$
0 \leq \sqrt{V[Z_n]} \leq K\sigma .
$$

However since the mixed filature technology is fundamentally given an opportunity for itself just when filament-end is feeding, the raw silk size deviation is never to be lead to the extent that it is greatly reduced. Thus $\rho_{i\ell} \doteq 0$, $i < \ell$ is generally established. That is to say, the deviation is given by

$$
V[Z] \doteq K\sigma^2 .
\tag{8.10}
$$

As is deduced from the above, the emphasis is, as the filature technology, placed upon faithfully preserving a combining method of raw material cocoons, multi-reeling, etc. in consideration of the following facts. (1) With multi-reeling filature, the target size of the raw silk is restricted to be the multiples of average cocoon filament size. (2) Likewise, the decrease of the size deviation is entrusted to the improvement of the silkworm breed allowing the cocoon filament size curve to take a smooth form. (3) Others.

8.2.2 Size Control of Fixed Size Raw Silk

As one of the measures to control aggressively the raw silk size characteristics, the fixed size reeling method allowing the size irregularity of the raw silk by changing the

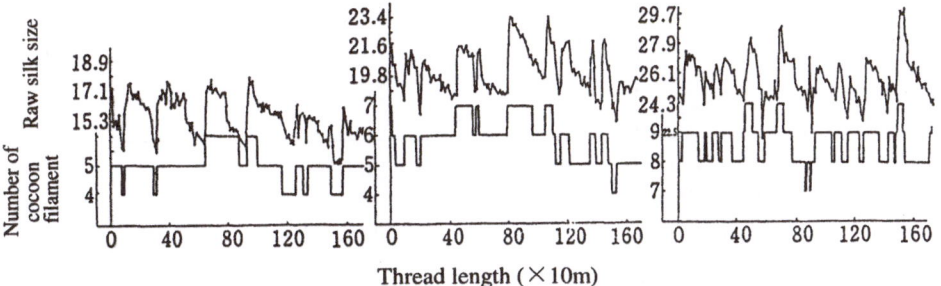

Figure 8.11 Size time series of fixed size raw silk and number of cocoon filaments.

number of the cocoon filament composing the raw silk has gradually been taken up on a trial basis since the early period of the 1950s. However a problem has arisen with the matter where the limit size for cocoon feeding value of the raw silk commanding the filament number change should be settled in a relation with the material characteristics. After the success of the application of the gap process onto the multi-reeling filature, statistical analysis has been promoted also with this problematic point. Furthermore under Akaike's leadership, the problematic point is converted into a problem of the pattern to which a theory of the renewal process is applied and is solved (Shimazaki 1961). Thus all of the Japanese raw silk have successfully been produced in accordance with this filature method.

Size Control Criterion of Fixed Size Raw Silk   When the raw silk during reeling process is made thinner to the extent of the given boundary size, fixed size filature allowing an end of a piece of the cocoon filament to be fed regardless of the number of the cocoon filament composing the raw silk (feeding-end) is made. At that time, the raw silk size is made thicker by the thickness of the cocoon filament that is end-fed at the feeding-end point as is shown in Figure 8.11. However when the cocoon filament number exceeds 5, it is observed that the thickness is linearly decreased. Therefore upon the fundamental assumption that the size of the fixed size raw silk is increased by the increment of the cocoon filament size and later is decreased at a constant gradient the same as the one of the increment, let it be supposed that distribution of the interval ℓ between the feeding-end and the next feeding-end is given by probability density $p(\ell)$. Now, let it be supposed that a position as arbitrary T is taken on the raw silk with the feeding-end point as an origin. Also let it be supposed that the length between the T and the next feeding-end is a random variable X and the probability density of X taking a value x is given by $P_T(x)$. Now, the probability density of the feeding-end of the first time being $T + x$ is $p(T + x)$. On the other hand, the probability of the feeding-end of the first time occurs at a length ζ on this side of the T and the feeding-end of the second time occurs on a place $T + x$ is given by $\int_0^T p_1(T - \zeta)p(x + \zeta)\,d\zeta$. In the meantime, the probability of the feeding-end of the third time occurs on a point $T + x$ is given by $\int_0^T p_2(T - \zeta)p(\zeta + x)\,d\zeta$. Here $p_k(T - \zeta)$ is a probability of the feeding-end of the k-th time occurs at the point $T - \zeta$.

Likewise when T takes a value large enough, the probability density of X taking a

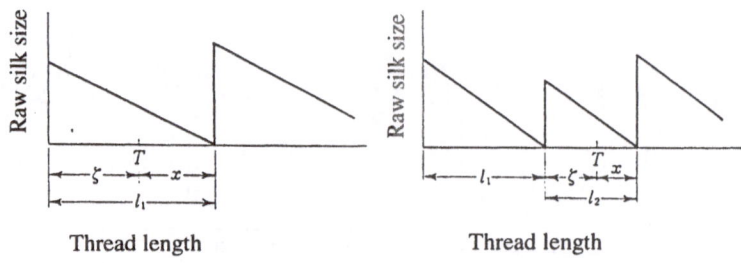

Figure 8.12   Model of fixed size reeling method.

vale x is given from the renewal theory (Cox 1962) as

$$P_T(x) = p(T + x) + \sum_{k=1}^{\infty} \int_0^T p(x + \zeta) p_k(T - \zeta) \, d\zeta \, .$$

Here, assuming that $T \to \infty$,

$$\lim_{T \to \infty} \sum_{k=1}^{\infty} p_k(T - \zeta) = \frac{1}{\mathrm{E}[\ell]} \, ,$$

where by putting $x + \zeta = \ell$, we have

$$\lim_{T \to \infty} P_T(x) = P(x) = \frac{1}{\mathrm{E}[\ell]} \int_x^{\infty} p(\ell) \, d\ell \, .$$

Now, by putting the descending angle of the raw silk size as θ, $\tan \theta = k$, the size of the feeding-end cocoon filament as Z, its probability density as $f(z)$, $z = k\ell$,

$$\Pr\{\ell \le x\} = \Pr\{k\ell \le kx\} = \Pr\{Z \le kx\}$$
$$= \int_0^{kx} f(z) \, dz \, .$$

Thus the boundary size indicating the feeding-end of a single-string cocoon filament is called the limit size for cocoon feeding and is expressed by C. Here by putting the value of the raw silk size of the fixed size raw silk exceeding C as Y and the probability element of Y taking the value y as $g(y)dy$,

$$g(y) \, dy = \frac{1}{\mathrm{E}[Z]} \int_y^{\infty} f(z) \, dz \, dy \, . \tag{8.11}$$

Also the n-th moment of Y is given by

$$\mathrm{E}[Y^n] = \int_0^{\infty} y^n g(y) \, dy = \frac{1}{(n + 1)\mathrm{E}[Z]} \int_0^{\infty} y^{n+1} f(y) \, dy$$
$$= \frac{1}{n + 1} \frac{\mathrm{E}[Z^{n+1}]}{\mathrm{E}[Z]} \, .$$

Denote the average size of the feeding-end cocoon filament as μ_Z and its variance as σ_Z^2. Also denote the average size of the fixed size raw silk as μ_Y and its variance

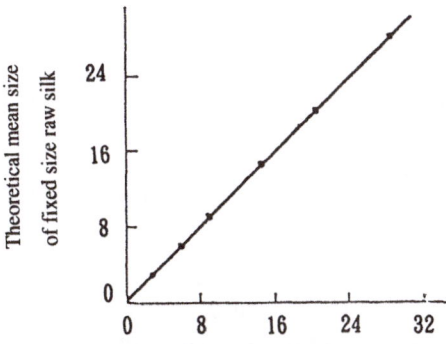

Figure 8.13 Observed mean size and estimate mean size of fixed size raw silk.

σ_Y^2, then we have

$$\mu_Y = C + \frac{\mu_Z}{2}\left\{1+\left(\frac{\sigma_Z}{\mu_Z}\right)^2\right\} \tag{8.12}$$

$$\sigma_Y^2 = \frac{\mathrm{E}[Z^3]}{3\mathrm{E}[Z]} - \frac{\mathrm{E}[Z^2]^2}{4\mathrm{E}[Z]^2}. \tag{8.13}$$

Since the size variance of the feeding-end cocoon filament is approximated by the log normal distribution, (8.13) is approximately given by

$$\sigma_Y^2 \doteqdot \frac{\mu_Z^2}{12}\left\{1+\left(\frac{\sigma_Z}{\mu_Z}\right)^2\right\}^2. \tag{8.14}$$

Thus the criterion of the raw silk size control, C, is given by

$$C = \mu_Y - \frac{\mu_Z}{2}\left\{1+\left(\frac{\sigma_Z}{\mu_Z}\right)^2\right\}, \tag{8.15}$$

from the target size μ_Y and the average size μ_Z of the cocoon filament, and the size variance σ_Z^2.

An example is shown in Figure 8.13. On the other hand, since the average of the cocoon filament size is around 3 deniers and the standard deviation is around 0.5 denier, σ_Z^2/μ_Z^2 is small enough to be 0.03. Accordingly, it is made known that the limit size for cocoon feeding should be settled no less 1/2 of the cocoon filament size than the target raw silk size. In accordance with the same procedure, the size deviation of the raw silk is given as

$$\sigma_Y \doteqdot \frac{\mu_Z}{\sqrt{12}}. \tag{8.16}$$

By comparing with the size standard deviation (8.10) of the multi-reeling raw silk, the size standard deviation of the fixed size raw silk is suppressed to be the value of

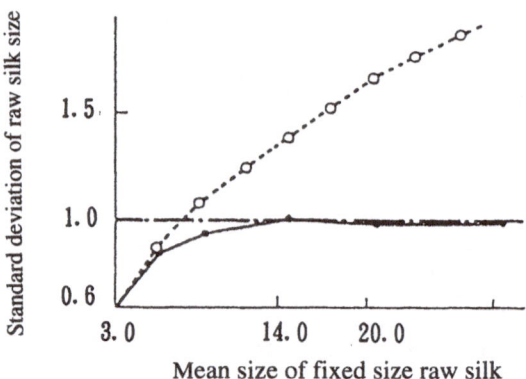

Figure 8.14 Size standard deviation of fixed size raw silk and fixed number of cocoon filament raw silk.

approximately 1/3 of the average cocoon filament size regardless of the thickness of the target size. An example is shown in Figure 8.14. Here, the raw silk detection criterion is detected by means of the detection thread length of 450mm. Since (8.16) has a value given in connection with continuous raw silk, the value takes another value with which the standard deviation obtained in (8.16) is furthermore approximately 0.2 is multiplied when the value for which the detection thread length is considered.

Since the size wave of the multi-reeling raw silk is subject to the undulation in thinness in a manner of multiplication with the overlap of the thick parts of the cocoon filament size curve, the variance is increased in proportion to the number of the cocoon filament. On the other hand, the fineness wave of the fixed size raw silk never becomes larger than the size of the cocoon filament fed in the end at the time of feeding-end and the fineness side is controlled with limit size for cocoon feeding. Therefore the fineness wave of the raw silk just reveals zigzag movement similar in a style to rectangular triangles arranged on a plane.

Production of the raw silk in Japan since 1960 conforms to this size control method in a unified fashion. In this connection, more than half of the raw silk produced in the world is still the multi-reeling raw silk. Figure 8.15 shows how the size standard deviation of Japanese raw silk is decreased because the production is converted into the fixed size reeling method.

8.3   Dwell Time in a Black Box

Cocoons thrown into the groping-end part that looks for the end of the cocoon circulate around the process. Then they are successively sent onto the next process, when they are changed into ended cocoons with the end being caught. However the cocoons once thrown cannot be identified with the ones already thrown. As a disposition to make the dwell time shorter is taken, the amount of brushing waste is increased. In the meanwhile as the dwell time becomes longer, the correct-end cocoons become short of supply resulting in irregularity in production. Thus a relation between the treatment condition and the dwell time has become a problem. However these can be known from the number of the cocoons that have been thrown and the time series analysis. As a result, the production of a series of correct-end cocoons to obtain a

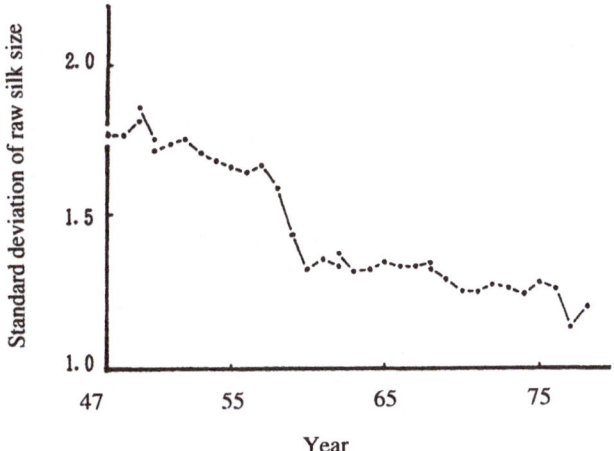

Figure 8.15 Transition of Japanese raw silk size deviation (21 denier raw silk).

single string of the correct end and an adequate control system to the supply process have been settled. This is based on Akaike's concept to estimate the responsiveness of the system to introduce the delay at random by bivariate the gap process with the utilization of the cross correlation between the input and the output.

8.3.1 Number of Ended Cocoons Conveyed

To analyze the behavior of the cocoons circulating around the groping-end part, the individual cocoons are identified as shown below.

I_n : Number of cocoons thrown into the groping-end part at the time point n.

$i_n(k)$: $i_n(k) = 1$ if the code of the cocoon taken out is k and $i_n(k) = 0$ otherwise by attaching formally the numbers on the individual cocoon.

$i_{n-\nu}(k_\nu)$: Taking k_ν as the k-th cocoon thrown at the time point $n - \nu$, $i_{n-\nu}(k_\nu) = 1$ provided the cocoon conveyed is k_ν and $i_{n-\nu}(k_\nu) = 0$ otherwise.

$\delta_{\ell \cdot \nu}(k_\nu)$: Supposing that the k_ν-th cocoon is conveyed ℓ hours after it is thrown in, $\delta_{\ell \cdot \ell}(k_\ell) = 1$, and $\delta_{\ell \cdot \nu}(k_\nu) = 0$ if $\ell \neq \nu$.

With such notations, $i_{n-\nu}(k\nu)\delta_{\ell \cdot \nu}(k_\nu)$ takes a value 1 if the cocoon of k_ν thrown at the time of n_ν is conveyed after ν hours and takes a value 0 otherwise. Accordingly the number of ended cocoons O_n is given as

$$O_n = \sum_{\nu=0}^{n} \sum_{k_\nu=1}^{N_{n-\nu}} i_{n-\nu}(k_\nu)\delta_{\ell \cdot \nu}(k_\nu). \tag{8.17}$$

8.3.2   Dwell Time Distribution

Let us consider the number of the cocoons thrown at the time of $n - m$, I_{n-m} and the covariance $\mathrm{Cov}(I_{n-m}, O_n)$ of the number of the ended cocoons O_n. First of all,

$$
\begin{aligned}
O_n I_{n-m} &= \left[\sum_{\nu=0}^{\infty} \sum_{k_\nu=1}^{N_{n-\nu}} i_{n-\nu}(k_\nu)\delta_{\ell\cdot\nu}(k_\nu)\right]\left[\sum_{k_m=1}^{N_{n-m}} i_{n-m}(k_m)\right] \\
&= \sum_{k_m=1}^{N_{n-m}} i_{n-m}(k_m)i_{n-m}(k_m)\delta_{\ell\cdot m}(k_m) \\
&\quad + \sum_{k_m \neq k'_m} \sum i_{n-m}(k_m)i_{n-m}(k'_m)\delta_{\ell\cdot m}(k_m) \\
&\quad + \sum_{\nu \neq m} \sum_{k_\nu=1}^{N_{n-\nu}} \sum_{k_m=1}^{N_{m-n}} i_{n-\nu}(k_\nu)i_{n-m}(k_m)\delta_{\ell\cdot\nu}(k_\nu),
\end{aligned}
$$

where $i_{n-m}(k_m)$ is a value taking 0 or 1, and therefore $\{i_{n-m}(k_m)\}^2 = i_{n-m}(k_m)$. In the meantime, $\sum_{k_m \neq k_m} i_{n-m}(k_m)i_{n-m}(k'_m)$ express the number of the permutation taking out 2 pieces from N_{n-m} and thus is given by $(I_{n-m})^2 - I_{n-m}$. Hence when the expected value of $O_n I_{n-m}$ is obtained by setting the probability that the thrown cocoon dwells for m hours as p_m,

$$
\mathrm{E}[O_n \cdot I_{n-m}] = \left\{\mathrm{E}\left[I_{n-m}^2\right] - \mathrm{E}[I_{n-m}]^2\right\} p_m + \sum_{\nu=0}^{\infty} \mathrm{E}[I_{n-m}]\,\mathrm{E}[I_{n-\nu}]\,p_\nu. \tag{8.18}
$$

In this case,

$$
\mathrm{E}\left[\sum_{k_m=1}^{N_{n-m}} i_{n-m}^2(k_m)\delta_{\ell\cdot m}(k_m)\right] = \mathrm{E}\left[\sum_{k_m=1}^{N_{n-m}} i_{n-m}(k_m)\delta_{\ell\cdot m}(k_m)\right] = \mathrm{E}[I_{n-m}]\,p_m
$$

$$
\mathrm{E}\left[\sum_{k_m \neq k'_m} \sum i_{n-m}(k_m)\delta_{\ell\cdot m}(k_m)i_{n-m}(k'_m)\right] = \mathrm{E}\left[\sum_{k_m \neq k'_m} \sum i_{n-m}(k_m)i_{n-m}(k'_m)\right] p_m
$$

$$
= \mathrm{E}[I_{n-m}(I_{n-m} - 1)]\,p_m
$$

$$
\mathrm{E}\left[\sum_{\nu \neq m} \sum_{k_\nu=1}^{N_{n-\nu}} \sum_{k_m=1}^{N_{n-m}} i_{n-\nu}(k_\nu)i_{n-m}(k_m)\delta_{\ell\cdot\nu}(k_\nu)\right] = \sum_{\nu \neq m} \mathrm{E}[I_{n-\nu}]\,\mathrm{E}[I_{n-m}]\,p_\nu.
$$

Therefore by setting $\mathrm{E}[I_{n-m}] = \mathrm{E}[I_{n-\nu}] = \mathrm{E}[I]$, $\mathrm{V}[I] = \mathrm{E}\left[I_{n-m}^2\right] - \mathrm{E}[I_{n-m}]^2$, we have

$$
\mathrm{E}[O_n I_{n-m}] = \mathrm{V}[I]\,p_m + \mathrm{E}[I]^2.
$$

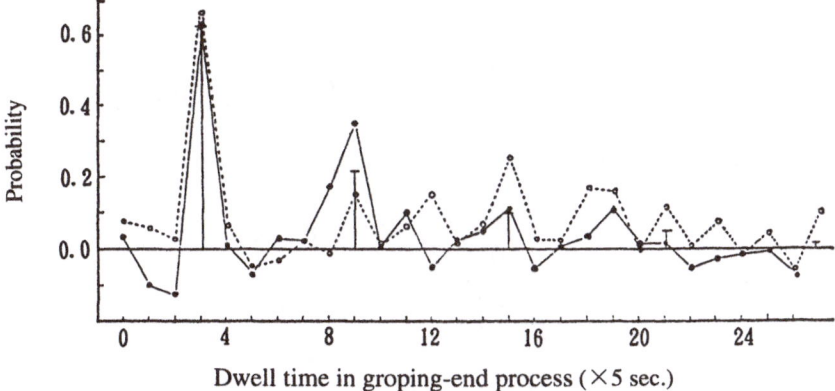

Figure 8.16   Estimation of dwell time frequency in groping-end process.

On the other hand, let the expected value of the cocoons thrown be identical to that of the cocoons taken out, and the covariance of these values is given as

$$\mathrm{Cov}(I_{n-m} \cdot O_n) = \mathrm{E}[I_{n-m} \cdot O_n] - \mathrm{E}[I_{n-m}]\,\mathrm{E}[O_n]$$
$$= \mathrm{V}[I]\,p_m \, .$$

Therefore the probability p_m of the cocoons dwelling on the groping-end part for m hours is given by

$$p_m = \frac{\mathrm{Cov}(I_{n-m} \cdot O_n)}{\mathrm{V}[I]} \, . \qquad (8.19)$$

By obtaining time series data $\{I_n\}$ of the cocoons thrown from the factory survey and by conducting a simulation experiment to which a dwell time distribution during ended cocoons are formed in the groping-end process is given, the ended cocoon time series $\{O_n\}$ conveyed is obtained. In the next stage, an estimated value of p_m obtained by means of (8.19) after the covariance and variance are gained from the time series data of $\{I_n\}$, $\{O_n\}$ is exemplified in Figure 8.16. From this, it is explained that the dwell time distribution can be estimated by (8.19).

8.3.3   Variance of the Number of the Ended Cocoons

Time-dependent variation of the number of the ended cocoons conveyed under the control of the transport process of the cocoons in the raw silk thread manufacturing process has become a controversial problem. On a sphere where the analysis of the previous section is a little different, the analysis of the problem is intended to be made (Bai, Shimazaki 1988).

With respect to the number $O_{n \cdot n - \nu}$ of the cocoons conveyed as ended cocoons at the time n among the cocoons thrown into the groping-end part at the time $n - \nu$, the probability that Z pieces of the cocoons are conveyed is given by

$$\mathrm{Pr}\{O_{n \cdot n - \nu} = Z \mid I_{n - \nu} = k\} = \binom{k}{Z} p_\nu^Z (1 - p_\nu)^{k - Z}$$

provided that the probability of the dwell time being set as p_ν under the condition that $I_{n-\nu}$ is k pieces. When the probability of the number of the cocoons thrown is set as g_k, we have

$$\Pr\{O_{n \cdot n - \nu} = Z\} = \sum_{k=Z}^{\infty} \binom{k}{\nu} p_\nu^Z (1 - p_\nu)^{k-Z} g_k, \quad Z = 0, 1, 2, \ldots$$

Accordingly, the mean and the variance of $O_{n \cdot n - \nu}$ are given by

$$
\begin{aligned}
\mathrm{E}[O_{n \cdot n - \nu}] &= \sum_{Z=0}^{\infty} Z \sum_{k=Z}^{\infty} \binom{k}{Z} p_\nu^Z (1 - p_\nu)^{k-Z} g_k \\
&= \sum_{k=0}^{\infty} \sum_{Z=0}^{k} \binom{k}{Z} p_\nu^Z (1 - p_\nu)^{K-Z} Z g_k \\
&= \sum_{k=0}^{\infty} k p_\nu g_k = p_\nu \mathrm{E}[I] \quad\quad (8.20)
\end{aligned}
$$

$$
\begin{aligned}
\mathrm{V}[O_{n \cdot n - \nu}] &= \sum_{z=0}^{\infty} [Z - \mathrm{E}[I] \cdot p_\nu]^2 \cdot \sum_{k=Z}^{\infty} \binom{k}{Z} p_\nu^Z (1 - p_\nu)^{k-Z} g_k \\
&= \mathrm{E}[I] \cdot p_\nu (1 - p_\nu) + \mathrm{V}[I] \cdot p_\nu^2 . \quad\quad (8.21)
\end{aligned}
$$

Thus, we have

$$\mathrm{V}[O_n] = \mathrm{V}\left[\sum_{\nu=1}^{\infty} O_{n \cdot n - \nu}\right] = \sum_{\nu=1}^{\infty} \mathrm{V}[O_{n \cdot n - \nu}] .$$

By substituting (8.21) to the above equation, we obtain

$$\mathrm{V}[O_n] = \mathrm{E}[I] \left(1 - \sum_{\nu=1}^{\infty} p_\nu^2\right) + \mathrm{V}[I] \sum_{\nu=1}^{\infty} p_\nu^2 . \quad\quad (8.22)$$

From this, it is comprehended that the following procedure is required in order to reduce the variation of the number of the cocoons conveyed.

1) The amount of the cocoons thrown is to be continuously a little adjusted. ($E[I]$ is to be reduced.)
2) The cocoons are to be thrown equally in amount as far as possible. ($V[I]$ is to be reduced.)
3) The groping-end mechanism is to be remodeled so that variance of the dwell time distribution will be reduced. $\left(\sum p_i^2 \to 1\right)$.

In the production/supply process of the correct-ended cocoons in the filature process, the 3 black boxes as indicated in the above are connected and the above-mentioned analytical results can be applied to a flow of the cocoons shifting them.

8.3.4 Several Problematic Point

The estimation accuracy of the dwell time distribution is unexpectedly bad. With respect to this matter, several attempts for observation are made.

Dispersion of $\hat{p}_\nu$ From the time series $\{x_n; n = 1, 2, \ldots, M\}$ of the cocoons thrown that are measured at the same time and the time series $\{y_n; n = 1, 2, \ldots, M\}$ of the cocoons conveyed, the covariance is at first obtained as

$$C_\nu(x \cdot y) = \frac{1}{M} \sum_{n=1}^{M-\nu} (x_n - \bar{x})(y_{n+\nu} - \bar{y}), \quad\quad (8.23)$$

where

$$\bar{x} = \frac{1}{M}\sum_{n=1}^{M} x_n, \quad \bar{y} = \frac{1}{M}\sum_{n=1}^{M} y_n.$$

Secondly, the estimate $\bar{p}_\nu$ of p_ν is obtained by

$$\bar{p}_\nu = \frac{C_\nu(x \cdot y)}{s^2(x)}, \tag{8.24}$$

where

$$s^2(x) = \frac{1}{M}\sum_{n=1}^{M}(x_n - \bar{x}_0)^2.$$

At that time, the variance of $C_\nu(x,y)$ is approximately given by (Bartlett 1968)

$$V[C_\nu(x \cdot y)] \doteq \frac{1}{M}V[I]\,V[O],$$

provided that the number of the thrown/conveyed cocoons is in accordance with the stationary process. From this

$$
\begin{aligned}
V[\bar{p}_\nu] &\doteq V\left[\frac{C_\nu(x \cdot y)}{V[I]}\right] = \frac{1}{V[I]^2}V[C_\nu(x \cdot y)] \\
&\doteq \frac{1}{M}\frac{V[O]}{V[I]} \tag{8.25} \\
&= \frac{1}{M}\left\{\frac{E[I]}{V[I]}\left(1 - \sum_{\nu=1}^{\infty} p_\nu^2\right)\right\} + \frac{1}{M}\sum_{\nu=1}^{\infty} p_\nu^2. \tag{8.26}
\end{aligned}
$$

In order to increase the estimation accuracy of $\bar{p}_\nu$, it is recommended to enlarge tentatively the variance by ceasing throwing or by increasing the amount of throwing.

Research by Division   The estimation accuracy of $\bar{p}_\nu$ is bad as is seen (8.25), and the M requires 1,500 points in order for the accuracy to stay within the range of ± 0.05 with the reliability of 95% even if $V[O]$ and $V[I]$ are the same. When such long-term data are required, attention has to be paid to stationarity. At that time, the time series is divided into N pieces of time series with the magnitude m and p_ν is recommended with respect to each time series. Then $\tilde{p}_\nu$ is estimated as the mean of $p_{\nu i}$'s $(i = 1, \cdots, N)$,

$$\tilde{p}_\nu = \frac{1}{N}\sum_{i=1}^{N} p_{\nu i}. \tag{8.27}$$

The expected value and variance of $\tilde{p}_\nu$ at that time are given as

$$E[\tilde{p}_\nu] = \frac{1}{N}\sum_{i=1}^{N} E[\bar{p}_{\nu i}] = p_\nu \tag{8.28}$$

$$V[\tilde{p}_\nu] = \frac{1}{N^2}\sum_{i=1}^{N} V[\bar{p}_{\nu i}] \doteq \frac{1}{Nm}\frac{V[O]}{V[I]}. \tag{8.29}$$

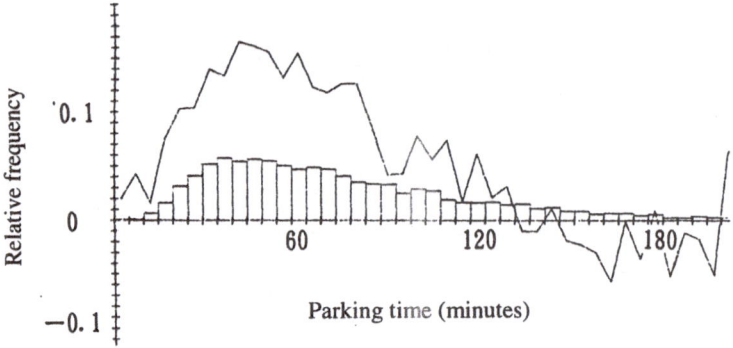

Figure 8.17 Parking time distribution in parking zone of supermarket (1).

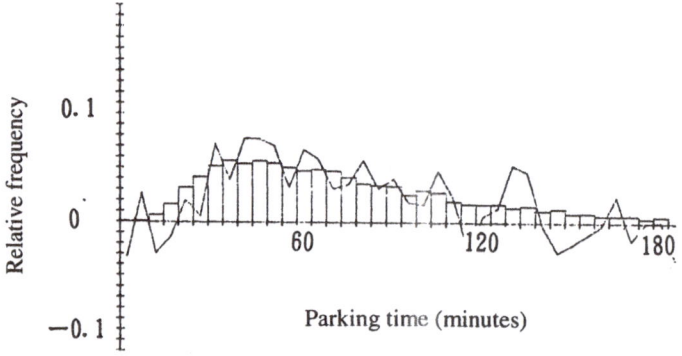

Figure 8.18 Parking time distribution in parking zone of supermarket (2).

Estimation of the Distribution   What result can be obtained is investigated when stationarity is not sufficed. As an example, parking time distribution is considered. The time when a car enters and leaves a parking zone of a supermarket and the body number of the car are measured at the unit of the second, and parking time distribution is obtained from the data. On the other hand, the parking time distribution $\{\bar{p}_\nu\}$ is obtained after the time series of the number of the cars entering and leaving the parking zone within 5 minutes $\{x_n; n = 1, 2, \ldots, M\}$, $\{y_n; n = 1, 2, \ldots, M\}$ are obtained. Since the data number M a day is around 100 points, a result of $\bar{p}_\nu$ estimated taking up the data for 6 days as a piece of the time series is shown in Figure 8.17. The number of the cars is subject to weekdays, and varies even in a day depending on the time band. Therefore, no stationarity cannot be guaranteed. This compels the probability of the individual dwell time estimated to take a relation $\bar{p}_\nu \gg p_\nu$ or to take a minus value. Figure 8.18 shows the result obtained by averaging $\bar{p}_\nu$ six times estimated each day. From this it can almost be observed how parking time distribution is made.

References

Akaike, H. (1956), "On a zero-one process some of its applications," *Annals of the Institute of Statistical Mathematics*, Vol. 8, No. 2, 87–94.

Akaike, H. (1959), "On the statistical control of the gap process," *Annals of the Institute of Statistical Mathematics*, Vol. 10, No. 3, 233–259.

Shimazaki, A. (1956), "On the distribution properties of dropping-ends of cocoon filaments (III), Distribution of non broken filament length," *Journal of Sericultural Science of Japan*, Vol. 25, No. 1, 65–70.

Shimazaki, A. (1961), "Studies on the statistical control of the raw silk production process (I)," *The Bulletin of the Sericultural Experiment Station*, Vol. 16, No. 6, 403–529.

Cox, D. R. (1962), *Renewal theory*, Methuen & Co. LTD., London, 61–70.

Shimazaki, A. and H. Akaike (1966), "Studies on the statistical control of the raw silk production process (IV), "The Bulletin of the Sericultural Experiment Station, Vol. 20, No. 2, 71–186.

Bai, L. and A. Shimazaki (1988), "Estimation of the staying time distribution of a cooked cocoon in the raw silk reeling machine," *Journal of Sericultural Science of Japan*, Vol. 57, No. 5, 369–378.

Chapter 9

Application to Pharmacokinetic Analysis

Akifumi Yafune
Bio-Iatric Center
The Kitasato Institute
5-9-1, Shirokane Minato-ku, Tokyo 108-8642, Japan

9.1 Introduction

Once a drug is administered to the human body, it undergoes the processes of absorption, distribution, metabolism and excretion. The area to study the kinetics of these processes is called pharmacokinetics.

The response magnitude of a drug depends on the drug concentration at the site of action. In practice, however, the drug concentration at the site of action can rarely be measured directly. Instead, the drug concentration is measured at a more accessible site, the blood (or plasma). Repeated blood samplings are made on subjects to estimate their pharmacokinetic profiles in blood, which are quite informative for evaluating the effect and safety of the drug (Rowland and Tozer 1995).

Even when the same dose is administered, the pharmacokinetic profiles generally show large inter-individual variations. These variations have to be taken into account in analyzing the pharmacokinetic profiles, that is, pharmacokinetic analysis.

In the phamacokinetic analysis, it is common to use pharmacokinetic models to describe pharmacokinetic profiles in blood. The models contains some exponential functions of elapsed time after the beginning of administration (Wagner 1975). In the pharmacokinetic model approach, the inter-individual variations are regarded as those of pharmacokinetic parameters included in the pharmacokinetic models.

Recently, there have often been performed Bayesian framework approaches in which the inter-individual variations of pharmacokinetic parameters are directly introduced into a Bayesian pharmacokinetic model as the prior distribution of pharmacokinetic parameters (Sheiner *et al.* 1977; Yamaoka *et al.* 1986). In the approaches, the parameters specifying the prior distribution are estimated by maximizing the likelihood function for the Bayesian pharmacokinetic model. The likelihood function contains a multiple integral, which is analytically intractable since pharmacokinetic models are nonlinear in pharmacokinetic parameters. Hence, the nonlinear models are approximated linearly by the Taylor expansion to calculate the integral analytically. As pointed out in the previous paper (Yafune and Ishiguro 1992), however, the accuracy of this linear approximation deteriorates according to the increase of the number of pharmacokinetic parameters, which may lead to incorrect model selections with AIC (Akaike Information Criterion) (Akaike 1973).

This chapter describes a Monte Carlo estimation of maximum log likelihood for Bayesian pharmacokinetic models and presents an example of application to pharmacokinetic analysis.

9.2   Pharmacokinetic Model

In clinical practice, drugs are administered by several routes, such as oral or intravenous administration. This section considers the case of oral administration.

As mentioned in the previous section, pharmacokinetic profiles in blood is commonly described by pharmacokinetic models consisting of some exponential functions of elapsed time after the beginning of administration. It is supposed that following an oral dose of a certain drug, blood samplings are made on n subjects at m measurement points. The pharmacokinetic model for the drug is supposed to consist of k exponential functions of elapsed time after the administration. Hence, the blood concentration at the j-th measurement point on the i-th subject is expressed by the following model

$$C_{ij} = \sum_{h=1}^{k} \alpha_{kih} \exp(-\beta_{kih}(t_j - t_{\ell ag(ki)})) + \varepsilon_{kij} \quad (i = 1, \ldots, n; \, j = 1, \ldots, m), \quad (9.1)$$

where C_{ij} is the observation at the j-th measurement point on the i-th subject; t_j is the elapsed time from the administration to the j-th measurement point; $t_{\ell ag(ki)}$ is the lag-time for the beginning of absorption; α_{kih} and β_{kih} are pharmacokinetic parameters; ε_{kij} is the intra-individual error term.

For oral administration, it is common to use a 2- or 3-exponential model, among which the better one has to be selected for each drug. In the selection, statistical approaches with AIC are quite useful.

The pharmacokinetic parameters in the model (9.1) have inter-individual variations. To express the inter-individual variations, α_{kih}, β_{kih} and $t_{\ell ag(ki)}$ are assumed to be mutually independent and normally distributed as

$$
\begin{aligned}
\alpha_{kih} &\sim N(\mu_{\alpha_{kh}}, \sigma^2_{\alpha_{kh}}), \\
\beta_{kih} &\sim N(\mu_{\beta_{kh}}, \sigma^2_{\beta_{kh}}), \\
t_{\ell ag(ki)} &\sim N(\mu_{t_{\ell ag(k)}}, \sigma^2_{t_{\ell ag(k)}}).
\end{aligned}
\qquad (9.2)
$$

The inter-individual variations of the parameters are introduced into the analysis by incorporating the normal distributions (9.2) as the prior distributions of the parameters.

For the intra-individual error term ε_{kij}, each term is assumed to be mutually independent and normally distributed as

$$\varepsilon_{kij} \sim N(0, \sigma^2_{\varepsilon_k}). \qquad (9.3)$$

Under the normal assumptions (9.2) and (9.3), by applying the model (9.1) to a real data set, $\mu_{\alpha_{kh}}, \sigma^2_{\alpha_{kh}}, \mu_{\beta_{kh}}, \sigma^2_{\beta_{kh}}, \mu_{t_{\ell ag(k)}}, \sigma_{t_{\ell ag(k)}}$ and $\sigma^2_{\varepsilon_k}$ can be estimated by the maximum likelihood method.

The model (9.1) is, however, nonlinear in pharmacokitenic parameters and the normal distributions (9.2) are incorporated as the prior distributions of the parameters. Hence, the maximum log likelihood cannot be analytically obtained and a numerical approach is needed. As mentioned in Introduction, the linear approximation by Taylor expansion has been generally used so far. The accuracy of this linear approximation, however, deteriorates according to the increase of the number of pharmacokinetic parameters.

To circumvent this problem, a Monte Carlo estimation of maximum log likelihood function is described in the next section.

9.3 Monte Carlo Estimation of Maximum Log Likelihood

k-exponential model (9.1) is rewritten as

$$C_{ij} = f(t_j \mid \boldsymbol{\theta}_{ki}) + \varepsilon_{kij}, \tag{9.4}$$

where C_{ij} is the observation at the j-th measurement point on the i-th subject; $f(t_j \mid \boldsymbol{\theta}_{ki})$ is the predicted concentration by k-exponential model; $\boldsymbol{\theta}_{ki}$ is the vector of pharmacokinetic parameters of the i-th subject; t_j is the elapsed time from the administration to the j-th measurement point; ε_{kij} is the intra-individual error term. As shown in (9.3), ε_{kij} is assumed to be mutually independent and normally distributed as

$$\varepsilon_{kij} \sim N(0, \sigma_{\varepsilon_k}^2).$$

For k-exponential model, the number of pharmacokinetic parameters is $2k + 1$ and $\boldsymbol{\theta}_{ki} = (\theta_{ki1}, \theta_{ki2}, \dots, \theta_{ki(2k+1)})$ is specified as

$$\boldsymbol{\theta}_{ki} = (\alpha_{ki1}, \dots, \alpha_{kik}, \beta_{ki1}, \dots, \beta_{kik}, t_{lag(ki)}). \tag{9.5}$$

From the prior distributions (9.2), each component of $\boldsymbol{\theta}_{ki}$ is assumed to be mutually independent and normally distributed as

$$\theta_{kir} \sim N(\mu_{kr}, \sigma_{kr}^2), \quad (r = 1, \dots, 2k + 1). \tag{9.6}$$

Given $\boldsymbol{\theta}_{ki}$, the simultaneous distribution of C_{ij}, $(j = 1, \dots, m)$ is given by

$$g(C_{i1}, \dots, C_{im} \mid \boldsymbol{\theta}_{ki}) \equiv \prod_{j=1}^{m} \psi(C_{ij} \mid f(t_j \mid \boldsymbol{\theta}_{ki}), \sigma_{\varepsilon_k}^2), \tag{9.7}$$

where $\psi(\cdot \mid \mu, \sigma^2)$ denotes the probability density function of the normal distribution with mean μ and variance σ^2. From the prior distributions (9.6), the simultaneous distribution of $\boldsymbol{\theta}_{ki}$ is specified by the density function

$$\pi(\boldsymbol{\theta}_{ki} \mid \boldsymbol{\omega}_k) \equiv \prod_{r=1}^{2k+1} \psi(\theta_{kir} \mid \mu_r, \sigma_r^2), \tag{9.8}$$

where $\boldsymbol{\omega}_k = (\mu_1, \sigma_1^2, \mu_2, \sigma_2^2, \dots, \mu_{(2k+1)}, \sigma_{(2k+1)}^2)$; μ_r and σ_r^2 denote the mean and variance specifying the prior distribution of the r-th component of $\boldsymbol{\theta}_{ki}$ in (9.5).

Thus, given $\boldsymbol{\omega}_k$ and $\sigma_{\varepsilon_k}^2$, the marginal distribution for the i-th subject is specified by the density function

$$
\begin{aligned}
h(C_{i1},\ldots,C_{im} \mid \boldsymbol{\omega}_k) &\equiv \int g(C_{i1},\ldots,C_{im} \mid \boldsymbol{\theta}_{ki})\pi(\boldsymbol{\theta}_{ki} \mid \boldsymbol{\omega}_k)d\boldsymbol{\theta}_{ki} \\
&= \int \prod_{j=1}^{m} \psi(C_{ij} \mid f(t_j \mid \boldsymbol{\theta}_{ki}), \sigma_{\varepsilon_k}^2)\pi(\boldsymbol{\theta}_{ki} \mid \boldsymbol{\omega}_k)d\boldsymbol{\theta}_{ki}. \quad (9.9)
\end{aligned}
$$

The set of observations for each subject is mutually independent. The log likelihood for the whole data is hence given by

$$
\begin{aligned}
\ell(\boldsymbol{\omega}_k) &\equiv \sum_{i=1}^{n} \log h(C_{i1},\ldots,C_{im} \mid \boldsymbol{\omega}_k) \\
&= \sum_{i=1}^{n} \log \left\{ \int \prod_{j=1}^{m} \psi(C_{ij} \mid f(t_j \mid \boldsymbol{\theta}_{ki}), \sigma_{\varepsilon_k}^2)\pi(\boldsymbol{\theta}_{ki} \mid \boldsymbol{\omega}_k)d\boldsymbol{\theta}_{ki} \right\}. \quad (9.10)
\end{aligned}
$$

The multiple integral in the log likelihood (9.10), which is analytically intractable, is calculated by a Monte Carlo method whose procedures are as follows.

First, $(2k+1)$-dimensional vectors are generated,

$$
\boldsymbol{\lambda}_\ell = (\lambda_{\ell 1}, \lambda_{\ell 2}, \ldots, \lambda_{\ell(2k+1)}), \quad (\ell = 1, 2, \ldots, M),
$$

where M is the number of the generated vectors, and $\lambda_{\ell r}$, $(r = 1,\ldots,2k+1)$ is a normal random number with mean μ_r and variance σ_r^2, which is generated from a standard normal random number $\nu_{\ell r}$ by

$$
\lambda_{\ell r} = \mu_r + \sigma_r \nu_{\ell r}, \quad (r = 1, 2, \ldots, 2k+1).
$$

$\boldsymbol{\lambda}_\ell$ is a realization from the distribution specified by $\pi(\boldsymbol{\theta}_{ki} \mid \boldsymbol{\omega}_k)$ in (9.8).

With $\boldsymbol{\lambda}_\ell$, $(\ell = 1,\ldots,M)$, the density function (9.9) is estimated by

$$
\hat{h}(C_{i1},\ldots,C_{im} \mid \boldsymbol{\omega}_k) = \frac{1}{M}\sum_{\ell=1}^{M} \left\{ \prod_{j=1}^{m} \psi(C_{ij} \mid f(t_j \mid \boldsymbol{\lambda}_\ell), \sigma_{\varepsilon_k}^2) \right\}. \quad (9.11)
$$

By taking M large enough, the estimate (9.11) gives a precise estimate for practical purposes.

Thus, the log likelihood (9.10) is estimated by

$$
\hat{\ell}(\boldsymbol{\omega}_k) = \sum_{i=1}^{n} \log \left[\frac{1}{M}\sum_{\ell=1}^{M} \left\{ \prod_{j=1}^{m} \psi(C_{ij} \mid f(t_j \mid \boldsymbol{\lambda}_\ell), \sigma_{\varepsilon_k}^2) \right\} \right]. \quad (9.12)
$$

By maximizing the log likelihood (9.12) numerically, the maximum likelihood estimate of $\boldsymbol{\omega}_k$ is obtained. With the maximum likelihood estimate $\hat{\boldsymbol{\omega}}_k$, AIC for k-exponential model is given by

$$
\text{AIC} = -2 \times \hat{\ell}(\hat{\boldsymbol{\omega}}_k) + 2 \times (4k+3), \quad (9.13)
$$

where $4k+3$ is the number of free parameters included in the Bayesian model.

Table 9.1   Blood concentrations in eight healthy male subjects $(ng/m\ell)$

No.	\multicolumn{11}{c}{Elapsed time after the administration (hr)}										
	0.25	0.5	0.75	1	2	3	4	6	8	10	24
1	2.5	5.1	14.3	17.1	20.5	21.6	23.3	16.2	14.6	12.0	4.7
2	1.9	10.8	21.1	28.5	30.9	30.9	30.2	20.0	17.3	13.5	3.5
3	2.4	11.5	18.1	31.2	35.7	36.4	30.6	24.2	20.0	12.2	2.4
4	4.9	9.8	18.2	20.1	31.2	27.5	26.0	22.6	15.5	13.4	2.7
5	3.3	14.7	25.6	38.0	35.6	29.8	29.7	24.6	20.5	16.0	2.5
6	3.9	13.5	17.8	21.0	22.8	23.0	19.1	14.4	11.8	9.8	1.1
7	0.5	9.0	15.5	20.9	24.7	24.3	22.8	18.0	14.4	8.9	3.2
8	2.0	5.3	16.2	21.7	25.4	28.0	23.0	21.0	13.6	10.0	1.2

The estimation error in (9.12) is evaluated as follows. By denoting $\hat{h}(C_{i1}, \ldots, C_{im} \mid \omega_k)$ in (9.11) by $\hat{h}_{ki}$, and its mean and variance by $\mu_{M_{ki}}$ and $\sigma^2_{M_{ki}}$, respectively, log $\hat{h}_{ki}$ is expanded around $\hat{h}_{ki} = \mu_{M_{ki}}$, by the Taylor expansion as

$$\log \hat{h}_{ki} = \log \mu_{M_{ki}} + \frac{1}{\mu_{M_{ki}}}(\hat{h}_{ki} - \mu_{M_{ki}}) + \cdots.$$

By ignoring the higher terms, the mean and the variance of log $\hat{h}_{ki}$ are estimated as $\log \mu_{M_{ki}}$ and $\sigma^2_{M_{ki}}/\mu^2_{M_{ki}}$, respectively. Thus, the estimation error in (9.12) is estimated as

$$\delta_{M_k} \equiv \left\{ \sum_{i=1}^{n} \left(\frac{\sigma^2_{M_{ki}}}{\mu^2_{M_{ki}}} \right) \right\}^{\frac{1}{2}}.$$

9.4   Example

Table 9.1 lists blood concentrations in eight healthy male subjects following an oral dose of a certain drug in a clinical phase I trial (Yafune and Cyong 1992). The original profiles are plotted in Figure 9.1. Samples were taken at $0.25, 0.5, 0.75, 1, 2, 3, 4, 6, 8, 10$ and 24 hours after the administration.

By applying 2- or 3-exponential model, the log likelihood function (9.12) is numerically maximized with $M = 8000$. Table 9.2 lists the estimated maximum log likelihood, its standard error and 95% confidence interval, and AIC of each model. Table 9.3 lists the maximum likelihood estimates of each component of ω_k in (9.8). In the present study, the numerical maximization was performed by Davidon method, one of quasi-Newton methods (Davidon 1968; Ishiguro and Akaike 1989).

As shown in Table 9.2, the AIC intervals for the 95% confidence intervals of maximum log likelihood do not overlap each other at all, which suggests that the value of M is large enough. The estimate of AIC for 2-exponential model is smaller than that for 3-exponential model, which suggests that for the present data set, 2-exponential model is more appropriate than 3-exponential model.

Figure 9.2 shows the estimated pharmacokinetic profile based on 1000 simulations from the distribution specified by the estimated parameters for 2-exponential model in Table 9.3. Among three curves in the figure, the central one is the estimated curve

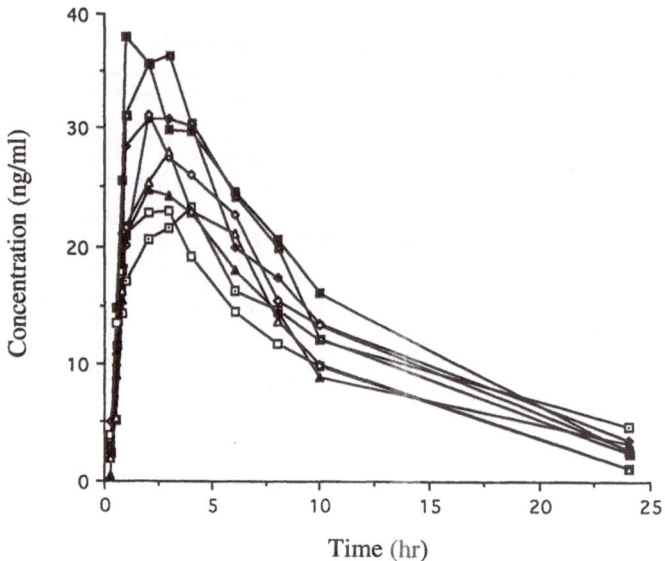

Figure 9.1 Blood concentrations in eight healthy male subjects following an oral dose of a certain drug in a clinical phase I trial

Table 9.2 Estimated maximum log likelihood, its standard error and 95% confidence interval, and AIC of each model ($M = 8000$)

	2-exponential model	3-exponential model
Estimate	−212.34	−212.33
Standard error	0.34	0.40
95% confidence interval	−213.01 ∼ −211.66	−213.11 ∼ −211.56
AIC	446.67	454.67
	(445.33 ∼ 448.01)*	(453.11 ∼ 456.22)*
Number of free parameters	11	15

* AIC interval for the 95% confidence interval of maximum log likelihood

with the means listed in Table 9.3; the upper and lower ones are the upper and lower limits of the 90% confidence interval at a given measurement point, which is defined as the interval including the central 90% of the 1000 simulated concentrations at the measurement point. The estimated profile well represents the original behavior in Figure 9.1 as a whole. Figure 9.2 indicates that the pharmacokinetic profiled have a large inter-individual variation. For example, the highest maximum concentration is about twice the lowest one, which suggest that the effect is suspected to have a large inter-individual variation also.

In clinical practice, the drug are commonly administered two or three times a day. Based on the estimated parameters for 2-exponential model in Table 9.3, pharmacokinetic profiles are estimated for the cases where the drug is administered two and three times a day. In the estimation, the drug is assumed to be administered 30 minutes

Table 9.3 Maximum likelihood estimates

<table>
<tr><td colspan="3" align="center">2-exponential model</td><td colspan="3" align="center">3-exponential model</td></tr>
<tr><td></td><td>μ</td><td>σ</td><td></td><td>μ</td><td>σ</td></tr>
<tr><td>α_{21}</td><td>40.006</td><td>6.961</td><td>α_{31}</td><td>−38.449</td><td>5.588</td></tr>
<tr><td>α_{22}</td><td>−35.051</td><td>4.688</td><td>α_{32}</td><td>29.174</td><td>3.608</td></tr>
<tr><td>β_{21}</td><td>0.121</td><td>0.003</td><td>α_{33}</td><td>13.075</td><td>2.345</td></tr>
<tr><td>β_{22}</td><td>1.371</td><td>0.280</td><td>β_{31}</td><td>1.276</td><td>0.271</td></tr>
<tr><td>$t_{\ell ag(2)}$</td><td>0.322</td><td>0.024</td><td>β_{32}</td><td>0.210</td><td>0.027</td></tr>
<tr><td>ε_2</td><td>****</td><td>1.919</td><td>β_{33}</td><td>0.105</td><td>0.003</td></tr>
<tr><td></td><td></td><td></td><td>$t_{\ell ag(3)}$</td><td>0.255</td><td>0.040</td></tr>
<tr><td></td><td></td><td></td><td>ε_3</td><td>****</td><td>1.920</td></tr>
</table>

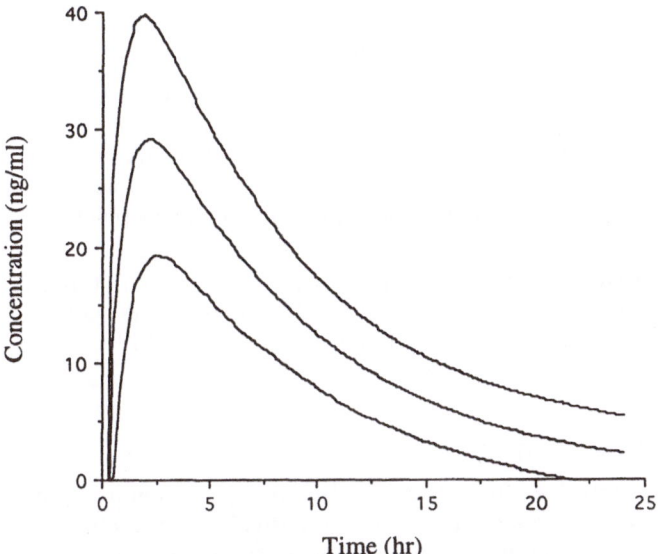

Figure 9.2 Estimated pharmacokinetic profile by 2-exponential model (Among three curves in the figure, the central one is the estimated curve with the means of the prior distributions; the upper and lower ones are the upper and lower limits of the 90% confidence interval at a given measurement point.)

before meals. Hence, the drug is assumed to be administered at 7:30 and 18:30 for two times a day, 7:30, 11:30 and 18:30 for three times a day. The estimated profiles for two times and three times a day are plotted in Figure 9.3 and 9.4, respectively. In both figures, the origin of time axis is the time point of the first administration. Among three curves in the figures, the central one is the estimated curve with the means of the prior distributions; the upper and lower ones are the upper and lower limits of the 90% confidence interval at a given measurement point.

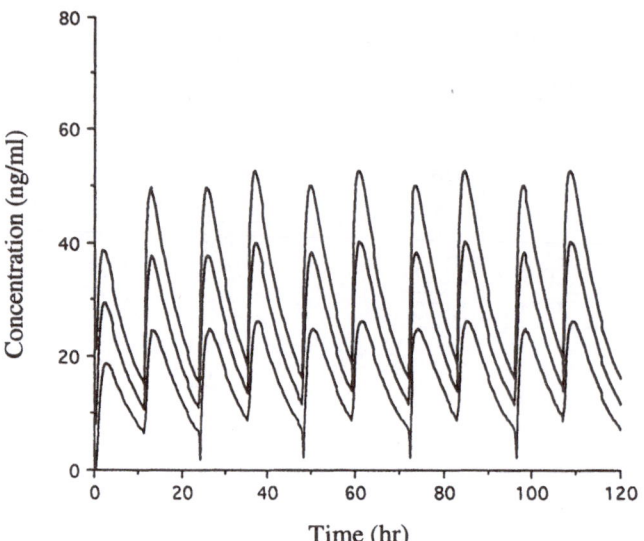

Figure 9.3 Estimated pharmacokinetic profile by 2-exponential model for the case where the drug is administered two times a day, at 7:30 and 18:30 (Among three curves in the figure, the central one is the estimated curve with the means of the prior distributions; the upper and lower ones are the upper and lower limits of the 90% confidence interval at a given measurement point.)

The figures show that although the pharmacokinetic profiles are stable on and after the second day, the estimated concentration for three times a day is obviously higher than that for two times a day. This result suggests that the effect is suspected to be quite different depending on the frequency of administration.

9.5   Concluding Remarks

This chapter described a Monte Carlo estimation of maximum log likelihood for Bayesian pharmacokinetic models in which inter-individual variations of pharmacokinetic parameters are directly introduced as normal prior distributions of the parameters.

In this chapter, an example of oral administration was presented. The proposed approach is, of course, readily applicable to the cases of administration other than oral.

In the present approach, however, some improvable points are left.

First, the prior distributions of pharmacokinetic parameters are not necessarily normal and parameters of a subject may have some correlations. For such cases, another appropriate distribution can be supposed for $\pi(\boldsymbol{\theta}_{ki} \mid \boldsymbol{\omega}_k)$ in the prior distribution (9.8). The number of parameters to be estimated increases according to the number of introduced correlations. If correlations are introduced too many, the estimates of parameters may not be reliable. Hence, correlations of clinical interest only should be introduced.

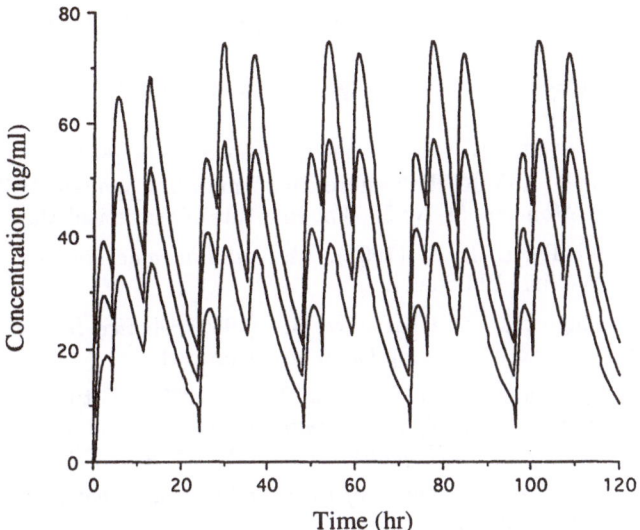

Figure 9.4 Estimated pharmacokinetic profile by 2-exponential model for the case where the drug is administered three times a day, at 7:30, 11:30 and 18:30 (Among three curves in the figure, the central one is the estimated curve with the means of the prior distributions; the upper and lower ones are the upper and lower limits of the 90% confidence interval at a given measurement point.)

Second, the variance of the error term in (9.3) may not be constant. For such a case, a log-normal model is commonly used in pharmacokinetic analysis.

Finally, a drug is often administered after meals, not before meals. In the administration after meals, the process of drug absorption is affected by meals, which has to be taken into account in pharmacokinetic modelings.

The model used in the present approach belongs to Bayesian models. One feature of the Bayesian models is that a pharmacokinetic profile for a new subject from the same population can be estimated based on a few or even a single observation made on the subject. Such Bayesian estimations are actually used in clinical practice for individual dosage adjustments.

Another feature is that various prior distributions can be supposed according to clinical data under investigation.

This chapter presented an example of pharmacokinetic application. Bayesian approaches with AIC are expected to have a wide range of application to clinical data analysis.

References

Akaike, H. (1973), "Information theory and an extension of the maximum likelihood principle," *2nd International Symposium on Information Theory* (Petrov, B. N. and Csaki, F. eds.), Akademiai Kiado, Budapest, 267–281. (Reproduced in *Break-*

throughs in Statistics, Volume 1, S. Kotz and N. L. Johnson, eds., Springer Verlag ag, New York, 1992.)

Davidon, W. C. (1968), "Variance algorithm for minimization," *Computer Journal*, Vol. 10, 406–410.

Ishiguro, M. and Akaike, H. (1989), "DALL: Davidon's algorithm for log likelihood maximization — A FORTRAN subroutine for statistical model builders—," *Computer Science Monographs*, No.25, The Institute of Statistical Mathematics, Tokyo.

Rowland, M. and Tozer, T. N. (1995), *Clinical Pharmacokinetics* (third edition), Williams &Wilkins, Philadelphia.

Sakamoto, Y., Ishiguro,M. and Kitagawa, G. (1986), *Akaike Information Criterion Statistics*, D. Reidel Publishing Company, Dordrecht.

Sheiner, L. B., Rosenberg, B. and Marathe, V. V. (1977), "Estimation of population characteristics of pharmacokinetic parameters from routine clinical data," *Journal of Pharmacokinetics and Biopharmaceutics*, Vol. 5, 445–479.

Wagner, J. G. (1975), *Fundamentals of Clinical Pharmacokinetics*, Drug Intelligence Publications, Inc., Illinois.

Yafune, A. and Ishiguro, M. (1992), "An exact application of maximum likelihood method to pharmacokinetic analysis," *Japanese Journal of Biometrics*, Vol. 13, 5–14.

Yafune, A. and Cyong, J. C. (1992), "Pharmacokinetics of ephedrine after oral administration of Sho-seiryu-to –Analysis in healthy male subjects–" *Japanese Journal of Oriental Medicine*, Vol. 43, 275–283.

Yamaoka, K., Tanaka, H., Okumura, K., Yasuhara, M. and Hori, R. (1986), "An analysis program MULTI(ELS) based on extended nonlinear least squares method for microcomputers," *Journal of Pharmacobio-Dynamics*, Vol. 9, 161–173.

Chapter 10

State Space Modeling of Switching Time Series

Fumiyasu Komaki
Department of Mathematical Engineering and Information Physics
School of Engineering, University of Tokyo
7-3-1 Hongo, Bunkyo-ku, Tokyo 113-8656, Japan
komaki@stat.t.u-tokyo.ac.jp

10.1   Introduction

In various fields of scientific research, it is possible to find time series data that cannot be adequately dealt with by using an ordinary linear Gaussian model. As seen in examples such as the concentration of hormone in the blood, electric potential of the nerve, and river-flow rate, such data exhibiting distinct pulse-shaped patterns can be taken as typical exemplification. Analysis using an input-output model can be made, provided that the input data including pulses is known and induces the pulses in the output. For example, Ozaki (1985) identified the changes in river-flow as the output of a non-linear system with the amount of rainfall as the input, and has been successful in predicting the river-flow rate. However, when the input data are not available, an alternative method is required.

In the field of endocrinology, hormonal time series data and a lot of related problems to be studied from statistical viewpoints have been discussed using various ad hoc methods (e.g. Rahe *et al.* 1980).

On the other hand, time series analysis using Bayesian procedures originated from Akaike (1980) has been highly developed. Here, the smoothness penalties for trends and other components are determined using an information criterion. This approach based on the Bayesian methods and the information criterion has been extended to non-Gaussian time series analysis; Kitagawa (1987) proposes to carry out smoothing and detecting discontinuous jumps in the trend by using the Pearson family of distributions for system innovation.

In this chapter, we introduce a model with stochastically switching modes and propose statistical methods based on non-Gaussian state-space modeling. Although we take hormonal time series data as an example, this approach has a wide range of applicability, not only for hormonal data or time series with pulses, but also for time series with stochastically switching modes. For details, refer to Komaki (1993).

10.2   Time Series Data with Pulses and the Existing Methods

The time series data considered here are illustrated in Figure 10.1. Each time series is composed of observed luteinizing hormone (LH) levels in the blood of a cow.

LH plays a pivotal role in an in-nerve endocrine system governing menstrual cycles. Explanation concerning the mechanism of LH secretion can be found in papers Diggle and Zeger (1989), Knobil and Hotchkis (1988), Lincoln *et al.* (1985), etc.

Hormonal time series are typical examples, where it is difficult to apply ordinary linear Gaussian models. Several types of statistical methods have been proposed up to the present and these methods are both endowed with advantages and disadvantages.

O'Sullivan and O'Sullivan (1988) analyzed hormonal data by introducing parameters to determine the shape, number, and position of the pulses. They assumed that the data are produced by adding observation errors to a superposition of irregularly positioned pulses of the same height. Their method aims to determine the locations of the peaks of the pulses. The problem of this formulation is that any kind of data can be interpolated exactly when a superposition of many small pulses is considered. Therefore, by minimizing a quantity called the generalized cross-validation, which is defined by combining the badness of the interpolation and the complexity of the model, the number of the pulses is determined. Since this optimization problem is accompanied with combinatorial complexity, it is quite difficult to obtain the optimal solution. Thus, an approximate solution is obtained by a procedure including increase/decrease of the number of the pulses. This method is inadequate for the purpose of simulation and prediction. We shall not follow this route here.

Diggle and Zeger (1989) introduced a new non-Gaussian time series model, taking the dynamics of the level change of the hormone in the blood into account. Their model is a first-order AR model including jumps and a feedback structure. The LH level y_i observed at the time i is generated by

$$y_i = \rho y_{i-1} + w_i,$$

where $0 < \rho < 1$. The innovation w_i is distributed according to the Gamma distribution with probability ϕ_i and the normal distribution with probability $1 - \phi_i$. Here, the Gamma distribution corresponds to jumps and the normal distribution corresponds to observation noises and random fluctuations with mean 0. The function ϕ_i is defined by

$$\phi_i = \phi(y_{i-1}; \beta_0, \beta_1) = \frac{1}{1 + \exp(-\beta_0 - \beta_1 y_{i-1})},$$

which is the logistic function dependent on y_{i-1}. The feedback structure they introduced is essential to analyze this kind of data, and the methods based on time series models are useful for prediction and simulation.

However, it can hardly be said that first-order Markovian models are appropriate for analysis of the data with pulses. While the pulse starts its elevation and reaches a peak, the ascent sometimes occurs more than twice. In this case, despite the fact that the pulse should intrinsically be considered a single pulse as Diggle and Zeger themselves state, it is liable to be construed on account of Markovian property that more than two pulses have continuously been produced. Furthermore, as seen in the second pulse on the first day of Cow A in Figure 10.1, the actual peak of the LH level appears to be placed between the two continuous observation points. This is due to

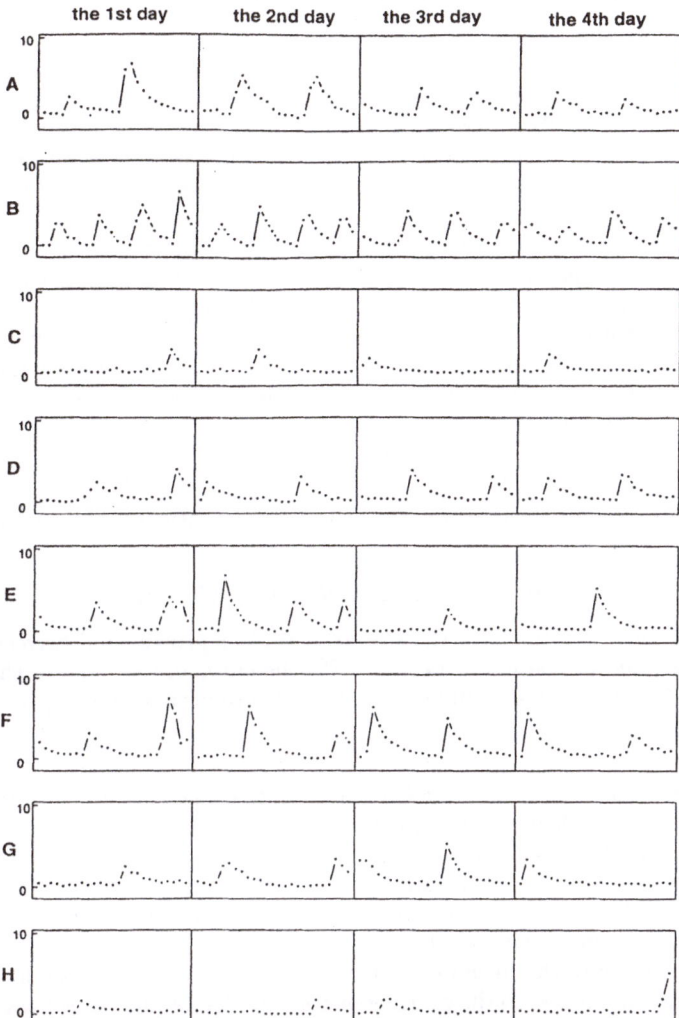

Figure 10.1 LH level time series for eight cows (A-H). For each cow, 25 observations are taken at 15-minute intervals over a six-hour period on four consecutive days.

the fact that the intervals of the individual points of the observation time are taken relatively roughly in comparison with the variation width of the LH level. Therefore to grasp the features of the dynamics in the change of the hormone, a model which takes the change of the continuous hormone level into account is required.

Because of the reasons above, it is necessary to introduce an analyzing method which is more faithful to the data.

10.3   The State Space Model for Time Series with Pulses

10.3.1   The Model

Let us consider the data of the eight cows. There are four time series for each cow and each series contains 25 values of LH in the blood. The dynamics of the LH level is comprised of two modes, i.e. rising mode and decreasing mode. Switching between the two modes occurs stochastically. The rising mode corresponds to, as referred to in O'Sullivan *et al.* (1984) and Knobil and Hotchkiss (1988), a state of the LH being released.

It is known that the level of the LH ascends when the activity of the pulse generator of the hypothalamus is high, and descends when the activity is low (e.g. Lincoln *et al.* 1985). The two high and low levels concerning the generation ratio of the spike of this pulse generator are distinct and correspond to the two modes of the model.

We consider that the level of the LH changes in accordance with the following stochastic differential equations. In the rising mode, a stochastic differential equation with linearly increasing drift

$$dx(t) = \alpha dt + \gamma dB(t) \qquad (\alpha > 0)$$

is supposed. In the decreasing mode, a stochastic differential equation with exponentially decreasing drift

$$dx(t) = -\beta x(t)dt + \gamma dB(t) \qquad (\beta > 0)$$

is considered. Here $B(t)$ denotes the standard Wiener process and corresponds to a relatively small fluctuation in comparison with the curve of the pulse of the LH. The data $\{y_\ell\}$ we observe at the sample times $\{\tau_\ell\}$ is produced by adding a observation noise to $\{x(\tau_\ell)\}$:

$$y_\ell = x(\tau_\ell) + w_\ell,$$

where the observation noise $\{w_\ell\}$ is a Gaussian white noise sequence with variance σ^2. As is described (iii) below, the two modes stochastically switch to each other. The probability of switching depends on $x(t)$.

Our interest is mainly concentrated on estimating the starting time of the two modes from the observed LH data. The peak of the pulse corresponds to the time of the initiation of the decreasing mode.

In order to compute the likelihood function, we explicitly formulate a state space model as an approximation to the continuous time model, by the following three steps.

(i) Time subdivision.

   If the individual sampling intervals are divided equally into n parts (here we put $n = 10$), then 241 time points $t_0, \ldots, t_{240}$ are obtained. The 25 time points that are sampled at the same interval for 6 hour are also included in these time points. The first time point t_0 is the first observation time, and the last time point t_{240} is the 25th observation time. The sample interval is taken as a unit of time. Here, the unit is 15 minutes. The step interval is Δt ($= t_{i+1} - t_i = 0.1$; $i = 0, \ldots, 239$) is 1.5 minutes. Thus $\tau_\ell = \ell - 1(\ell = 1, \ldots, 25)$, where τ_ℓ denotes ℓ-th observation time. The time points without observations are to be dealt with as missing observations. As is to be described later, the state at the time points where no observations are obtained can be estimated by smoothing. One of the advantages of the state space method is the fact that missing observations can very naturally be dealt with.

(ii) Shape of the pulses

The model operates in two modes, i.e. the rising mode and decreasing mode. In the rising mode, the level of the LH increases linearly with time. In the decreasing mode, the level is attenuated exponentially. Here, we introduce the time series $m(t)$ corresponding to the modes. The rising and decreasing modes correspond to $m(t_i) = 1$ and $m(t_i) = 0$, respectively. Thus, the state can be defined by the pair that cannot be directly observed

$$s_i = (m_i, x_i) \qquad (i = 0, \ldots, 240).$$

Here, m_i and x_i denote $m(t_i)$ and $x(t_i)$, respectively. We consider a parameter ξ of the increasing gradient and a parameter ρ of the decreasing ratio for the individual cows. These values are assumed to be constant during the whole observation period. Thus, when $m(t_i) = 1$,

$$x(t_{i+1}) = x(t_i) + \xi + v_i \qquad (\xi > 0),$$

and, when $m(t_i) = 0$,

$$x(t_{i+1}) = \rho x(t_i) + v_i \qquad (0 < \rho < 1).$$

Here, v_i denotes the system noise at the time t_i independently distributed according to $N(0, \lambda^2)$. We observe $x(t)$ with additional noise

$$y_\ell = x(t_i) + w_\ell, \tag{10.1}$$

where $i = 10(\ell - 1)$, and w_ℓ denotes the observation noise independently distributed according to $N(0, \sigma^2)$.

(iii) The switching between the two modes.

Diggle and Zeger (1989) modeled the fact that release of LH is controlled by the negative feedback from the LH level in blood. Here, we introduce the structure of similar feedback structure into our model. First, we assume the onset probability of LH release $\phi(x(t_i))$ is depending on the level of the hormone level $x(t_i)$ at t_i such that

$$\phi(x) = \frac{1}{1 + e^{-\beta_0 - \beta_1 x}}.$$

When $\beta_1 < 0$, the probability of the pulse occurring increases as $x(t)$ decreases. Second, to model the variation of the height of the pulse, we assume that the probability of switching from the rising mode to the decreasing mode is given by

$$\psi(x) = \frac{1}{1 + e^{-\gamma_0 - \gamma_1 x}} \qquad (\gamma_1 \geq 0).$$

In the decreasing mode, release of the LH into the blood starts with the probability $\phi(x_i)$ depending on the level of the LH. In the rising mode, the mode switches to the decreasing mode with probability $\psi(x_i)$ and the release stops.

From the procedure (i) – (iii) described above, the transition probability densities of the state $s_i = (m_i, x_i)$ are given by

$$
\begin{aligned}
\Pr\{(0, dx_{i+1}) \mid (0, x_i)\} &= \{1 - \phi(x_i)\}\, n(x_{i+1}; \rho x_i, \lambda^2)\, dx_{i+1} \\
\Pr\{(1, dx_{i+1}) \mid (0, x_i)\} &= \phi(x_i)\, n(x_{i+1}; x_i + \xi, \lambda^2)\, dx_{i+1} \\
\Pr\{(0, dx_{i+1}) \mid (1, x_i)\} &= \{1 - \psi(x_i)\}\, n(x_{i+1}; \rho x_i, \lambda^2)\, dx_{i+1} \\
\Pr\{(1, dx_{i+1}) \mid (1, x_i)\} &= \psi(x_i)\, n(x_{i+1}; x_i + \xi, \lambda^2)\, dx_{i+1},
\end{aligned}
\tag{10.2}
$$

where $n(x_{i+1}; \mu, \lambda^2)$ denotes the normal density with mean μ and variance λ^2.

In addition to the state $(m_0, x_0), \ldots, (m_{240}, x_{240})$ that cannot directly be observed, the model contains eight parameters ξ, ρ, β_0, β_1, γ_0, γ_1, λ^2, and σ^2. We put $\theta = (\xi, \rho, \beta_0, \beta_1, \gamma_0, \gamma_1, \lambda^2, \sigma^2)$.

It can be proved by using the theorem by Tweedie (1983) that the Markovian state space model defined above has a stationary distribution. Therefore in a strict meaning, the likelihood can be defined.

10.3.2   Calculation of Log Likelihoods

Let $y_1, \ldots, y_{25}$ be the observed time series. The likelihood function is given by

$$q_\theta(y_1) \prod_{l=1}^{24} q_\theta(y_{l+1} \mid y_1, \ldots, y_l). \qquad (10.3)$$

Here $q_\theta(y_1)$ is obtained by

$$q_\theta(y_1) = \int q_\theta(y_1 \mid s_0) p_\theta(s_0) ds_0, \qquad (10.4)$$

where $p_\theta(s)$ is the density function of the stationary distribution of the Markovian process $\{s_i\} = \{(m_i, x_i)\}$. The conditional density $q_\theta(y_{l+1} \mid y_1, \ldots, y_l)$ is successively obtained by

$$q_\theta(y_{l+1} \mid y_1, \ldots, y_l) = \int q_\theta(y_{l+1} \mid s_{10l}) p_\theta(s_{10l} \mid y_1, \ldots, y_l) ds_{10l} \qquad (\ell = 1, \ldots, 24),$$

where the integral concerning the state s_i is defined by the summation of the integrals with respect to x_i with $m_i = 0$ and 1. The conditional density of the integrated function in (10.4) is determined by the relations (10.1) and (10.2).

The likelihood (10.3) is calculated by the numerical integration shown in Kitagawa (1987). First, to obtain the stationary initial distribution $p_\theta(s)$, the transition of (10.2) is repeated until the distribution of the state converges. Second, the convolution

$$q_\theta(y) = \int p_\theta(x) n(y - x; 0, \sigma^2) dx$$

is obtained by using (10.1). Here $p_\theta(x) = p_\theta\{(0, x)\} + p_\theta\{(1, x)\}$. In this way, the contribution $q_\theta(y_1)$ of the initial value to the likelihood (10.3) can be evaluated.

Next, the posterior density of the s_0 given the observation y_1 is obtained by Bayes' theorem

$$p_\theta(s_0 \mid y_1) = \frac{q_\theta(y_1 \mid s_0) p_\theta(s_0)}{q_\theta(y_1)}.$$

Then, $p_\theta(s_{10} \mid y_1)$ can be obtained from $p_\theta(s_0 \mid y_1)$ by repeating 10 times the following transition step

$$p_\theta(s_{i+1} \mid y_1) = \int p_\theta(s_{i+1} \mid s_i) p_\theta(s_i \mid y_1) ds_i.$$

The predictive distributions $q_\theta(y_2 \mid y_1)$ can be obtained by

$$q_\theta(y_2 \mid y_1) = \int q_\theta(y_2 \mid s_{10}) p_\theta(s_{10} \mid y_1) ds_{10}.$$

The other predictive distributions $q_\theta(y_3 \mid y_1, y_2), \ldots, q_\theta(y_{25} \mid y_1, \ldots, y_{24})$ to calculate the likelihood (10.3) can also be calculated in the same way.

We assume that the four time series obtained from the same cow are independent of each other. The sum of the four log likelihoods with the same parameters is used for inference. By maximizing it using a quasi-Newton method, the maximum likelihood estimates for each cow can be obtained.

10.3.3 Posterior Distributions of the States

The states that are not observed directly can be estimated by Bayesian smoothing procedure with the estimated parameters $\hat{\theta}$. This method corresponds to the fixed internal smoothing in Kalman filtering theory. Let Y_m be the set of the observations that have been obtained until time m. Since the observations can only be obtained every 10th point, $Y_m = Y_{m+1} = \cdots = Y_{m+9}$ holds when $m = 10i$.

First, we calculate the predictive distributions $p(s_1 \mid Y_0), \ldots, p(s_{240} \mid Y_{239})$ and filtered distributions $p(s_0 \mid Y_0), \ldots, p(s_{240} \mid Y_{240})$ by the procedure described in the previous section. At the points where the observations are not obtained, the relation $p(s_m \mid Y_{m-1}) = p(s_m \mid Y_m)$ holds. Next, the conditional distributions $p(s_i \mid Y_{240})$ $(i = 0, \ldots, 240)$ are calculated successively in the inverse direction with respect to i by

$$p(s_i \mid Y_{240}) = p(s_i \mid Y_i) \int \frac{p(s_{i+1} \mid s_i)}{p(s_{i+1} \mid Y_i)} p(s_{i+1} \mid Y_{240}) ds_{i+1}.$$

This relation can be obtained from the equation

$$p(s_i, s_{i+1} \mid Y_{240}) = \frac{p(s_{i+1} \mid Y_{240}) p(s_{i+1} \mid s_i) p(s_i \mid Y_i)}{p(s_{i+1} \mid Y_i)}. \tag{10.5}$$

From the posterior density of this states, various kinds of information can be obtained. For example, a marginal posterior probability $\Pr(m_i = 1 \mid y_1, \ldots, y_{25})$ where the state is in the rising mode can be obtained by integrating $p(s_i \mid y_1, \ldots, y_{25})$ with respect to x_i (Figure 10.2). In Figure 10.2, nearly vertical lines crossing upwards stand for the onsets of the pulses and nearly vertical ones crossing downwards correspond to the location of the peaks of the pulses.

Figure 10.3 A–H show the posterior densities $p(x_i \mid y_1, \ldots, y_{25})$ for the individual cows. These are obtained by taking the summation of $p(s_i \mid y_1, \ldots, y_{25})$ with respect to $m_i = 0, 1$. In the figure, the dotted lines indicate the central 95% lines of the marginal posterior probability. The marginal posterior probability of the onset of the two modes is obtained by integrating (10.5) with respect to x_ℓ and $x_{\ell+1}$ (the figure is omitted). From these facts, it can be said that the model has been successful in catching the dynamics of the time series from relatively roughly sampled observations.

10.3.4 Model Selection and Simulations

We can evaluate the goodness of fit of the assumed model by comparing several different statistical models for the same data. Various kinds of modifications of the model can be considered. We compare several different models via their values of AIC, which is defined by

$$\text{AIC} = -2 \, (\text{the maximum log likelihood}) + 2 \, (\text{the number of parameters}).$$

The model that minimizes the value of AIC is selected.

Instead of the linear increase of the LH level supposed in the original model, we can consider the exponential increase given by

$$x(t_{i+1}) = \max\{\eta x(t_i) + v_i, 0\} \qquad (\eta > 1).$$

In this model, the shape of the pulses is similar to the two-sided exponential function assumed by O'Sullivan and O'Sullivan (1988). AIC selects the model with the linear LH level increase for the seven cows except cow E. This result shows that the model

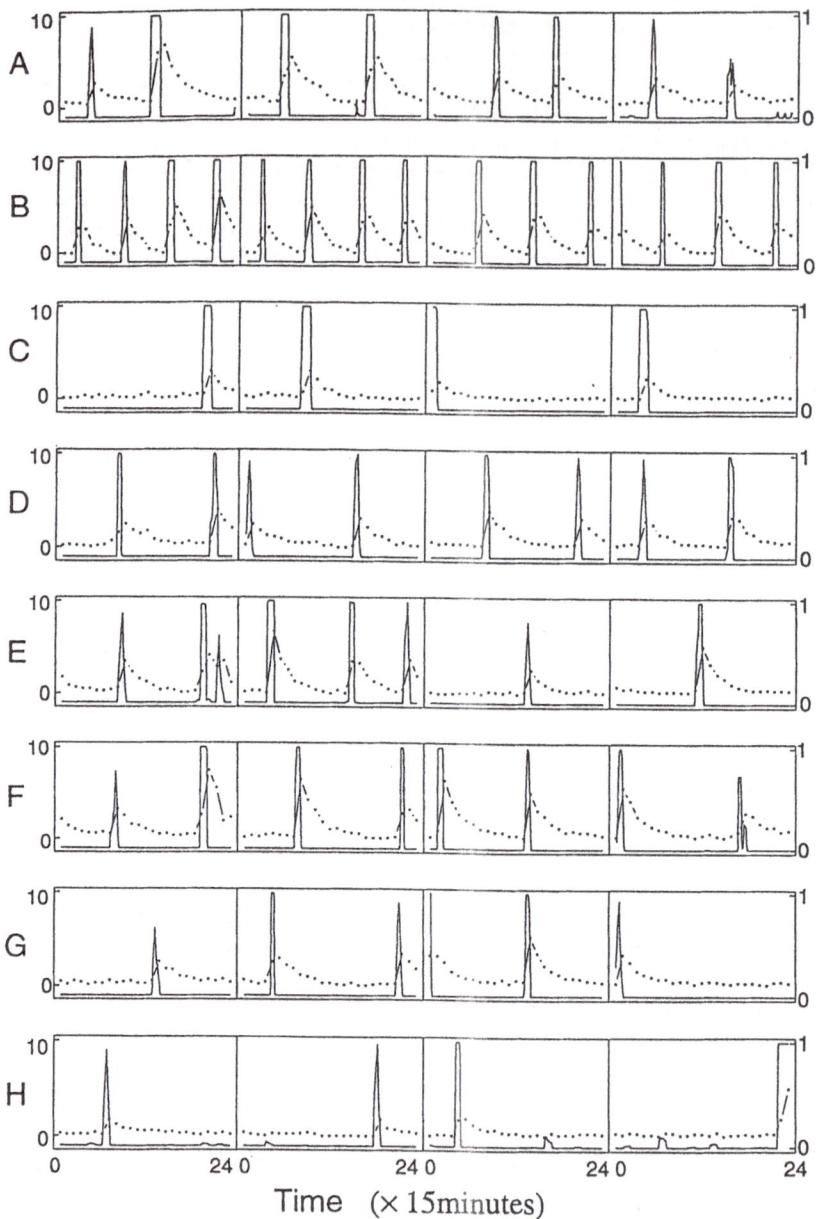

Figure 10.2   Posterior probability of the the rising mode ($m = 1$).

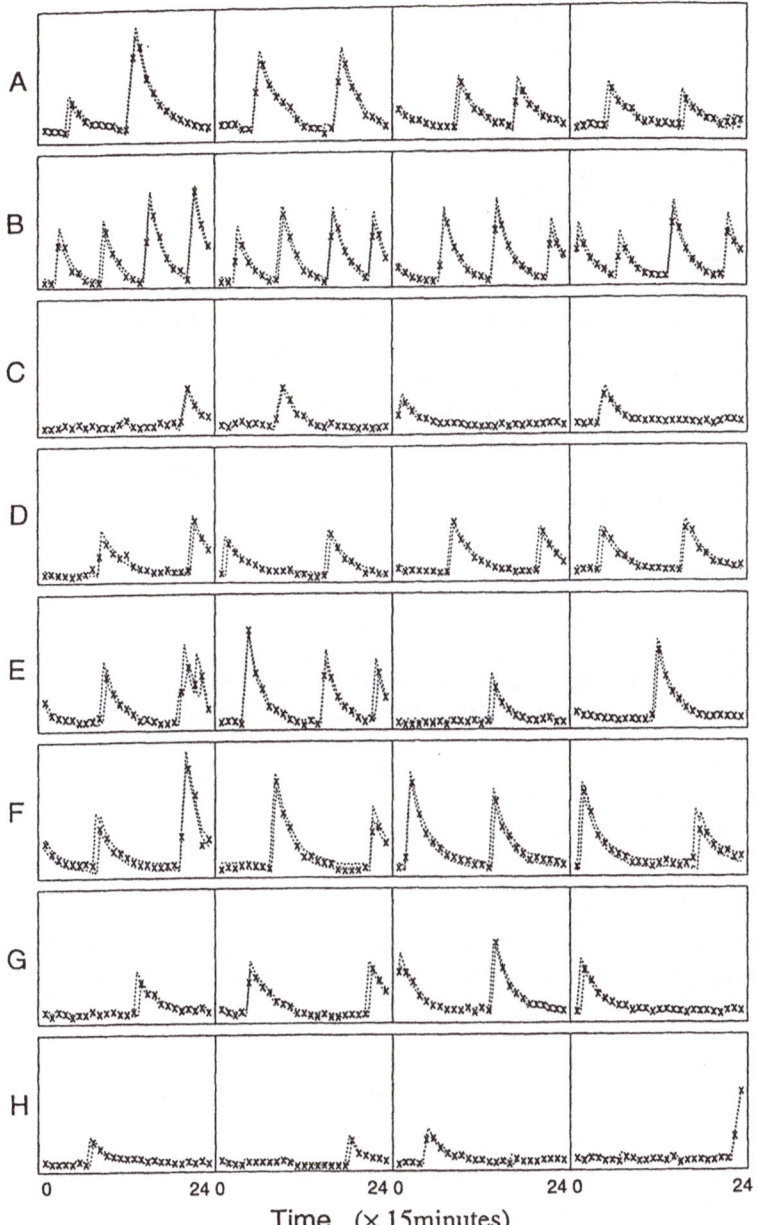

Figure 10.3 Luteinizing hormone levels, shown by crosses, and their 95 % confidence lines, dotted lines, estimated by the Bayes procedure.

with the linear increase fits the present data better. The result of the comparison with several kinds of modified versions reveals that the fit of the original model is better than the modified ones. This conclusion depends on the observation number and observation time interval.

The simulated data obtained by using the model with the estimated parameters displays similar features to the original data. The model can generate a variety of patterns using various different parameter values. We re-estimated the parameters using the simulated data to check the performance of the estimation procedure. Bayesian smoothing procedure was applied to the simulated data in which the occurrence of the pulses is frequent and the LH levels at the onset times of the pulses are relatively in fluctuation. In such a case, it is difficult to estimate the mode m and LH level x. The result has revealed that although the confidence interval becomes considerably wide, the original states can be estimated properly. When the observations are provided twice as densely, the confidence intervals become much narrower with successful acquisition of more accurate recovery of the position and level of the peaks. This is a reasonable result because more detailed information about the original level x is provided. These results show that our method can extract the information concerning the state from the observations.

10.4   Conclusions

Various kinds of non-Gaussian time series with pulses can be modeled appropriately using the state space representation. This approach has a wide range of application for time series with switching modes. By using the state space formulation, we can calculate the likelihood and apply Bayesian inference procedure. Furthermore, the estimated model can be used for simulation and prediction.

By dividing the time intervals between observations into sub-intervals, the unobserved states can be estimated by filtering and smoothing. Therefore, even when the original observation points are considerably sparse, it is possible to extract suitable information from the data, provided that the steps of filtering and smoothing are taken finely enough. This method can also be applied to the irregularly-sampled data by considering an appropriate time subdivision.

Although non-Gaussian state space methods require a large amount of calculation, the required amount of the calculation is in proportional to the data size.

References

Akaike, H. (1980), "Likelihood and the Bayes procedure," in *Bayesian Statistics*, eds. J. M. Bernardo, M. H. De Groot, D. V. Lindley, and A. F. M. Smith, Valencia, University Press, 143–166.

Diggle, P. and Zeger, S. (1989), "A non-Gaussian model for time series with pulses," *Journal of the American Statistical Association*, Vol. 84, 354–359.

Kitagawa, G. (1987), "Non-Gaussian state-space modeling of nonstationary time series," *Journal of the American Statistical Association*, Vol. 82, 1032–1041.

Knobil, E. and Hotchkiss, J. (1988), "The menstrual cycle and its neuroendocrine control, in *The Physiology of Reproduction*, eds. E. Knobil, J. Neill, L. Ewing, G. Greenwald, C. Markert, and D. Pfoff, New York: Raven Press, 1971–1994.

Komaki, F. (1993), "State space modeling of time series sampled from continuous processes with pulses," *Biometrika*, Vol. 80, 417–429.

Lincoln, D. W., Fraser, H. M., Lincoln, G. A., Martin, G. M. and McNeilly, A. S. (1985), "Hypothalamic pulse generators," *Recent Progress in Hormone Research*, Vol. 41, 369–419.

O'Sullivan, F., Whitney, P., Hinselwood, M. M. and Houser E. R. (1984), "The analysis of repeated measures experiments in endocrinology," *Journal of Animal Science*, Vol. 59, 1070–1079.

O'Sullivan, F. and O'Sullivan, J. (1988), "Deconvolution of episodic hormone data: An analysis of the role of season on the onset of puberty in cows," *Biometrics*, Vol. 44, 339–353.

Ozaki, T. (1985), "Statistical identification of storage models with application to stochastic hydrology," *Water Resources Bulletin*, Vol. 21, 663–675.

Rahe, C. H., Owens, R. E., Fleeger, J. L., Newton, H. J. and Harms, P. G. (1980), "Pattern of plasma luteinizing hormone in the cyclic cow: Dependence upon the period of the cycle," *Endocrinology*, Vol. 107, 498–503.

Tweedie, R. L. (1983), "Criteria for rates of convergence of markov chains, with application to queuing and storage theory," in *Probability, Statistics and Analysis*, eds., J. F. C. Kingman and G. E. H. Reuter, Cambridge University Press, 260–276.

Wilson, R. C., Kesner, J. S., Kaufman, J. M., Uemura, T., Akema, T. and Knobil, E. (1984), "Central electrophysiologic correlates of pulsatile luteinizing hormone secretion in the rhesus monkey," *Neuroendocrinology*, Vol. 39, 256–260.

The author of this article wishes to express their wholehearted gratitude to Professors Y. Ogata, T. Ozaki, and T. Matsunawa who have been kind enough to provide the author with various kinds of guidance throughout this study. The author are very grateful to Professor Diggle who kindly offered the LH data to the author. The author's appreciation is extended to such distinguished researchers of the Institute of Statistical Mathematics including Professors H. Akaike, G. Kitagawa, and K. Tanabe who have kindly encouraged the author by furnishing them with precious advice and constructive suggestions. Acknowledgment would also like to be given to Biometrika Trustees for their kind permission for the reproduction of Figure 1 (p. 419) and Figure 2(p. 424) from Komaki (1993).

Chapter 11

Time Varying Coefficient AR and VAR Models

Xing-Qi Jiang
Department of Economics, Asahikawa University
3-jo 23-Chome, Nagayama, Asahikawa, Hokkaido, 079–8501 Japan
jiang@asahikawa-u.ac.jp

11.1 Introduction

Autoregressive (AR) models are very useful for time series analysis. As shown in Figure 11.1, there is correspondence between the AR model and the autocovariance function and the power spectrum of an univariate stationary time series. Therefore, if an AR model is estimated from a time series, then the estimates of the autocovariance function and the power spectrum are obtained immediately. For the analysis of multivariate time series, Figure 11.2 shows the relation between the vector autoregressive (VAR) model and the cross-covariance function, the cross-spectrum, and the relative power contribution. Akaike and Nakagawa (1988), and Kitagawa (1993) developed procedures and programs for the analysis of stationary time series by using the AR and VAR models.

Since 1980, there are significant developments in the analysis of nonstationary time series. The nonstationary time series may be classified into two types: the time series with nonstationary mean (e.g., economic data), and the time series with nonstationary covariance (e.g., seismic data). For analyzing the time series with nonstationary mean, the seasonal adjustment models and the ARIMA models are often applied. On the other hand, various kinds of method are also proposed for analyzing the time series with nonstationary covariance. Priestley (1965) developed an approach to the spectral analysis of univariate nonstationary processes by introducing the notion of evolutionary spectrum. Ozaki and Tong (1975), and Kitagawa and Akaike (1978) developed a locally stationary AR modeling technique.

Kitagawa (1983) proposed a procedure for estimating the time varying coefficient AR models by introducing the state space model based on the Bayesian smoothness priors. In his approach, the estimates of the time varying AR coefficients and the likelihood of the hyperparameters of the models are obtained by the Kalman filter, and then the hyperparameters are estimated by the method of maximum likelihood. To determining the AR order, the minimum AIC procedure is used. Kitagawa and

175

Gersch (1985) extended Kitagawa's procedure to the case that the hyperparameters are also time varying and applied their extended procedure to the analysis of seismic data. Kitagawa (1986, 1993) are detailed explanations with programs for estimating the time varying coefficient AR models and the time varying spectrum.

The procedures for estimating the time varying coefficient AR models can be extended to the VAR models. Jiang (1992), and Jiang and Kitagawa (1993) realized this extension and developed a procedure for estimating the time varying coefficient VAR models. Furthermore, they proposed methods for estimating the time varying cross-covariance function, the time varying cross-spectrum, the time varying relative power contribution, and the time varying frequency-wavenumber spectrum. In Jiang (1992), and Jiang and Kitagawa (1993), analyses of seismic data were also shown to exemplify their time varying coefficient VAR modeling.

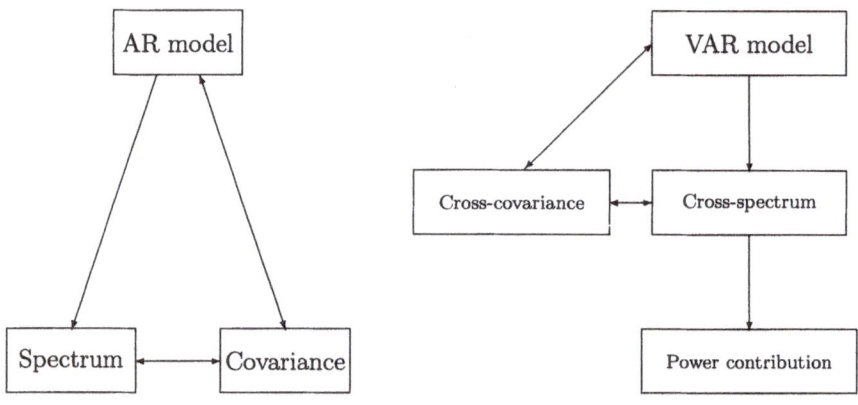

Figure 11.1 AR model, autocovariance and power spectrum

Figure 11.2 VAR model, cross-covariance,
cross-spectrum and power contribution

11.2   Time Varying Coefficient AR Models

11.2.1   The Models

For an univariate time series y_n, the time varying coefficient AR model can be expressed by

$$y_n = \sum_{\ell=1}^{p} a_\ell(n) y_{n-\ell} + \varepsilon_n, \qquad (11.1)$$

where p is the AR order, and $\{a_\ell(n); \ \ell = 1, \ldots, p\}$ the AR coefficients at time n. Suppose that ε_n is a Gaussian white noise sequence with zero mean and unknown variance σ^2. It is also supposed that ε_n and $y_{n-\ell}$ are independent of each other for $\ell > 0$.

In the Bayesian modeling, the time varying coefficients, $\{a_\ell(n); \ \ell = 1, \ldots, p\}$, are regarded as random variables and smoothness priors are introduced by assuming that the AR coefficients vary smoothly. The smoothness priors can be considered as the stochastic constrains on the smoothness of the changes of the AR coefficients and are expressed by the stochastic difference equations

$$\Delta^q a_\ell(n) = \nu_{n\ell}, \tag{11.2}$$

where Δ^q is the qth order difference operator and is defined by $\Delta a_\ell(n) = a_\ell(n) - a_\ell(n-1)$, $\Delta^q a_\ell(n) = \Delta^{q-1}(\Delta a_\ell(n))$. Here, it is supposed that $\nu_{n\ell}$ is a Gaussian white noise sequence with zero mean and unknown variance τ_ℓ^2. It is also supposed that $\nu_{n\ell}$ and ε_n are independent of each other, and ν_{ni} and ν_{nj} are independent of each other for $i \neq j$. Under the assumption that the spectrum of the time series changes smoothly with frequency, we can simplify the models by putting $\tau_1^2 = \tau_2^2 = \cdots = \tau_p^2 = \tau^2$. By this simplification the number of the unknown hyperparameters can be reduced and hence the estimation of the models become more easy (see, Kitagawa 1993, Kitagawa and Gersch 1996).

Behavior of the time varying AR coefficients depends on the value of the difference order q. For example, when $q = 1$, the stochastic difference equations in (11.2) represent the random walk models

$$a_\ell(n) = a_\ell(n-1) + \nu_{n\ell}.$$

When $q = 2$, they are given by

$$a_\ell(n) = 2a_\ell(n-1) - a_\ell(n-2) + \nu_{n\ell},$$

and indicate that the change of each AR coefficient is approximated locally by a straight line and is disturbed by a Gaussian white noise sequence. We will show how to determine the appropriate value of the difference order q with the AR order p in Subsection 11.2.3. In practice, we only need to consider lower order cases such as $q = 1$ or 2.

11.2.2   State Space Representation of the Models

In the previous section, it was explained that the time varying coefficient AR models can be generally defined by the Bayesian models as shown in (11.1) and (11.2). The models can be also expressed by the state space representation:

$$\begin{aligned} x_n &= F x_{n-1} + G w_n, \\ y_n &= H_n x_n + \varepsilon_n. \end{aligned} \tag{11.3}$$

Firstly, when $q = 1$, the state space model is given by the following expressions:

$$\begin{aligned} x_n &= (a_1(n), \ldots, a_p(n))^t, \quad w_n = (\nu_{n1}, \ldots, \nu_{np})^t, \\ F &= I_p, \quad G = I_p, \quad H_n = (y_{n-1}, \ldots, y_{n-p}), \end{aligned}$$

where $(*)^t$ denotes the transposition, I_p the $p \times p$ identity matrix. On the other hand, when $q = 2$, it is given by

$$\begin{aligned} x_n &= (a_1(n), \ldots, a_p(n), a_1(n-1), \ldots, a_p(n-1))^t, \\ w_n &= (\nu_{n1}, \ldots, \nu_{np})^t, \\ F &= \begin{bmatrix} 2I_p & -I_p \\ I_p & 0 \end{bmatrix}, \quad G = \begin{bmatrix} I_p \\ 0 \end{bmatrix}, \\ H_n &= (y_{n-1}, \ldots, y_{n-p}, 0, \ldots, 0). \end{aligned}$$

Also as explained in the previous section, for both $q = 1$ and 2, the covariance matrices of the system noise w_n and observation noise ε_n are respectively given by

$$Q = \mathrm{E}\{w_n w_n^t\} = \tau^2 I_p, \quad R = \mathrm{E}\{\varepsilon_n^2\} = \sigma^2.$$

Assuming that the initial conditions $x_{0|0}$ and $V_{0|0}$, and the observations $\{y_n; n = 1, \ldots, N\}$ are given, we can obtain the conditional mean and the covariance matrix of the state vector at each time by the successive application of the following Kalman filter for $n = 1, \ldots, N$.

[**Prediction**]

$$
\begin{aligned}
x_{n|n-1} &= F x_{n-1|n-1}, \\
V_{n|n-1} &= F V_{n-1|n-1} F^t + \tau^2 G G^t.
\end{aligned}
\tag{11.4}
$$

[**Filter**]

$$
\begin{aligned}
K_n &= V_{n|n-1} H_n^t (H_n V_{n|n-1} H_n^t + \sigma^2)^{-1}, \\
x_{n|n} &= x_{n|n-1} + K_n(y_n - H_n x_{n|n-1}), \\
V_{n|n} &= (I_m - K_n H_n) V_{n|n-1}.
\end{aligned}
\tag{11.5}
$$

Here, $x_{\ell|n}$ and $V_{\ell|n}$ denote respectively the mean and the covariance matrix of the state vector x_ℓ given the data $\{y_1, \ldots, y_n\}$, and m represents the dimension of the state vector x_n. Note that the computational complexity of the Kalman filter is reduced to $O(mN)$ by using the sparse structure of the matrices F and G. On the other hand, the method of least squares generally requires $O(m^3 N^3)$ computational complexity.

Based on the results of the above computations, we can obtain the smoothed estimate $x_{n|N}$ of the state vector and the corresponding covariance matrix $V_{n|N}$ by the following backward computations:

[**Fixed Interval Smoothing**]

$$
\begin{aligned}
A_n &= V_{n|n} F^t V_{n+1|n}^{-1}, \\
x_{n|N} &= x_{n|n} + A_n(x_{n+1|N} - x_{n+1|n}), \\
V_{n|N} &= V_{n|n} + A_n(V_{n+1|N} - V_{n+1|n}) A_n^t.
\end{aligned}
\tag{11.6}
$$

Note that since the time varying AR coefficients $a_\ell(n)$ are included in the state vector x_n, the estimates $\hat{a}_\ell(n)$ of the time varying AR coefficients can be immediately obtained from $x_{n|N}$.

11.2.3   Estimation and Identification of the Models

By using the results of the Kalman filter, the conditional density function of y_n, given the data $Y_{n-1} = \{y_1, \ldots, y_{n-1}\}$, are expressed as follows:

$$f(y_n|Y_{n-1}; \sigma^2, \tau^2) = (2\pi v^2(n))^{-\frac{1}{2}} \exp\left\{-\frac{(y_n - H_n x_{n|n-1})^2}{2v^2(n)}\right\},$$

where $v^2(n) = H_n V_{n|n-1} H_n^t + \sigma^2$ expresses the variance of the prediction error of y_n based on the observations $Y_{n-1} = \{y_1, \ldots, y_{n-1}\}$. Since the joint density function of $Y_N = \{y_1, \ldots, y_N\}$ is expressed by

$$f(Y_N|\sigma^2, \tau^2) = \prod_{n=1}^{N} f(y_n|Y_{n-1}; \sigma^2, \tau^2),$$

given the model orders p and q, the log-likelihood for the hyperparameters σ^2 and τ^2 is approximately obtained as follows (see, Kitagawa 1993):

$$\ell(\sigma^2, \tau^2; p, q) = -\frac{1}{2}\left\{ N \log 2\pi + \sum_{n=1}^{N} \left(\log v^2(n) + \frac{(y_n - H_n x_{n|n-1})^2}{v^2(n)} \right) \right\}. \tag{11.7}$$

The hyperparameters σ^2 and τ^2 are estimated by the method of maximum likelihood, or equivalently by maximizing the log-likelihood (11.7) with respect to these hyperparameters. However, the problem is essentially the estimation of the hyperparameter $d^2 = \sigma^2/\tau^2$. As described previously, since σ^2 is the variance of the residual ε_n of the models in (11.1), it indicates the goodness of fit of the models to the data. On the other hand, τ^2 is the variance of the disturbance $\nu_{n\ell}$ in the stochastic difference equations which controls the smoothness of the time varying AR coefficients. Therefore, d^2 is a hyperparameter that expresses the ratio of σ^2 to τ^2, hence it is a parameter which controls the trade-off between the goodness of fit and the smoothness of the time varying coefficients. Estimation of the hyperparameters using the method of maximum likelihood is equivalent to the determination of an appropriate trade-off parameter. Since there are only two free parameters among σ^2, τ^2 and d^2, we have only to estimate two hyperparameters of interest by the method of maximum likelihood.

Let the maximum likelihood estimates of the hyperparameters σ^2 and τ^2 be $\hat{\sigma}^2$ and $\hat{\tau}^2$, respectively, the Akaike information criterion, AIC, for evaluating the models is defined by (see for example, Akaike 1974)

$$\text{AIC}(p, q) = -2\ell(\hat{\sigma}^2, \hat{\tau}^2; p, q) + 2 \times 2,$$

where p and q are the AR order and the difference order, respectively. Among all possible models with different values of the model orders p and q, we may choose one set which minimizes the AIC as the best by using the minimum AIC procedure.

11.2.4 Estimation of Time Varying Spectrum

For a stationary AR process with order p the coefficients $\{a_\ell; \ell = 1, \ldots, p\}$ and the innovation variance σ^2, the power spectrum is given by

$$s(f) = \frac{\sigma^2}{\left| 1 - \sum_{\ell=1}^{p} a_\ell \exp(-2\pi i \ell f) \right|^2}, \qquad -\frac{1}{2} \leq f \leq \frac{1}{2}. \tag{11.8}$$

Therefore by replacing the AR coefficients in (11.8) with the time varying AR coefficients, an instantaneous spectrum at time n for the nonstationary time series can be obtained by

$$s_n(f) = \frac{\sigma^2}{\left| 1 - \sum_{\ell=1}^{p} a_\ell(n) \exp(-2\pi i \ell f) \right|^2}, \qquad -\frac{1}{2} \leq f \leq \frac{1}{2}. \tag{11.9}$$

The time varying AR coefficients, $\{a_\ell(n); \ell = 1, \ldots, p\}$, can be estimated by the method explained previously, which permits the instantaneous spectrum $s_n(f)$ to be also obtained by (11.9) as a function of time. The spectrum computed by (11.9) is called the time varying spectrum. The details for estimating the time varying spectrum can be found in Kitagawa (1983, 1993), and Kitagawa and Gersch (1996).

11.3   Time Varying Coefficient VAR Models

11.3.1   The Models

The time varying coefficient VAR models are used for vector (or multivariate) AR models with the time varying coefficients. For a k-variate time series $z_n = (y_{n1}, \ldots, y_{nk})^t$, the time varying coefficient VAR models are expressed by

$$z_n = \sum_{\ell=1}^{p} A_\ell(n) z_{n-\ell} + u_n, \qquad (11.10)$$

where p is the order of the models, and $A_\ell(n)$ the VAR coefficient matrix with lag ℓ at time n. In (11.10), u_n is a k-variate Gaussian white noise sequence with mean zeros and covariance matrix Σ_n, and it is supposed that u_n and $z_{n-\ell}$ are independent of each other for $\ell > 0$.

However, if the method stated in Section 11.2 is extend to the multivariate case directly, the number of unknown parameters in the VAR coefficients is $k^2 p$, and hence we have to estimate simultaneously the parameters possibly with high dimension. This difficulty is alleviated by using VAR models with simultaneous response introduced by Kitagawa and Akaike (1981). The time varying coefficient VAR models with simultaneous response are expressed by

$$z_n = D(n) z_n + \sum_{\ell=1}^{p} B_\ell(n) z_{n-\ell} + v_n, \qquad (11.11)$$

where $v_n = (\varepsilon_{n1}, \ldots, \varepsilon_{nk})^t$ is a k-variate Gaussian white noise sequence with mean zeros and covariance matrix V. Here V is a diagonal matrix with the diagonal elements $\sigma_1^2, \ldots, \sigma_k^2$ as unknown hyperparameters. On the other hand, $D(n)$ and $B_\ell(n)$ ($\ell = 1, \ldots, p$) are the simultaneous response matrix and coefficient matrices defined as follows:

$$D(n) = \begin{bmatrix} 0 & 0 & \cdots & 0 \\ b_{210}(n) & 0 & \ddots & \vdots \\ \vdots & \ddots & \ddots & 0 \\ b_{k10}(n) & \cdots & b_{k(k-1)0}(n) & 0 \end{bmatrix},$$

$$B_\ell(n) = \begin{bmatrix} b_{11\ell}(n) & b_{12\ell}(n) & \cdots & b_{1k\ell}(n) \\ b_{21\ell}(n) & b_{22\ell}(n) & \cdots & b_{2k\ell}(n) \\ \vdots & \vdots & \ddots & \vdots \\ b_{k1\ell}(n) & b_{k2\ell}(n) & \cdots & b_{kk\ell}(n) \end{bmatrix}.$$

The time varying coefficient VAR models with simultaneous response have mutually independent innovations so that the models can be expressed by the following k models

$$y_{ni} = \sum_{j=1}^{k} \sum_{\ell=0}^{p} b_{ij\ell}(n) y_{(n-1)j} + \varepsilon_{ni}, \quad \varepsilon_{ni} \sim \mathrm{N}(0, \sigma_i^2) \quad (i = 1, \ldots, k), \qquad (11.12)$$

where $b_{ij0}(n) = 0$ for $j \geq i$ and both ε_{ni} and ε_{nj} are white noise sequences which are independent of each other for $i \neq j$. Hereafter, we call equation (11.12) the ith channel model for each $i = 1, \ldots, k$. A remarkable merit of the time varying coefficient

VAR models with simultaneous response is that we can estimate independently the models in (11.12) with respect to $i = 1, \ldots, k$.

As shown in Kitagawa and Akaike (1981), there are one-to-one correspondences between the models in (11.10) and (11.11) such that

$$
\begin{aligned}
A_\ell(n) &= (I - D(n))^{-1} B_\ell(n) \quad (\ell = 1, \ldots, p), \\
\Sigma_n &= (I - D(n))^{-1} V (I - D(n))^{-t}.
\end{aligned}
$$

Therefore if the models in (11.11), or equivalently (11.12), are estimated, then the models in (11.10) will naturally be obtained.

As the univariate case shown in Section 11.2, in order to obtain the reasonable estimates of the time varying VAR coefficients of the models in (11.12), the smoothness priors for the time varying coefficients are introduced as follows:

$$
\Delta^q b_{ij\ell}(n) = \nu_{ij\ell}(n), \tag{11.13}
$$

where $\nu_{ij\ell}(n)$ is a Gaussian white noise sequence with zero mean and unknown variance $\tau_{ij\ell}^2$. We assume that $\nu_{ij\ell}(n)$ and ε_{ni} are independent of each other, and that $\nu_{ij\ell}(n)$ and $\nu_{rst}(n)$ are independent of each other for $\{r, s, t\} \neq \{i, j, \ell\}$. In (11.13), Δ^q denotes the qth order difference operator with respect to time n such as $\Delta b_{ij\ell}(n) = b_{ij\ell}(n) - b_{ij\ell}(n-1)$. As shown in Jiang (1992), and Jiang and Kitagawa (1993), under the assumption that the spectrum of the time series is smooth, we have

$$
\tau_{ij\ell}^2 = \tau_i^2,
$$

and the number of the hyperparameters is reduced significantly. This mitigates the difficulty in the numerical optimization for estimating the hyperparameters.

11.3.2 Estimation and Identification of the Models

As stated in the previous subsection, by using the representation of the time varying coefficient VAR models with simultaneous response, the models can be expressed by k independent linear models as shown in (11.12). By combining the models in (11.12) with the smoothness priors in (11.13), the time varying VAR coefficients can be estimated successfully. Noted that the ith channel model contains two hyperparameters (σ_i^2, τ_i^2) to be estimated.

Here, we only give the state space representation of the ith channel model with the difference order $q = 1$. The details of the models with $q = 2$ can be found in Jiang and Kitagawa (1993). Let $m = kp + i - 1$, and

$$
F = I_m, \quad G = I_m,
$$

the state vector x_n and the vectors H_n and w_n be defined as follows:

$$
\begin{aligned}
x_n &= (\overbrace{b_{i10}(n), \ldots, b_{i(i-1)0}(n)}^{i-1}, \overbrace{b_{i11}(n), \ldots, b_{i1p}(n)}^{p}, \ldots, \overbrace{b_{ik1}(n), \ldots, b_{ikp}(n)}^{p})^t \\
H_n &= (y_{n1}, \ldots, y_{n(i-1)}, y_{(n-1)1}, \ldots, y_{(n-p)1}, \ldots, y_{(n-1)k}, \ldots, y_{(n-p)k}) \\
w_n &= (\nu_{i10}(n), \ldots, \nu_{i(i-1)0}(n), \nu_{i11}(n), \ldots, \nu_{i1p}(n), \ldots, \nu_{ik1}(n), \ldots, \nu_{ikp}(n))^t
\end{aligned}
$$

Furthermore, by replacing y_n, ε_n, σ^2, and τ^2 in the state space models (11.3) with y_{ni}, ε_{ni}, σ_i^2, and τ_i^2 respectively, the time varying coefficient VAR models can be expressed by the same state space models. Therefore, algorithms using the Kalman filter can be utilized without any modification.

Moreover, when the orders p and q of the models are given, the log-likelihood for the hyperparameters σ_i^2 and τ_i^2 is expressed by (11.7). By maximizing $\ell(\sigma_i^2, \tau_i^2; p, q)$, we obtain the maximum likelihood estimates $\hat{\sigma}_i^2$ and $\hat{\tau}_i^2$ of the hyperparameters, and the AIC of the ith channel model is defined by

$$\text{AIC}_i(p, q) = -2\ell(\hat{\sigma}_i^2, \hat{\tau}_i^2; p, q) + 2 \times 2.$$

Since the models of the individual channels are independent of each other, the AIC of the entire models is calculated by the sum of the AIC's of the models in the individual channels, i.e.,

$$\text{AIC}(p, q) = \sum_{i=1}^{k} \text{AIC}_i(p, q).$$

The values of the model orders p and q are determined by minimizing $\text{AIC}(p, q)$.

11.3.3   Time Series Analysis by the Models

In this section, several methods for the nonstationary time series analysis using the time varying coefficient VAR models are shown. A method for estimating time varying frequency-wavenumber spectrum is also proposed by Jiang (1992) and Jiang and Kitagawa (1993).

(1) Estimation of time varying covariance function   As the stationary case, the time varying Yule-Walker equations are defined based on the parameters of the time varying coefficient VAR models, i.e.,

$$
\begin{aligned}
C_{nn} &= \sum_{\ell=1}^{p} A_\ell(n) C_{(n-\ell),n} + \Sigma_n, \\
C_{n,(n-m)} &= \sum_{\ell=1}^{p} A_\ell(n) C_{(n-\ell),(n-m)} \quad (m = 1, 2, \ldots, p),
\end{aligned}
\tag{11.14}
$$

where $C_{n,(n-m)}$ is the time varying covariance function at time n defined by

$$C_{n,(n-m)} = \text{E}\{z_n z_{n-m}^t\}.$$

When the initial values of the time varying covariance function $\{C_{n,(n-m)}; n = 0, 1, \ldots, 1$ $p; m = 0, 1, \ldots, p - 1\}$ are given, the whole of the time varying covariance function of the time series can be recursively calculated by (11.14).

However, the initial values for the equations in (11.14) are usually unknown. Instead of this, under the assumption that the time series is locally stationary, we obtain the instantaneous covariance function $\tilde{C}_{nm}$ by solving the following instantaneous Yule-Walker equations:

$$
\begin{aligned}
\tilde{C}_{n0} &= \sum_{\ell=1}^{p} A_\ell(n) \tilde{C}_{n\ell} + \Sigma_n, \\
\tilde{C}_{nm} &= \sum_{\ell=1}^{p} A_\ell(n) \tilde{C}_{n,(\ell-m)} \quad (m = 1, 2, \ldots, p).
\end{aligned}
\tag{11.15}
$$

In (11.15), $\tilde{C}_{nm}$ is the covariance function at time n and lag m assuming that the models are locally stationary. Since the instantaneous Yule-Walker equations (11.15) are linear equations, they can be solved easily (see, Kitagawa 1993).

(2) Estimation of time varying cross-spectrum   By the same idea, the instantaneous cross-spectrum $P_n(f)$ at time n and frequency f is estimated from the time varying coefficient VAR models as follows:

$$P_n(f) = (A_n(f))^{-1} \Sigma_n (A_n^*(f))^{-t}, \quad -\frac{1}{2} \leq f \leq \frac{1}{2}. \tag{11.16}$$

Here, $A_n(f)$ is the instantaneous frequency response function of the models and is obtained by

$$A_n(f) = \sum_{\ell=0}^{p} A_\ell(n) \exp(-2\pi i\ell f)$$

with $A_0(n) = -I$, and $A_n^*(f)$ expresses the conjugate complex matrix of $A_n(f)$. By using the parameters in the time varying coefficient VAR models with simultaneous response, the instantaneous cross-spectrum is also estimated by

$$P_n(f) = (B_n(f))^{-1} V (B_n^*(f))^{-t}, \quad -\frac{1}{2} \leq f \leq \frac{1}{2}, \tag{11.17}$$

where $B_n(f)$ is calculated by

$$B_n(f) = \sum_{\ell=0}^{p} B_\ell(n) \exp(-2\pi i\ell f)$$

with $B_0(n) = D(n) - I$, and V is the covariance matrix of the models (11.11). From the instantaneous cross-spectrum, the instantaneous coherency spectrum between any two channels can also be calculated.

(3) Estimation of time varying relative power contribution   The relative power contribution developed by Akaike and Nakagawa (1988) is often applied as an useful measure for analyzing the feedback systems. Here, we extend the concept of relative power contribution to the time varying coefficient VAR processes and define the instantaneous relative power contribution as

$$r_{nij}(f) = \frac{|e_{nij}(f)|^2 \sigma_j^2}{\sum_{\ell=1}^{k} |e_{ni\ell}(f)|^2 \sigma_\ell^2}, \quad -\frac{1}{2} \leq f \leq \frac{1}{2}, \tag{11.18}$$

where $e_{nij}(f)$ is the (i,j)th element of the matrix $E_n(f) = B_n(f)^{-1}$, σ_j^2 the jth diagonal element of the covariance matrix V of the models (11.11). The instantaneous relative power contribution $r_{nij}(f)$ expresses the degree of contribution of the innovation in the jth channel model to the power of the ith one at frequency f, and is useful for the analysis of the time varying characteristics of the mutual influence of the multi-channel signals.

11.4   An Example of Seismic Data Analysis

In this section, we show an example of analyzing seismic data to illustrate how to analyze the time series with nonstationary covariance by using the time varying coefficient VAR models. The data are the three component seismograms, i.e., the seismograms in East-West (EW), North-South (NS), and Up-Down (UD) directions, of the 1982 Urakawa-oki earthquake data (Takanami, 1991). The code names of

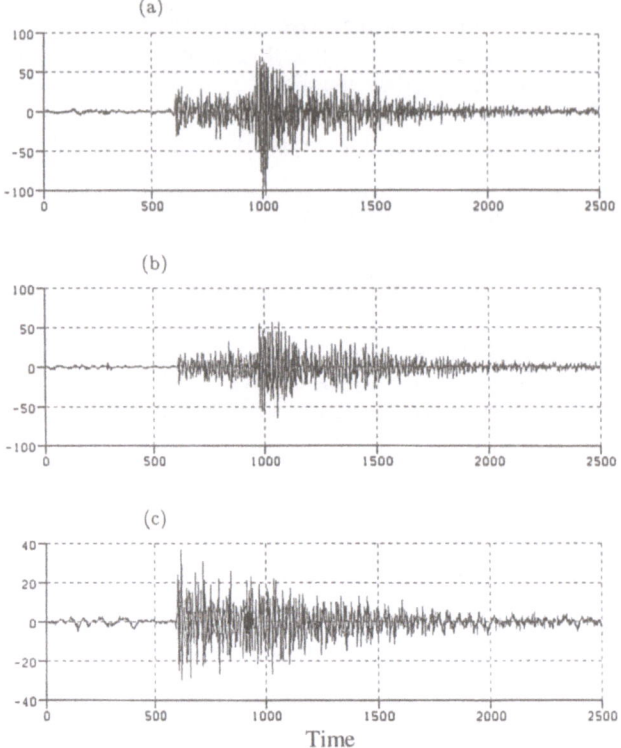

Figure 11.3 The seismograms: (a) EW, (b) NS, (c) UD

the data are MYE2F, MYN2F, and MYU2F, respectively. The sampling interval of the original data is approximately 0.01 second. The data analyzed here, however, are obtained by resampling every other observation from the original data (i.e., the sampling interval becomes approximately 0.02 second). Figure 11.3 shows a plot of the 3-variate seismic data with the sampling size $N = 2500$.

As shown in Figure 11.3, each seismogram is composed of the three parts, i.e., the background noise (before arrival of the earthquake), P-wave (after the first abrupt change of amplitude), and S-wave (after the second abrupt change of amplitude). The purpose of the analysis is to obtain information on variations in the direction of the seismic wave by observing the fluctuations of the power spectra and relative power contribution in the three dimensional space. However the three parts of the seismograms shown in Figure 11.3 have considerably different variances, and therefore we can not expect to obtain very good estimates of the models by using the data directly. Therefore the data are first transformed so that the series become approximately homogeneous in the variance. The transformation of the data is realized by using the program TVVAR developed by Kitagawa (1993). Figure 11.4 shows a plot of the normalized seismograms.

We show here the results obtained by fitting the time varying coefficient VAR models to the normalized data. In Table 11.1, AIC versus the values of the VAR order p and the difference order q are listed. As shown in the table, the minimum

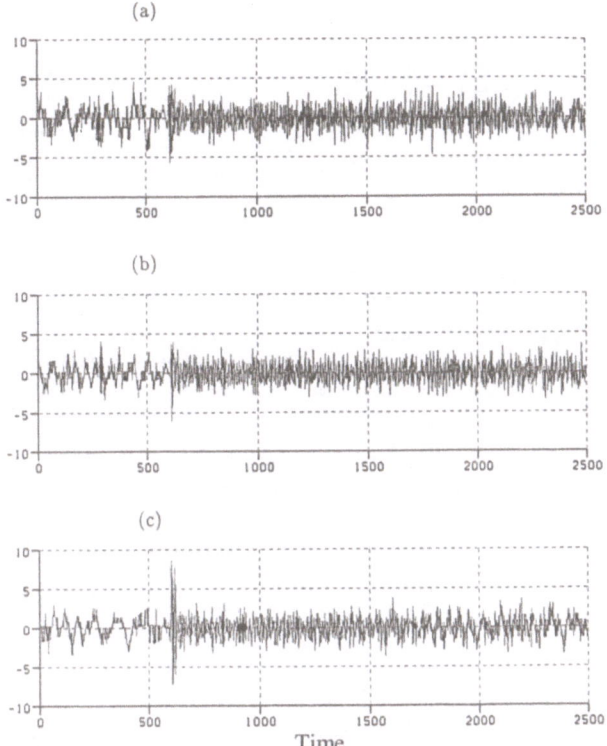

Figure 11.4   The normalized seismograms: (a) EW, (b) NS, (c) UD

Table 11.1   AIC of the models

q = 1				q = 2			
p	AIC	p	AIC	p	AIC	p	AIC
1	22131	6	19041	1	22475	6	20073
2	20201	7	19019	2	20745	7	20194
3	19231	8	19015	3	19909	8	20365
4	19071	9	*18988*	4	19788	9	20522
5	19019	10	19006	5	19907	10	20689

AIC is attained at $p = 9$ and $q = 1$. According to the minimum AIC procedure, the analysis of the seismic data is realized by applying the minimum AIC model. The maximum likelihood estimates of the hyperparameters are as follows: $\hat{\sigma}_1^2 = 0.594$, $\hat{\tau}_1^2 = 0.361 \times 10^{-3}$, $\hat{\sigma}_2^2 = 0.598$, $\hat{\tau}_2^2 = 0.152 \times 10^{-3}$, $\hat{\sigma}_3^2 = 0.714$, and $\hat{\tau}_3^2 = 0.349 \times 10^{-3}$.

Figure 11.5 shows plots of the estimates of several typical time varying coefficients, b_{iik} ($i = 1, 2, 3 \, ; k = 1, 2, 3$) and b_{ij0} ($1 \leq j < i \leq 3$). Most of the estimates of the coefficients show significant step changes corresponding to the arrival of the P-wave and S-wave, (at the time around $n = 600$ and $n = 960$), respectively. The arrival time of these waves is estimated by the procedure introduced by Kitagawa (1993). They

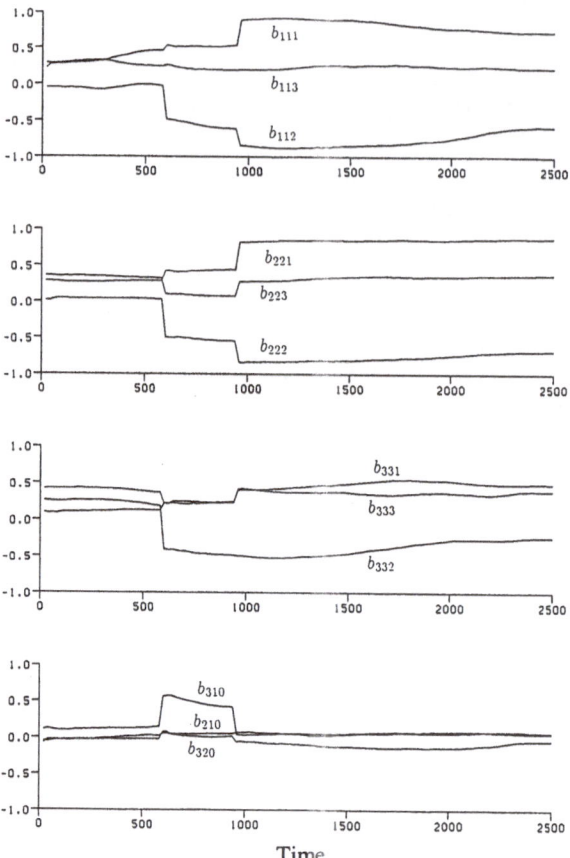

Figure 11.5   Typical examples of estimated time varying VAR coefficients

are used, through human intervention, to specify the positions of outliers having large noise variance in the smoothness priors. Therefore the prior constraints for smoothness of the VAR coefficients are relaxed so that the abrupt changes can be shown. As illustrated in Figure 11.5, the coefficients take quite different values in the three intervals (i.e., the background noise, the P-wave, and the S-wave). It is also shown that the coefficients related to the two horizontal waves behave similarly, while those related to the vertical wave are different especially in the S-wave mode.

Figure 11.6 shows the time evolution of the estimated instantaneous auto- and cross-correlation functions. The graphs on the diagonal part show the auto-correlation functions, and the rest are the cross-correlation functions. From the change in the instantaneous auto- and cross-correlation functions, the information about the correlation between the seismic signals in the individual directions and individual time points can be obtained.

As described earlier, the time varying power spectrum and cross-spectrum of the vector time series with nonstationary covariance can be estimated from the time varying coefficient VAR models. The time varying power spectra of the individual

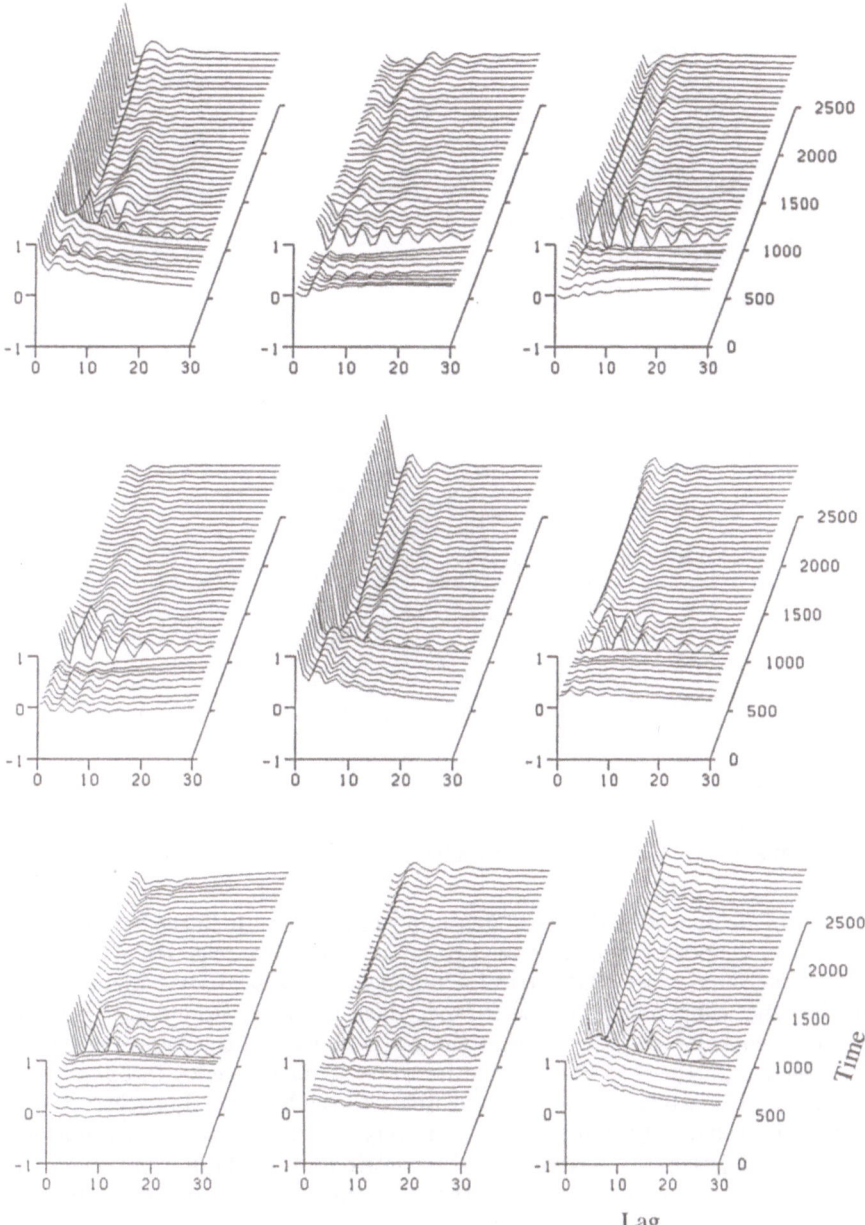

Figure 11.6   Estimated instantaneous auto- and cross-correlation functions

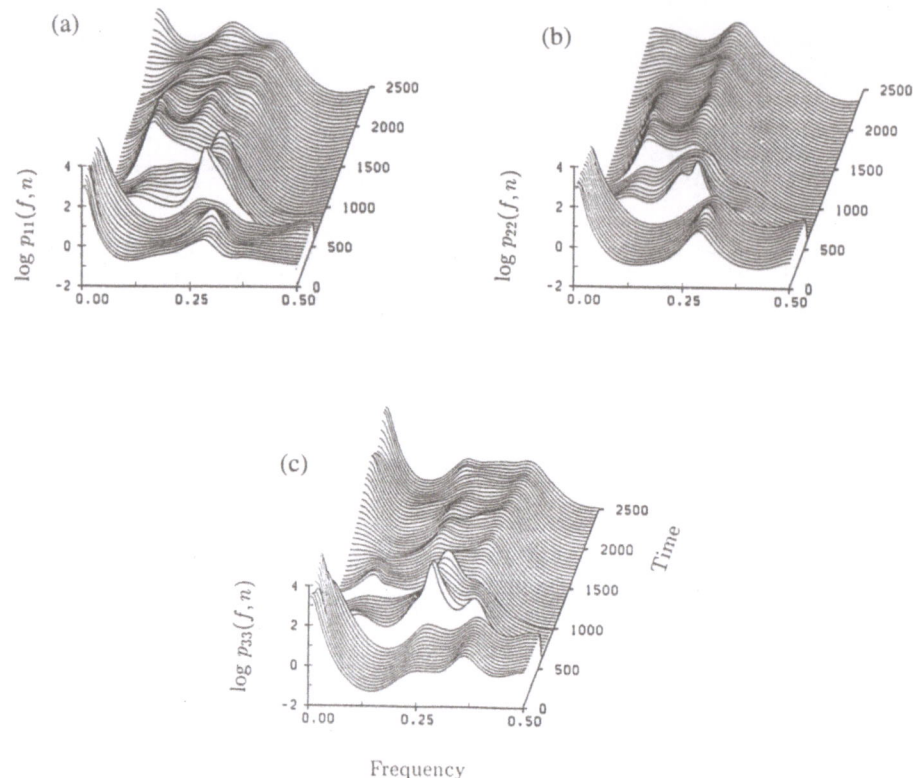

Figure 11.7 Estimated instantaneous spectra: (a) EW, (b) NS, (c) UD

components are plotted in Figure 11.7. As show in this figure, in the background noise and P-wave, the power spectra of the two horizontal component resemble each other. In the background noise, the vertical component has a peak at a higher frequency than the other components. In the P-wave, there is a peak around at $f = 0.2$ for all the components. We can see also that the power spectra of the S-wave are abundant in the change. In the S-wave, the spectrum at a lower frequency (around $f = 0.12$) dominates in its early stage, and then switches to the higher frequencies.

Figure 11.8 shows the graphs of the time varying relative power contribution. The plot on the (i, j)th position indicates the contribution of the innovation in the jth component to the power spectrum of the ith component. By observing these plots, we can obtain information on the mutual influence. For the P-wave and S-wave, significant power contribution found in the figure are shown in Table 11.2 and Table 11.3, respectively. In these tables, the (i, j)th element indicates that the ith component has significant power contribution from the jth one (shown by 1), and the number in each bracket shows its significant frequency.

From these observations, we would conclude as follows: (1) In the case of P-

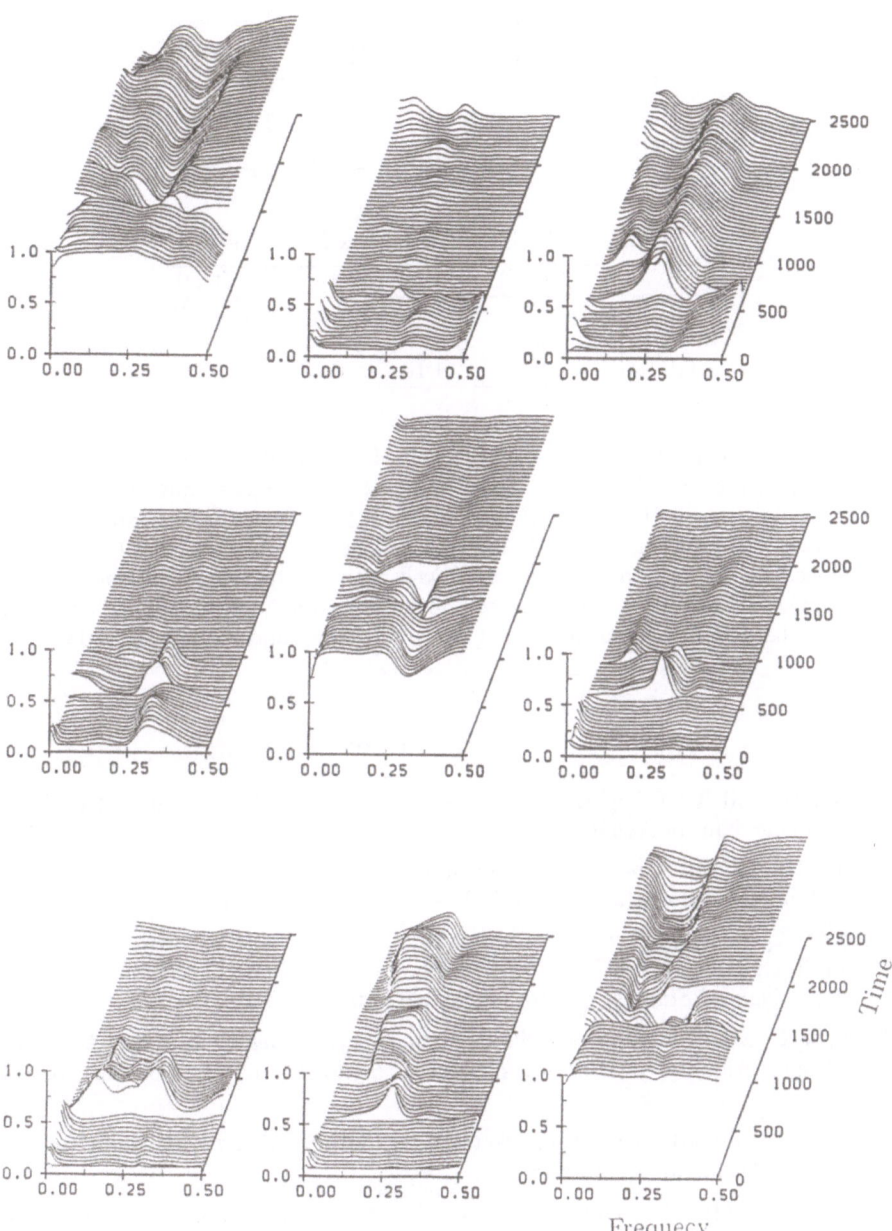

Figure 11.8   Estimated instantaneous relative power contribution

Table 11.2 Influence matrix for the P-wave

	1 (EW)	2 (NS)	3 (UD)
1 (WE)	—	0	1 (0.20)
2 (NS)	1 (0.25)	—	1 (0.20)
3 (UD)	1 (0.10~0.25)	0	—

Table 11.3 Influence matrix for the S-wave

	1 (EW)	2 (NS)	3 (UD)
1 (EW)	—	0	1 (0.20)
2 (NS)	0	—	0
3 (UD)	0	1 (0.10~0.20)	—

wave, the NS component is affected by the EW and UD ones, and the EW and UD components are affected by each other. This indicates that the EW and UD components contain the essential information. (2) On the other hand, in the case of S-wave, the EW component is induced by the UD one, and the UD component is induced by the NS one. These agree well with the fact that the epicenter of the earthquake is roughly in the west of the observatory. The details of the example analyzed here can be found in Jiang (1992) and Jiang and Kitagawa (1993).

References

Akaike, H. and T. Nakagawa (1988), *Statistical Analysis and Control of Dynamic Systems*, Kluwer Academic Publishers, Dordrecht.

Akaike, H. (1974), "A new look at the statistical model identification," *IEEE Transactions on Automatic Control*, Vol. AC–19, No. 6, 716–723.

Jiang, X. Q. (1992), "Bayesian Methods for Modeling, Identification and Estimation of Stochastic Systems," *unpublished Ph.D. dissertation, The Graduate University for Advanced Studies, Department of Statistical Science*, Tokyo.

Jiang, X. Q. and G. Kitagawa (1993), "A time varying coefficient vector AR modeling of nonstationary covariance time series," *Signal Processing*, Vol. 33, No. 3, 315–331.

Kitagawa, G. (1983), "Changing spectrum estimation," *Journal of Sound and Vibration*, Vol. 89, No. 1, 433–445.

Kitagawa, G. (1986), "Time varying coefficient AR modeling," *Proceedings of the Institute of Statistical Mathematics*, Vol. 34, No. 2, 273–283 (in Japanese).

Kitagawa, G. (1993), *FORTRAN77 Programming for Time Series Analysis*, Iwanami Shoten Publishers, Tokyo (in Japanese).

Kitagawa, G. and H. Akaike (1978), "A procedure for the modeling of non-stationary time series," *Annals of the Institute of Statistical Mathematics*, Vol. 30–B, No. 2, 351–363.

Kitagawa, G. and H. Akaike (1981), "On TIMSAC–78," in: D.F. Findley, ed., *Applied Time Series Analysis II*, Academic Press, 499–547.

Kitagawa, G. and W. Gersch (1985), "A smoothness priors time-varying AR coefficient modeling of nonstationary covariance time series," *IEEE Transactions on Automatic Control*, Vol. AC–30, No. 1, 48–56.

Kitagawa, G. and W. Gersch (1996), *Smoothness Priors Analysis of Time Series*, Lecture Notes in Statistics No.116, Springer-Verlag, New York.

Ozaki, T. and H. Tong (1975), "On the fitting of non-stationary autoregressive models analysis," *Proceedings of the 8th Hawaii International Conference on System Sciences*, 224–226.

Priestley, M. B. (1965), "Evolutionary spectra and non-stationary processes," *Journal of the Royal Statistical Society, Series B*, Vol. 27, No. 2, 204–237.

Takanami, T. (1991), "Seismograms of foreshocks of 1982 Urakawa-oki earthquake, AISM Data 43-3-01", *Annals of the Institute of Statistical Mathematics*, Vol. 43, No. 3, 605.

Chapter 12

Statistical Control of Cement Process

Yoshitaka Yagihara
System Sogo Kaihatsu Co. Ltd.
4-8-17 Hongo, Bunkyo-ku, Tokyo 113-0033 Japan

12.1   Introduction

One of the most important processes in a cement plant is a kiln process. The kiln process is a multiple input and output system having many internal noise sources and internal feedback loops, and the process is, viewing from the physical characteristics, a distributed parameter system which has an extension of the behavior over space and time. Its model is given by the partial differential equations concerning the gas temperature T_G and solid temperature T_M derived from the heat balance and material balance in the kiln. At the initial stage of the development (1962) when the first computer for process control was installed to Kumagaya Plant of Chichibu Cement Co. Ltd., this mathematical model was used for control. However the mathematical model was in reality not satisfactory enough to be used as the model for stabilizing control because of the fact that none of noise sources existing in the actual process were considered.

On the other hand, by aggregating a kiln process into a lumped parameter system, the conventional local PID control based on the single input and single output model was also tried. This feedback system is somewhat robust for noise, but the results were not necessarily successfully operated because no mutual interference (internal feedback loops) among the process variables could be taken account of. For the realization of the stabilizing controller, identification of the model for objective process which takes into account the mutual interference among the process variables and the noise characteristics is the key points and some kinds of statistical approach were required.

Taking advantage of the opportunity when the second computer for process control was introduced to the Kumagaya Plant in 1967, application of time series analysis to a raw-material blending mill process and rotary kiln process was started. First of all, controlled variables, manipulated variables, sampling interval, etc. were determined through a result of the spectrum analysis and the extensive knowledge yielded from experiences of the physical and chemical characteristics or from the problems arising during the measurement. After that, identification of the multivariate AR model of the

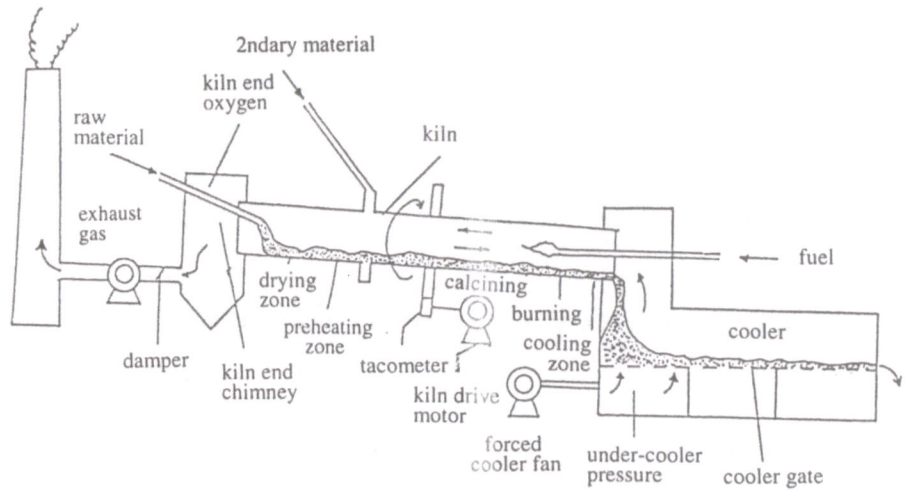

Figure 12.1 Wet rotary kiln

kiln process by the FPE criterion equivalent to AIC was executed. Thus computer control (the optimum control by means of the optimum regulator) of the wet-type cement kiln has been obtained with success.

The results of the study and development at this primary stage are described in detail in Otomo, Nakagawa and Akaike (1969, 1972), Akaike and Nakagawa (1972, 1988). The outcomes of the study and development contributed greatly to the field of the control, since it established the practical model identification method of the noisy industrial process and the systematic approach of the control system design for the first time.

Shortly after these outcomes were obtained, for the purpose of the improvement of the productivity and energy saving, the cement plant was changed in its style of production from the wet process (Figure 12.1) with which raw materials in a form of wet slurry (in a state of 32% moisture being contained) to the dry process (NSP kiln) (Figure 12.3) with which powder raw materials are used. In 1983 by applying the time series analysis also to this NSP kiln process, a practical AR models of the calcining process and clinker cooling process were obtained. Since that time on, the on-line control based on this model has been continued (Hagimura, Saitoh, Yagihara and Kominami 1986; Hagimura, Saitoh and Yagihara 1988). In this chapter, we present our experience of the kiln process control of this new type.

12.2   Cement Plant

For the readers who are unfamiliar with a cement manufacturing plant, an outline of the manufacturing process of cement is given. The cement manufacturing process is, as illustrated in Figure 12.2, comprised of a raw material process, burning process (NSP kiln process), finishing process and shipping process.

In the raw material process, 4 types of natural raw materials (lime stone, clay, calcium silicate, and pyrite cinder) containing CaO, SiO_2, Fe_2O_3, and Al_2O_3 are weighed,

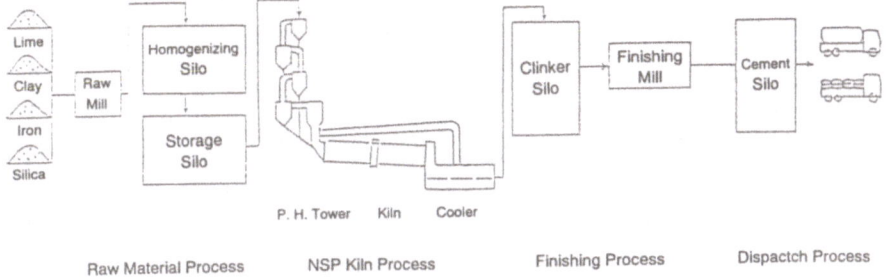

Figure 12.2 Cement manufacturing process

crashed and blended by mill. The blended raw materials are homogenized and stored in a silo. The main objective of the control at this stage is to control the constant feed weigher (CFW) of the individual raw materials so that the 3 component module of the blended raw materials (HM, SM, and IM) will maintain the required values. The irregularity in the component module will give rise to disturbance in the kiln process to be followed, resulting in deterioration of the products.

The NSP kiln process is composed of the following 3 subprocesses.

1) **Cyclone suspension preheater tower (P.H. tower)**
   Here, preheating and calcining (evolution of carbon dioxide from calcium carbonate) of the blended raw materials is made. If the blended materials enter the rotary kiln without a sufficient reaction then, a half-burned phenomenon of the clinker appears, giving rise to deterioration in the quality of the products.

2) **Rotary kiln process**
   In this process, the clinker, a half-finished product, is burned at a temperature of approximately 1450°C. Owing to wide variation of the heat that is generated and to be supplied in the burning zone, in where clinker mineral determining the quality of the cement is combined, half-burned clinker is prone to be produced. Also here, internal disturbance such as muddling failure (phenomenon of removal of the coating adhered to the inside wall of the kiln) or so-called long flame, short flame, etc. shifts the distribution of temperature in kiln, and as the result a wide range of variation of the heat quantity is induced.

3) **Clinker quenching process**
   Here, the clinker fed from the kiln onto cooler grate is rapidly cooled by the cooling air. The air heated to high temperature by exchanging the heat with the clinker is recovered as the secondary air for burning in the kiln and preheater tower (formation of a feedback loop). Accordingly the variation of the brought-in heat quantity of the secondary air causes the disturbance influencing the kiln and preheater tower.

In the finishing process, the clinker is crashed together with gypsum in the ball mill to be formed as cement, i.e. a final product.

As described above, the kiln process is the most important subprocess to maintain good quality of the cement among the manufacturing sub-processes, and stabilized calcining of the blended raw materials is indispensable and steady burning and steady quenching of the clinker are required. These subprocesses, which are always suscep-

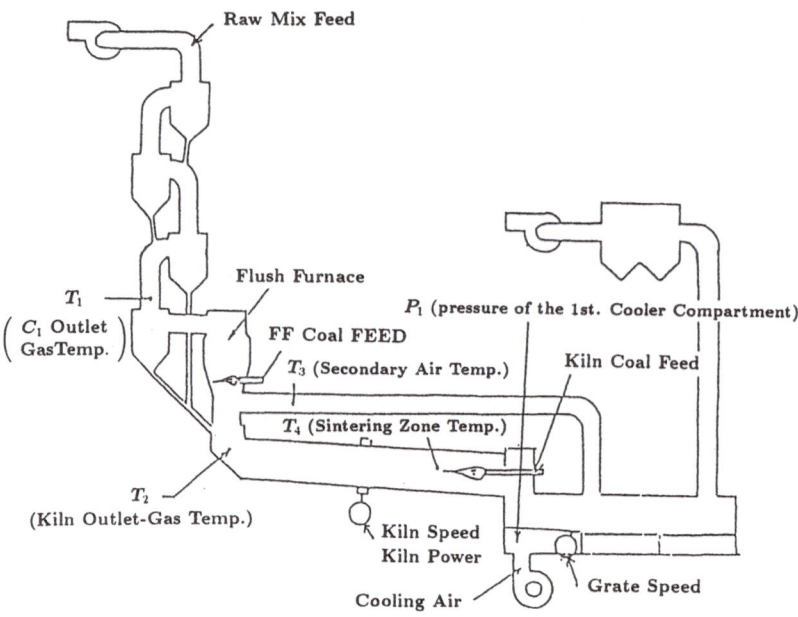

Figure 12.3 Dry NSP kiln

tible to the trouble of a variety of the internal disturbance and contains internal feedback loops, are hard to be controlled. Therefore it is required to develop a practical procedure of model identification based on the statistical approach and the control system design by the modern control theory.

12.3 Identification and Control of the Kiln Process

12.3.1 Calcining Ratio Control

As illustrated in Figure 12.3, the calcining process exists, in the FF (Flush Furnace). Here, the chemical reaction (evolution of carbon dioxide from calcium carbonate)

$$CaCO_3 \rightarrow CaO + CO_2 \uparrow \qquad (12.1)$$

is in progress at a temperature of approximately 850°C. The above reaction is an endothermic reaction of approximately 420kcal/kg, and the required heat is given by the purified coal of the FF together with the exhaust gas from the rotary kiln and the secondary air from the clinker cooler. Meanwhile variation of the exhaust gas from the rotary kiln and the brought-in heat quantity from the secondary air is reacted as disturbance on the calcining process.

Big turbulence of the outlet gas temperature T_1 of the C_4 cyclone where calcining reaction is in progress, disturbs the kiln process. Therefore to maintain a stabilized calcining reaction, the C_4 cyclone outlet gas temperature T_1 should be retained in a narrow range around a specified value by the operation of the FF consumption of purified coal. On the other hand, the time lag between T_1 and the secondary air

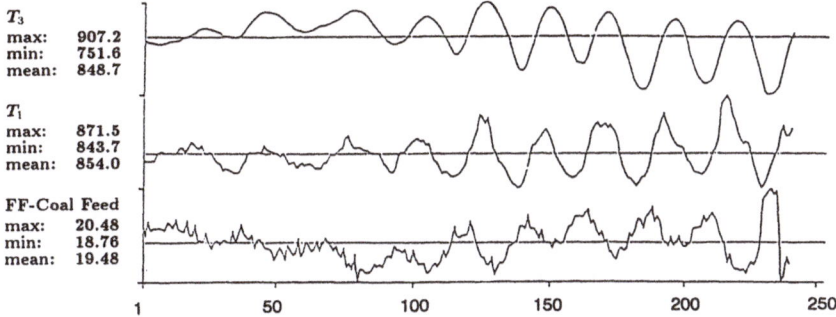

T_3
max: 907.2
min: 751.6
mean: 848.7

T_1
max: 871.5
min: 843.7
mean: 854.0

FF-Coal Feed
max: 20.48
min: 18.76
mean: 19.48

1   50   100   150   200   250

Figure 12.4   PID control and lag oscillation of the C_4 outlet gas temperature T_1

temperature T_3 is approximately 25min. Therefore under specific disturbance, lag oscillation caused by the delay in time occur. This fact is shown in Figure 12.4. The classical PID control was not only unable to reduce the oscillation, but often induced it, because of the lack of the consideration of the interaction between T_1 and T_3.

The purpose of the control is to stabilize T_1 by adjusting the FF consumption of purified coal. To analyze the thermal behavior of the calcining process, the 3 input variables of the kiln consumption of purified coal, FF consumption of purified coal and feed amount of the blended raw materials, and the 3 output variables of T_1, T_2 (exhaust temperature from rotary kiln), and T_3 were chosen (Figure 12.3). Thus the multivariate AR model

$$X_n = \sum_{m=1}^{M} A_m X_{n-m} + U_n \tag{12.2}$$

was used, where X_n is the 6-dimensional process variable vector, A_m is the coefficient matrix and U_n is the white noise vector.

The individual variables are expressed by the deviation from the individual set-point values (target value of the control). At the next stage, the minimum AIC procedure was applied to the data of the sampling interval 1min and the length 150, and the selected model order M was 2. Figure 12.5 shows the power contribution ratio of the other variable to T_1. The power contribution ratio indicates that while the contribution of the kiln consumption of fine coal (K-COAL) and T_2 to T_1 are restricted in the low-frequency region close to the zero frequency, the feeding amount of the blended materials (RAW-FEED) and T_3 significantly contribute to T_1 on the intermediate frequency band.

The great contribution of K-COAL and T_2 in the low-frequency region suggests that the variables of these factors govern the fundamental thermal behavior to retain the whole process of the kiln under an adequate environment. The results coincide with experiences of the authors of this article. The primary objective here is to suppress the 25min delay oscillation, and it is advisable to pay attention exclusively to the intermediate frequency band of 20–60min. Thus a 4-variable model (T_3, T_1; FF-COAL, RAW-FEED) excluding K-COAL and T_2 was selected. Meanwhile from the assumption that the first 6-variable model is the model (1) and from the comparison of the models (2), (3), ... obtained by taking up or throwing out and choosing the variables as listed in Table 12.1 with AIC, it is explained that the model giving

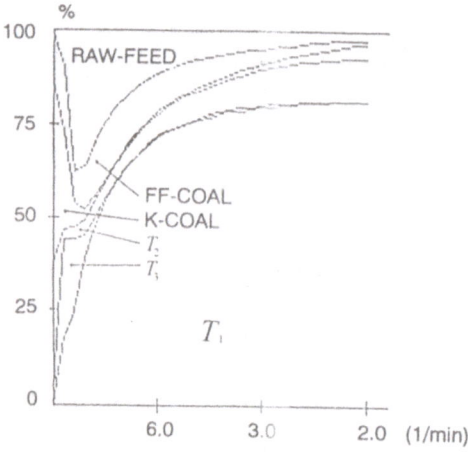

Figure 12.5   Power contribution ratio of the C_4 outlet gas temperature T_1

Table 12.1   Comparison of the models by AIC

	Model(1) $M = 2$	Model(2) $M = 3$	Model(3) $M = 3$	Model(4) $M = 3$	Model(5) $M = 3$
T_2	◯	◯	◯		
T_3	◯	◯	◯	◯	◯
T_1	◯	◯	◯	◯	◯
Kiln Coal	◯				
FF-Coal	◯	◯	◯	◯	◯
Raw Mix Feed	◯	◯			◯
Min. AIC(M)	1529.4	1530.5	1539.4	924.9	914.5

the minimum AIC coincides with the 4-variable model (5) obtained from the above-mentioned empirical judgment.

ARX model of the 4-variable (T_3, T_1; FF-COAL, RAW-FEED) was modeled by

$$x_n = \sum_{m=1}^{M} a_m x_{n-m} + \sum_{m=1}^{M} b_m y_{n-m} + w_n , \qquad (12.3)$$

where x_n is the 2-dimensional controlled variable vector (output), and y_n is the 2-dimensional manipulated variable vector. Meanwhile w_n is the white noise vector and a_m and b_m are the coefficient matrices. By the minimum AIC procedure, the order 3 ARX model was selected. The variance covariance matrix of the prediction error w_n was very close to the diagonal matrix (Table 12.2), and the relative contribution of FF-COAL T_1, RAW-FEED, and T_3 to T_1 were almost the same as the ones in Figure 12.5.

From the author's experiences, it can be judged that the brevity of this model (T_1, T_3, FF-COAL, RAW-FEED) is the most adequate factor for the present objective of the control. However the amount of production in proportion to the feed rate

Table 12.2 Normalized covariant matrix of the innovation

	T_3	T_1	FF-Coal	Raw Mix Fffd
T_3	1.00	−0.05	0.03	0.02
T_1	−0.05	1.00	−0.07	−0.12
FF-Coal	0.03	−0.07	1.00	−0.06
Raw Mix Feed	0.02	−0.12	0.06	1.00

[system outputs] T_3, T_1
[system input] FF-Coal, Raw Mix Feed

of the blended raw material is determined at the production management level, and therefore no actual amount of production coincides with the amount of production determined at the management level if RAW-FEED is frequently operated by the control. Hence with the actual control, we adopted the model (4) shows in Table 12.1 from which RAW-REED is eliminated. With this model, no consideration can be made even if RAW-FEED happens to be operated for some reason or other. Thus there is a possibility that offset of T_1 may be produced. From this it can be said that no attempt should have been made with the operation of RAW-FEED by the control gain.

To design the optimal controller, the process model is represented by the state space model

$$Z_n = \Phi Z_{n-1} + \Gamma y_{n-1} + v_n \tag{12.4}$$

$$x_n = H Z_n, \tag{12.5}$$

where Z_n is the 6-dimentional state variable vector, y_n is the manipulated variable vector, and v_n is the 6-dimentional white noise vector. Meanwhile Φ, Γ, and H are the coefficient matrices defined below using the coefficient matrices a_m and b_m ($m = 1, 2, 3$),

$$\Phi = \begin{bmatrix} a_1 & I & 0 \\ a_2 & 0 & I \\ a_3 & 0 & 0 \end{bmatrix}, \quad \Gamma = \begin{bmatrix} b_1 \\ b_2 \\ b_3 \end{bmatrix}, \quad H = \begin{bmatrix} I & 0 & 0 \end{bmatrix}.$$

Next, to determine the amount of control of the manipulated variable y_n, the optimal controller is designed so as to minimize the performance index J_I defined by

$$J_I = E\{K_I\} \tag{12.6}$$

$$K_I = \sum_{n=1}^{I} \{Z_{n-1}^T Q Z_{n-1} + y_{n-1}^T R y_{n-1}\}, \tag{12.7}$$

where E is the expectation and Q is the positive semi-definite weighting matrix. Meanwhile R is the positive definite weighting matrix, and I was put equal to 10. From the optimal control gain G which was calculated by using the dynamic programming method, the control rule was determined as

$$y_n = G Z_n. \tag{12.8}$$

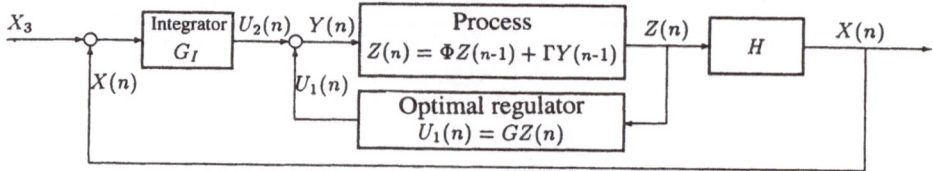

Figure 12.6 LQ control + integral control

The weighting matrices Q and R determining the control gain G is tuned through the simulation after various values are determined for the diagonal elements of the matrices. As the primary value of the tuning, the inverse of the variance of the prediction error of the corresponding controlled variable is usually designated to Q. In the meantime, the inverse of the variance of the raw data of the corresponding manipulated variable is designated for R. However the control gain designed by this Q is usually too strong, and this control tend to cause the hunting oscillation in the actual process which is also under the influence of the variables not included in the model. Therefore it is necessary to make adjustment of the Q matrix referring to the result of the simulation.

The optimal controller considered in the above is a so-called optimal regulator (a linear controller minimizing the variation around the set-point under the quadratic evaluation function: LQ controller), and an idea of the alteration of the set-point as a target value is out of consideration. Therefore with the set-point value alteration or the continuous disturbance, offset (stationary deviation) is produced. This problem is solved by designing the optimum regulator with integral compensation (LQI controller) to which a controller performing an integral action is added. However since the practical LQ controller has already been obtained at this stage, a feedback control system by the integral control is constituted (Figure 12.6) by regarding the feedback system composed of the original process and LQ controller as an objective process.

Supposing that the noise is neglected in (12.4), the model of the objective process can be expressed by

$$Z_n = (\Phi + \Gamma G)Z_{n-1} + \Gamma u_{2,n-1} \tag{12.9}$$
$$x_n = HZ_n. \tag{12.10}$$

On the other hand, assuming that the target value of the controlled variable is x_s, the output $u_{2,n}$ of the integral controller is given as

$$u_{2,n} = u_{2,n-1} + G_I(x_s - x_n), \tag{12.11}$$

where G_I is the gain of the integral controller. It is revealed from (12.9) and (12.10) that the steady state gain K_P of the objective process is given as

$$K_P = H(I - \Phi - \Gamma G)^{-1}\Gamma. \tag{12.12}$$

Referring to the above equation, G_I was adjusted by simulation.

The results of the on-line control by the classical PID controller and the optimal controller adopted here is illustrated in Figure 12.7. (The optimal control is realized

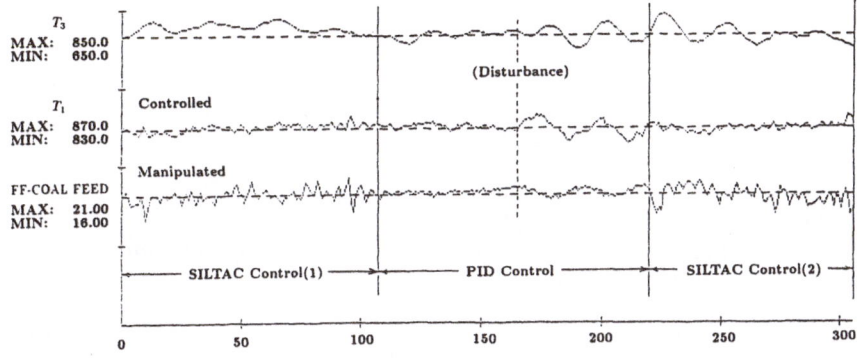

Figure 12.7 Comparison between PID control and SILTAC control of the calcining process

Table 12.3 Mean value, variance, standard deviation at PID control and at SILTAC control

		SILTAC(1)	PID	SILTAC(2)
T_3^{*1}	Mean	805.50	784.90	794.20
	Variance	169.00	234.60	1152.00
	S. D.	13.00	15.32	33.94
T_1^{*2}	Mean	849.90	849.30	839.10
	Variance	3.52	9.98	3.85
	S. D.	1.88	3.16	1.96
FF-Coal	Mean	18.53	18.67	18.45
	Variance	0.105	0.026	0.175
	S. D.	0.324	0.160	0.145

[*1] Setpoint was free.
[*2] Setpoint was 850.0°C.

by using the control system SILTAC shall hereafter be called the SILTAC control. Here, SILTAC is developed by System Sogo Kaihatsu Co. Ltd.)

No information of T_3 is used with the PID control in determining FF-COAL, whereas the information of T_3 is actually utilized with the SILTAC control in predicting T_1. Since T_3 cannot be controlled directly with FF-COAL, the control gain of T_3 is set to almost 0 by adjusting Q. From Figure 12.7, it is perceived that the process is exceedingly stabilized under the SILTAC control (1), whereas delay oscillation is caused with the PID control owing to muddling failure resulting in instability. This lag oscillation is rapidly attenuated by the SILTAC control (2) to recover the stability.

Table 12.3 compare the mean and the variance of both the controls, and Figure 12.8 shows the comparison between and the power spectra at both controls. From these table and figure, it is comprehended that although T_3 is not controlled, T_1 is controlled in a smaller range around the set-point value (850°C) by the SILTAC control than by the PID control. SILTAC control also performed better in the low frequency variation

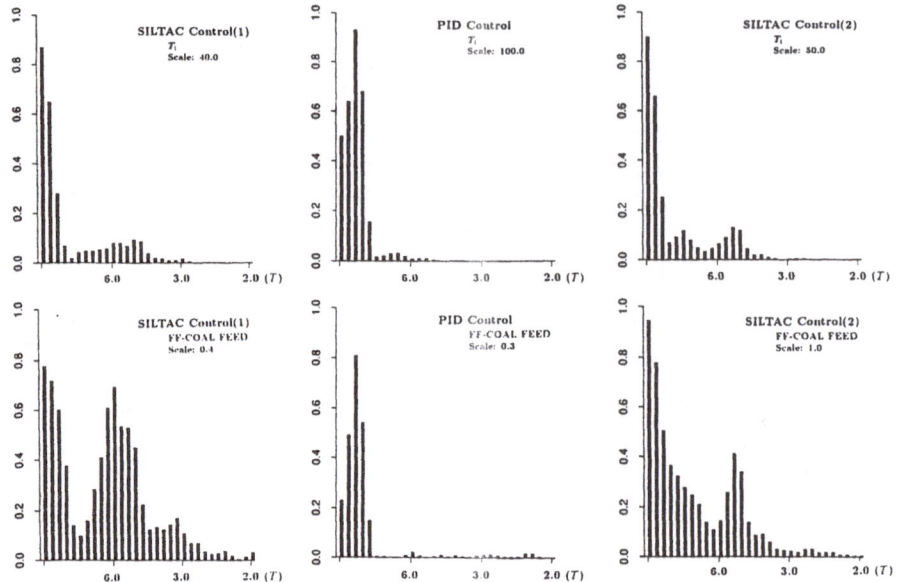

Figure 12.8 Comparison of the power spectrum at between PID control and SILTAC control

closer to the zero frequency. (Note the difference among the individual scales in Figure 12.8.)

Concerning the variation with the period of approximately 5min, the PID control is better. However this is found to be due to bad adjustment of the FF-COAL set-point value control (PID control) in executing the SILTAC control.

12.3.2 Clinker Cooler Control

The high-temperature clinker fed from a kiln is rapidly cooled by the heat exchange with the cooling air, transferring at a settled layer thickness over the grate in reciprocal motion. The hot air produced by the heat exchange with the clinker is recovered as the secondary air for combustion in the kiln and FF when the temperature reaches 750–850°C. The big variation of this heat exchange disarrays the kiln and the combustion of the purified coal of FF, resulting in variation of the sintering zone temperature T_4 and C_4 outlet gas temperature T_1 together with deterioration of the fuel productivity (a fuel amount required to burn 1 ton of the clinker). Concurrently with this, the clinker quality is also badly influenced. A principal cause of the heat quantity variation of the secondary air is the variation of the depth of clinker on the grate, which can be indirectly perceived by the pressure in the compartment under the grate. In case of the conventional control, the grate speed was regulated so that the pressure would be retained at a set-point value. However even if the pressure is retained at a set-point value, the secondary air temperature is not necessarily kept constant. The secondary air temperature is influenced by the temperature of the clinker fed from the kiln.

Using the 5 variables given in Figure 12.9, that is, the power of the kiln driving motor (kiln power), burning zone temperature (T_4), secondary air temperature (T_3), cooler pressure (P_1), and grate speed, time series analysis of the cooler process was

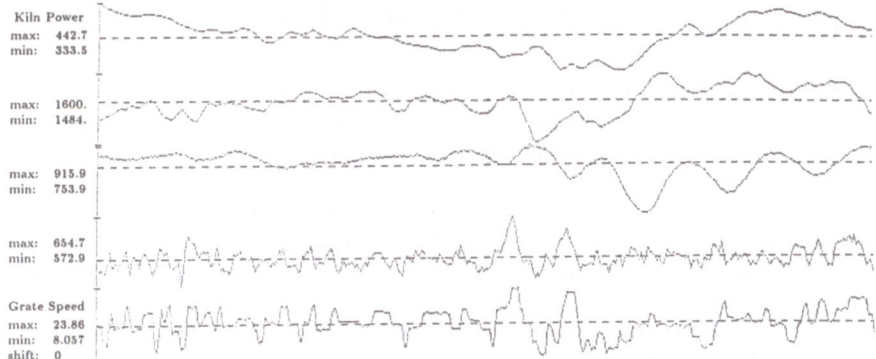

Figure 12.9 Raw data records of the cooler process

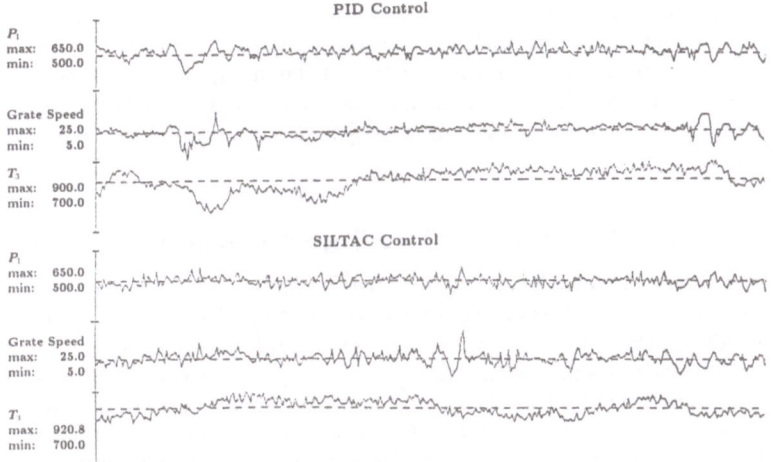

Figure 12.10 Comparison between PID control and SILTAC control of the cooler process

attempted. As can be deduced from Figure 12.9, it is evident that the kiln power, T_4 and T_3 are slow in response comparing with the other P_1 and grate speed. Thus it is explained that none of good model can be obtained even if identification is made by using the combination of such variables whose time constants are extremely different. If it is intended that the dynamic characteristics of P_1 and grate speed be made clear, then rapid sampling will be required. However this compels the kiln power and T_4, T_3 to do the same rapid sampling, and therefore something extremely closer to a unit circle appears in the characteristic root of the identified model. Thus the model is changed into quite an instabilized one. From this reason, the authors of this article considered a model in which the kiln power and T_4 are removed. Figure 12.10 gives results by the PID control and SILTAC control. The process under these controls was stationary and stabilized. Thus both controls achieved nearly similar results.

12.4   Collection and Identification of the Data under the On-line Control

In on-line control for a long time, the dynamic characteristics of the process occasionally become incompatible with the original identified model, and sometimes the conventional model is even inadequate when a special kind of the products is switched to another. Even in such an occasion, it is desirable to modify the model without stopping the on-line control. When the model is inadequate, the characteristics of the prediction error of the model change and a mean value of the prediction error, for example, drifts in either of the directions. Therefore by observing the prediction error of the model continuously, it is possible to detect the inadequacy of the model. As input and output data under on-line control tend to cause the multiple collinearity, a white noise with adequate variance is added to the control signals of the manipulated variables during on-line control, and then identification was performed off-line using the observed data. According to the author's experience, the identification by this method is successful in almost all the occasions provided that the white noises are independent of each other and their magnitude is adequate (variation with twice to 3 times of the standard deviation of the raw data of the manipulated variables is added). An important problem is how to determine the control gain, but theoretical solution was not found yet. Therefore, the tuning of Q and R was performed heuristically by the simulation.

12.5   Optimal Production Level and Pursuit Control

Although the production level is determined by managers of the factory, the operators at work on the spot of production have to determine the set-point values in good balance for all the variables of the control system complying with the production level. The operators' performance indexes at that time are production costs and quality of the products. With the multiple-input and multiple-output system, how to determine the set-point values is an exceedingly difficult but very important problem.

The linear programming method can be applied to determine the optimum set-point values. The constraint that is imposed by the process is given by a process model, but composition of an objective function is furthermore given as another problem. For the process model with constraints, the sampling interval of the set-point value is enough to be far longer than in case of the stabilized control owing to the fact that the set-point values are not changed so frequently. Supposing that the dynamic process model of the set-point value model is given with (12.3) under the condition, the stationary model shown below is obtained from the final value theorem of the sampled-value control theory (Jury 1958)

$$x_s = K_P y_s \tag{12.13}$$

$$K_P = \left(I - \sum_{m=1}^{M} a_m\right)^{-1} \left(\sum_{m=1}^{M} b_m\right), \tag{12.14}$$

where x_s, y_s are the components of the balanced set-point value vectors (that are obtained by solving LP problem), and K_P is a steady state gain matrix of the process. Here, supposing that m_x and m_y are the mean vectors of x_n, y_n being used for identification and x_L, y_L and x_U, y_U are the vectors of the lower and upper limits of the individual variables, formulation of LP problem here is described as

Table 12.4 Steady state gain K_p of the process and the restricted value $E_s(m_x, m_y)$

		$-K_p$			E_s
$.13881E+1$	$.58279E-1$	$.46885E-1$	$-.57690E-1$	$-.51754E-2$	$.29373E+0$
$.42974E+3$	$-.97157E+0$	$-.58943E+0$	$.21355E+0$	$.23971E+1$	$.28986E+1$
$.51408E+2$	$.23601E+0$	$-.24734E+1$	$-.81482E+0$	$.99792E+0$	$.40000E+1$
$.61488E+1$	$.58970E+0$	$.62358E+0$	$-.17899E+0$	$-.85463E-1$	$.13453E+1$
$-.13041E+4$	$-.13533E+2$	$-.61773E+0$	$.73241E+1$	$-.45758E+1$	$.12858E+3$

$$\text{object function} \quad ; \quad J = \alpha^T x_s + \beta^T y_s$$

$$\text{constraints} \quad ; \quad \begin{cases} x_s - K_P y_s = E_s(m_x, m_y) \\ x_L \le x_s \le x_U \\ y_L \le y_s \le y_U \end{cases} \quad ,$$

where $E_s(m_x, m_y) = (m_x - x_L) - K_P(m_y - y_L)$, and therefore $E_s(m_x, m_y)$ is also changed if m_x, m_y are changed during the on-line period.

Although a linear function can be adopted the objective function J from the judgment concerning the problem settlement, the problem is found in the determination of α and β. From the fact that the purpose resides in acquisition of the lowest cost and keeping of the high quality, application of the principal component analysis or canonical correlation analysis is used in determining the coefficients. However here using the time series data comprised of the data measured from the mean value per every 8 hours in Figure 12.11 (the sampling interval = 8 hours), application of the principal component analysis was attempted.

First of all, obtaining K_P and $E_s(m_x, m_y)$ with the aid of the time series data gives the information in Table 12.4. Judging from the polarity and magnitude, it is explained that adequate steady state gain is noted.

Secondly when principal component analysis was applied to the same time series data, it is explained from the sign of the factor loading and the roughly estimated magnitude of the value that the first principal component expresses the 2 categories, i.e. the cost and quality. Therefore taking up the eigen vectors of this principal component as α and β, the objective function of LP is determined as

$$\begin{aligned} J = \ & 0.260x_{s1} + 0.043x_{s2} - 0.070x_{s3} + 0.497x_{s4} - 0.384x_{s5} \\ & + 0.404y_{s1} + 0.348y_{s2} + 0.343y_{s3} + 0.196y_{s4} - 0.303y_{s5} \, . \end{aligned}$$

Solving the LP problem referred to above gives the result in Table 12.5.

The result obtained as above expresses the object of the retention of the high quality at the lowest cost with concrete set-point values of the individual variables, and is quite a reasonable one. After the set-point values of the individual variables are determined, it is noted that the pursuit control (LQI control) directed to the set-point values starts its operation (Figure 12.12).

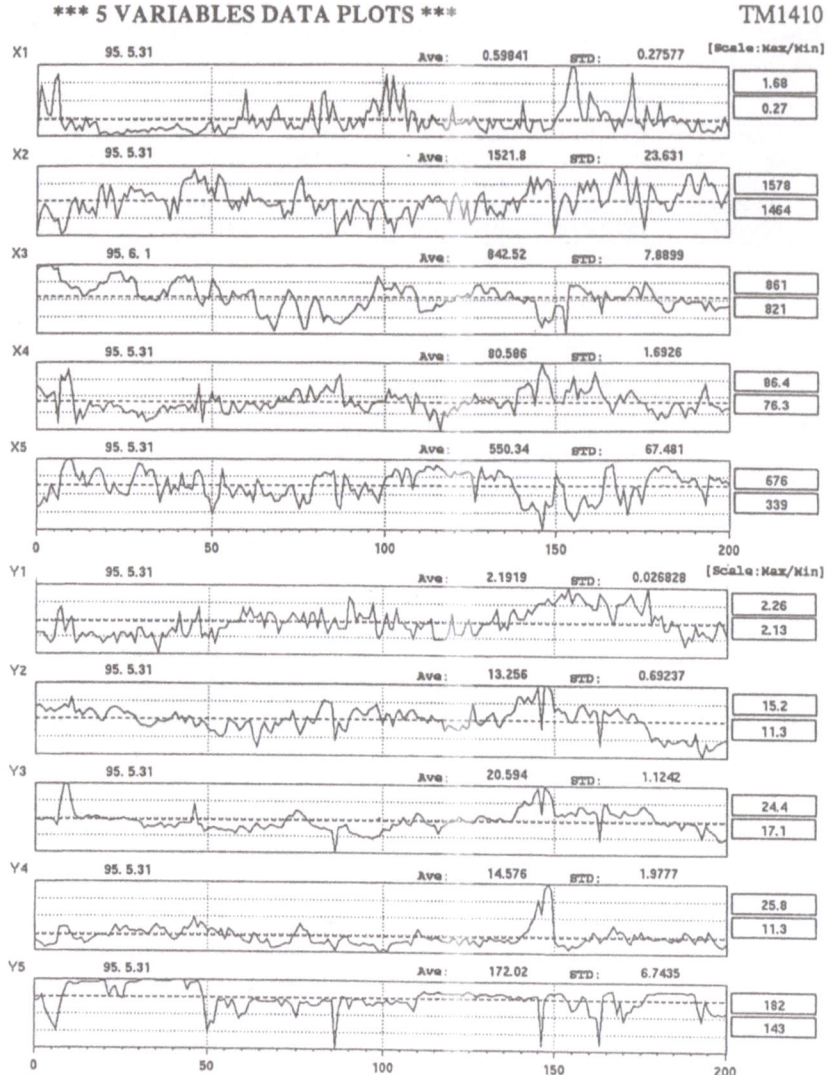

Figure 12.11 Kiln data for the analysis of the optimum production level (Δt=8hr.)

12.6 Conclusions

Taking up the 2 processes that are the most important among the cement manufacturing processes, i.e. calcining process of the NSP kiln process and cooling process, an outline of the successful examples of the identification and control is shown. In the meanwhile, obtaining the sets of the well-balanced set-point values is a task of extreme difficulty for operators. In consideration of this matter, some trials to the problem to determine the optimum production level has been referred to.

Table 12.5 The optimum production level by LP

Variable	Mean[1] (A)	LP solution[2] (B)	Difference (A)−(B)
X_1 ; fcao[3]	0.5984	0.3636	−0.2384
X_2 ; T201	1521.80	1513.41	−7.89
X_3 ; T32	842.52	840.42	−2.10
X_4 ; OIL-l/t[4]	80.59	79.11	−1.48
X_5 ; W200	550.34	617.63	67.29
Y_1 ; Kiln-HM	2.1919	2.1651	−0.0268
Y_2 ; W252	13.26	12.57	−0.69
Y_3 ; W262	20.59	19.47	−1.12
Y_4 ; N215-1	14.58	16.55	1.97
Y_5 ; N200[5]	172.02	170.28	−1.74

[1] Mean value (m_x, m_y) of the data used for the identification

[2] The optimum set-point value (X_s, Y_s) obtained by the linear programming method

[3] fcao is the free lime in the clinker, and is related with the quality of the cement.

[4] OIL-l/t means fuel per unit (litter / clinker ton).

[5] N200 (kiln rotation speed) is related with the output of the cement.

[6] Evaluation function is composed of the eigen vectors of the first component.

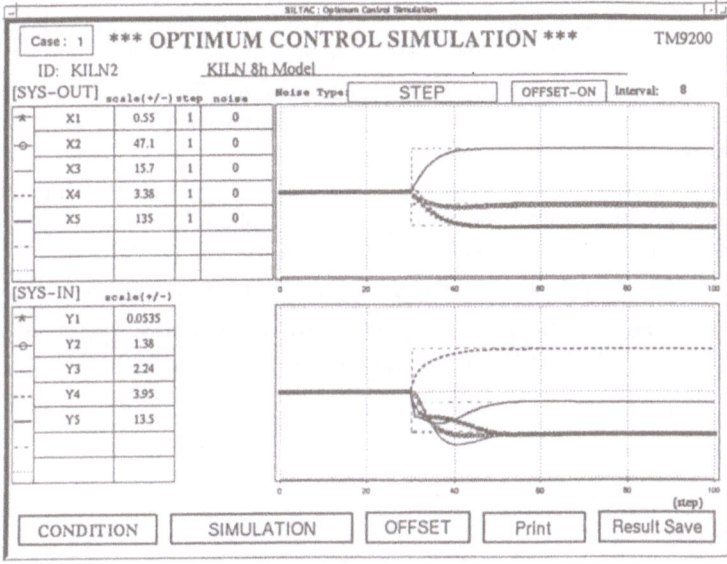

Figure 12.12 Change and pursuit control of the optimum production level

Actual industrial processes are subjected to much of trouble by noise, and many of the variables interfere in each other through the internal feedback loop, resulting in complicated behavior. To explain the structure of this complicated behavior, it seems that the statistical approach is indispensable. The important factor in identifying the process from the actual data and in designing control is to observe carefully the real data and to make clear the objective of the control, and furthermore to endow the model obtained as a result of the identification with the background supported by theory and experience.

There is no guarantee that the stationary characteristics of the process is certain to be retained for a long period, and a nonlinear phenomenon may occur in some occasions. To allow the control to be continued for a long time, construction of the tough control system with a feedback system covering the affairs as referred to above might be required, and furthermore, the control focused on the trendy variation of running data must be considered.

References

Akaike, H. and T. Nakagawa (1972), Statistical analysis and control of dynamic systems, Saiensu-sha (in Japanese). (English translation was published from Kluwer Academic Publishers in 1988.)

Otomo, T., Nakagawa, T. and Akaike, H. (1969), "Implementation of computer control of a cement rotary kiln through data analysis," *Technical Session 66, IFAC 4th World Congress*, Warsaw, 115–140.

Otomo, T., Nakagawa, T. and Akaike, H. (1972), "Statistical approach to computer control of cement rotary kilns," *Automatica*, Vol. 8, 35–48.

Hagimura, S., Saitoh, T., Yagihara, Y. and Kominami, T. (1986), "The hierarchical control with stability and production level control of cement NSPkilns," *IFAC 5th MMM Congress*, Tokyo, 77–78.

Hagimura, S., Saitoh, T. and Yagihara, Y. (1988), "Application of time series analysis and modern control theory to the cement plant," *Annals of the Institute of Statistical Mathematics*, Vol. 40, No. 3, 419–438.

Jury, E. I. (1958), Sampled-data control systems, John Wiley & Sons Inc., 30–31

Nakagawa, T. and Y. (1986), "The approach to the design of optimum production level and control", *J. SICE* (in Japanese), Vol. 24, 77–83.

Chapter 13

Analysis of a Human/2-wheeled-Vehicle System by ARdock

Makio Ishiguro[1] and Takio Oya[2]
[1]The Institute of Statistical Mathematics
4-6-7 Minami-Azabu, Minato-ku, Tokyo 106-8569, Japan
ishiguro@ism.ac.jp
[2]Meiji University
1-1-1 Higasi-Mita, Tama-ku, Kawasaki-shi 214-8571, Japan

13.1 Introduction

For bicycles or motorcycles that are running straight, the following conditions should be satisfied. (1) Balancing stability should be retained. (2) Directional stability (route retention) should be retained. Why can a 2-wheeled vehicle run without capsizing? This has long been a problem stimulating people's keen interest, and the cause was once said to be gyroscopic effect. However computation reveals instantly that the gyroscopic effect is almost negligible. A 2-wheeled vehicle is intrinsically of static instability, and is liable to be overturned without control. This is because the running bicycle is continuously under the influence of many kinds of disturbances. Even the bicycle's rider him/herself is often a cause of the disturbance. Hereafter, the disturbance shall be just called noise.

For directional and balancing stability the ground contact points of a 2-wheeled vehicle has to be shifted transversely through lateral ground forces. The mechanism of generation of lateral ground forces is as follows: When the handlebar is suddenly turned to the right, for example, making a slip angle of the tire, a force from the road toward the right-side direction is generated forcing the wheel to be shifted to the right. (When the handlebar is turned to the left, generation of the force is in a left-side direction. No wording such as "to the right for example" shall never be employed hereafter. All the 2-wheeled vehicles are of the symmetric structure.) The said force is called a cornering force. In addition to the above, a force called camber thrust is generated and the ground contact point is shifted to the right when the wheel is tilted to the right side. This tilting angle is called as camber angle. The rider can tilt the bicycle's body to the right with tilting the upper half of his/her body to the

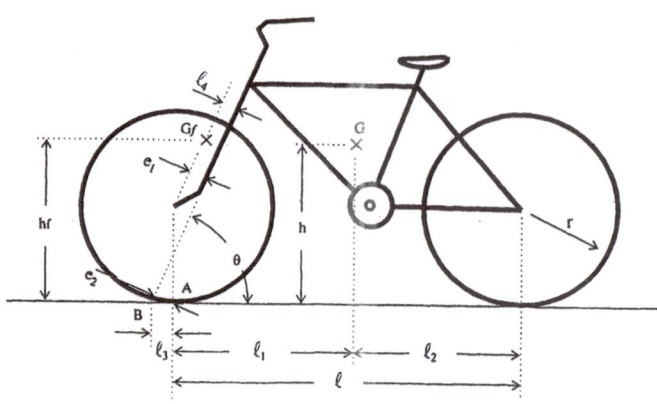

Figure 13.1   Alignment of steering system of a bicycle, caster angle θ, and trail ℓ_3

left. In short, balancing stability and route keeping of a 2-wheeled vehicle are realized through the cornering force and camber thrust generated by the rider's handlebar steering and body leaning in a lateral direction.

To secure the balancing stability, the straight line connecting the grounding points of the front and rear wheels is controlled to come to a spot directly under the center of gravity of total mass through the operation such as firstly turning the handlebar to the right or tilting the rider's body leftward when the total mass is off to the right. To recover the original direction of the advancement of the bicycle which is deviated to the right side, adjustment of the direction should be made retaining the balancing stability. This is realized through the operation such as firstly tilting the bicycle's body to the left by turning furthermore the handlebar to the right, then turning the handlebar to the left. A bicycle is prevented from being capsized thanks to such kinds of operation and manipulation. In addition to this, the bicycle itself provides the dynamic feedback mechanism prevent the bicycle from being exposed to danger. The bicycle's body is designed so that the handlebar turns itself naturally to the right whenever the bicycle's body is tilted to the right, even if the bicycle's rider makes no handlebar steering. This is the performance dynamically produced in accordance with the alignment of the steering system (Figure 13.1), i.e. the geometric condition such as caster angle θ (inclination of steering axis), trail ℓ_3 (distance between the point where the extension line of the handlebar axis crosses the ground and the front wheel ground contact point), position of the center of gravity of the steering system G_f. In case of hands-free riding, the operation and process for generating lateral ground forces are as follows. If the rider tilts the upper half of his body to the left, the bicycle's body is tilted to the right side with the reaction torque, then the handlebar automatically turned to the right by the front wheel system alignment producing the cornering force toward the right direction. Concurrently with this, camber thrust in the same direction is produced. In general, the cornering force is more dominant than the latter.

Almost all of the control manipulation is made by the rider's skillful operation, but the real nature of the control or details of the control circuits have not yet been explained because the manipulation in question is made unconsciously. Investigation has been made in various manners with the control mechanism or control manipulation of actual 2-wheeled vehicle running, but no explanation has been made until now. The greatest difficulty was found in the following 2 points. (1) The roll angle of the bicycle leaning from side to side was difficult to measure since its amplitude is as minute as of the standard deviation 1 degree. (2) The bicycle is under the feedback control resulted from the rider's unconscious handlebar steering and body leaning entangled together with dynamic response of the 2-wheeled vehicle itself, and consequently the estimation of the significant characteristics has been almost impossible until now even if the individual variates can be measured.

Owing to the reasons described as above, earlier study has been restricted in the theoretical or experimental researches on the stability of the human/2-wheeled-vehicle system or on the transfer function of the control systems by setting control mechanism model (handlebar steering or shift of the center of gravity) (Iguchi 1962; Tsukada and Oya 1981), human technological approach concerning the properties such as human body's response (Kageyama and Kogo 1984), researches to identify various characteristics of the human/2-wheeled-vehicle by applying specific relatively great pulse input in actual running (Sharp 1971; Hattori et al. 1975; Zellner and Weir 1979; Aoki 1979; Nishimi et al. 1985; Nagai 1986), and researches on the statistical characteristics of running motorbicycle (Yokomori et al. 1991). No study has been done with the matters stepping into the control issues of the actual human/2-wheeled-vehicle at straight running. Use of precision gyroscopes has become generally possible since about 20 years ago, and the problem (1) has now been solved. However at that time, some technical books and specific documents even stated clearly that identifying the control circuit characteristics of feedback system from the observation of the stationary irregular variation process was hard to be made, and no solution was obtained with the problem (2). Nevertheless, introduction of Akaike's TIMSAC (Akaike and Nakagawa 1972) was a breakthrough to the solution of the problem, and the analysis using TIMSAC has been made in the Mechanical Dynamics Laboratory of Oya, one of the authors, since about 15 years ago (Oya et al. 1991).

The object of this study is not only to be acquainted with the rider's control characteristics in running 2-wheeled vehicle, but also to compare the transfer function of the 2-wheeled vehicle itself obtained simultaneously with the mechanically obtained bicycle body's transfer function. In this chapter, an attempt is made with the identification of the control circuits by fitting the multidimensional autoregressive model to the time series of the measurement of various variables recorded when a bicycle is running straight. The results of the said attempt using the software ARdock (Ishiguro 1994) for the system analysis of the autoregressive model developed by Ishiguro are hereby introduced.

13.2   Data

Among the physical quantities concerned to the control actions of the human/2-wheeled-vehicle system, the two variates as the controlled quantities; (1) the roll angle (RO) and (2) steering angle of handlebar (HA), and the three variates as the operation quantities; (3) the handlebar torque (HT), (4) saddle torque (ST: the torque which is added to the saddle by the leaning body and is expected to be a measure of the power of control with leaning body), and (5) pedal part torque (PT: the torque which is added with leaning body to the foot-rests fixed to the frame) are measured. The

Table 13.1   Measured variables

Variable	Symbole	Unit	Method of measurement
Roll angle	RO	deg	fixing a free-gyroscope to the frame
Steering angle	HA	deg	fixing a large pulley to the handle axis coaxially, whose rotation is transmitted to a small pulley fixed to a potentiometer
Handlebar torque	HT	kgf·cm	setting a subhandlebar to the handle axis, and measuring the torque with strain gage method
Saddle torque	ST	kgf·cm	applying strain gages on the base of saddle, and measuring the torque, which corresponds to the leaning of the rider's body
Pedal part torque	PT	kgf·cm	detaching pedals and cranks, fixing foot-rests to the frame for measuring the torque, which corresponds to the leaning of the rider's body

Table 13.2   Dimensions of the bicycle

Total mass of bicycle	m_0	28.1kg
Mass of steering system	m_f	7.5kg
Mass of wheel	m_w	2.5kg
Wheel radius	r	0.327m
Height of center gravity of bicycle	h	0.585m
Height of center gravity of steering system	h_f	0.785m
Wheel base	ℓ	1.140m
Horizontal distance between center gravity of bicycle and front wheel axle	ℓ_1	0.630m
Trail	ℓ_3	0.031m
Perpendicular distance from center gravity of steering system to steering axis	ℓ_4	0.030m
Caster angle	θ	70deg
Fork off-set	e_1	0.083m
Caster length	e_2	0.029m
Moment of inertia of bicycle around x axis at center gravity	J_x	1.83kg·m^2
Moment of inertia of bicycle around z axis at center gravity	J_z	3.53kg·m^2
Moment of inertia of wheel around axle	J_w	0.20kg·m^2

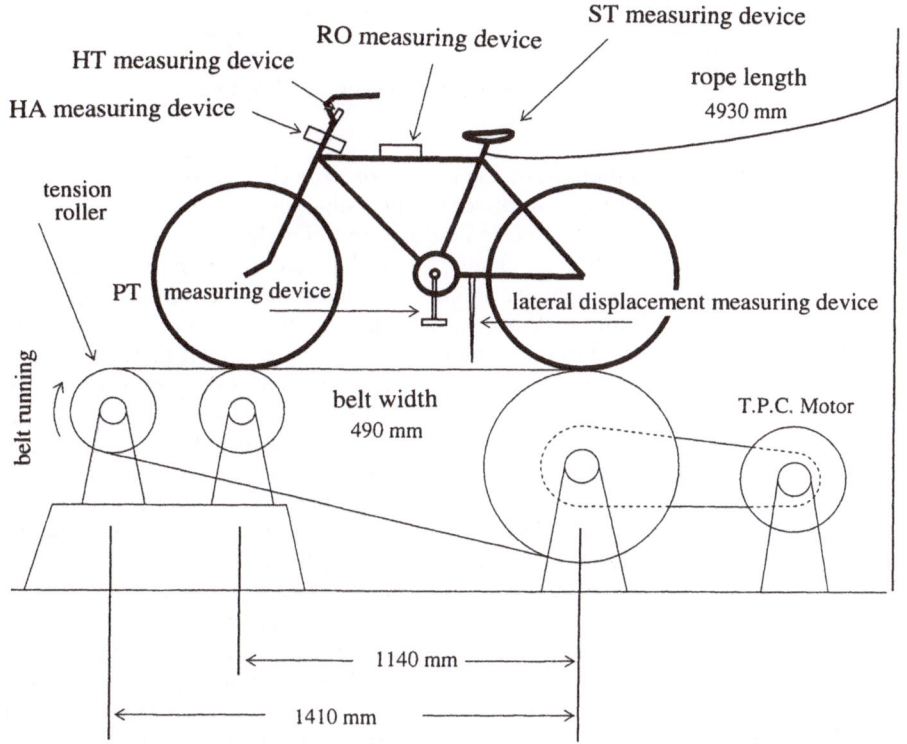

Figure 13.2 Apparatus of the indoor running experiment

measuring methods and units concerning the five variates referred to above are listed in Table 13.1.

In every cases, a leftward direction was determined as plus. The bicycle offered for the experiment was the remodeled unit of a 26" (outside diameter of wheel shown in inches) bicycle on the market, whose dimensions are listed in Table 13.2.

To avoid the disturbance caused by pedalling, a method with traction running was used outdoors and a method with running on the moving belt was used indoors. The road width was 5m in case of the road experiment, and instructions were given to the rider so as not to especially be conscious of the route retention.

The indoor experiment is conducted on a running belt of width 490mm. The rider has to make running especially keeping in mind that the bicycle will neither be capsized nor be deviated from the belt width. Longitudinal position restriction is made with a rope freely connected to the bicycle's body in the vicinity of the center of gravity as illustrated in Figure 13.2 in consideration of easy riding.

The front wheel is placed approximately 15mm ahead of the axis of the smaller roller (diameter 200mm). Owing to the moving belt tension, it can hardly be concluded that the reaction force point of the front wheel ground contact is a 15mm point, but it might be permissible to state from estimation that the ground contact point will

Table 13.3 Standard deviation of measured values

RO	HA	HT	ST	PT
0.94deg	1.6deg	23kgf·cm	14kgf·cm	24kgf·cm

be found approximately 7mm ahead of the roller axis. For this reason, a forward component of the ground contact force, approximately 3–4kgf, of the front wheel is yielded and a good balance with the rope tension is acquired. Furthermore as the influence exercised on the steering stability, an elongation of the trail is seen in some occasions. The effective ground contact point is, as can be judged from the description made above, found at $8(= 15 - 7)$mm distance toward the rear from the point immediately under the front wheel shaft. From the fact that the intrinsic trail of the bicycle is 31mm, it is considered under the condition in question that $31 + 8 = 39$mm will be the effective trail length. In the case the trail is substantially lengthened, tendencies are noted that: (1) Shimmy (spontaneous steering turning shake, vibration, or wobble of several Hz caused often mostly with a 2-wheeled vehicle during running at high speed) is suppressed even at high speed (more than 40km/h). (2) Natural rotational force of the steering is strengthened when the speed is lowered. As a conclusion, it is assumed that no abnormality is produced especially with this restriction method.

Each of four riders with the weight approximately 60kg was engaged in ordinary running at the speed of 10, 15, and 20km/h, and the five variates in Table 13.1 were recorded. On the laboratory apparatus for running test, the displacement Y of the position in a lateral direction was also measured. However, under the consideration of the three points shown below, analyses in this chapter have been made with the 5 variables of RO, HA, HT, ST, and PT. (1) From a point of view of the running dynamics, it is judged that the variable Y is not important one as explanatory variable for the other variables. (2) According to the preparatory power contribution analysis by TIMSAC, almost none of the contribution of Y to the other variables was found. (3) No values of Y are measured in the road running experiment.

After passing through the low-path filter of 10Hz in consideration of aliasing, AD-converted variables at the sampling interval of 50ms were inputted to a computer. As an example, a record of the road running of 20km/h by Mr. T is depicted in Figure 13.3. The standard deviation of the variation of the individual measured values are given in Table 13.3.

13.3   AR Model and ARdock

ARdock is an instrument for fitting a multidimensional AR model shown below to the data and for comprehension of the physical meaning of the fitted AR model. It is named ARdock after a dock to inspect and repair ships (Ishiguro 1989; Ishiguro 1994). The instrument allows the interactive use of strengthened version functions of TIMSAC (Akaike and Nakagawa 1972).

$$\begin{cases} \boldsymbol{x}_t = \sum_{m=1}^{M} A_m \boldsymbol{x}_{t-m} + L\omega_t \\ \omega_t \sim N(0, I) \end{cases} \qquad (t = \dots, -1, 0, 1, 2, \dots) \qquad (13.1)$$

For fitting models revised versions of TIMSAC's FPEC and MULMAR by Akaike *et al.* are used, as shown in Table 13.4. INPR and STPR enable easy computation of the impulse response and step response from a variable to another variable, respectively. With SPEC calculation of the multivariable power spectrum is done, the power contribution is computed simultaneously. Using PBP, stability of the system can be examined. It is very important to examine beforehand the stability of the system by PBP because even if the system is quite unstable, SPEC returns the results. The system, even if it is a stable one, becomes in some occasions unstable when the circuit is cut with CUT.

By CUT operations, it is possible to conduct experiments to cut a stream of the information inside the system. Furthermore, ARdock has a function to replace some part of the system with the optimum control system for the remaining part.

13.4 Numerical Results

The estimated values of the variance/covariance matrix $\Sigma(= LL^T)$ of the residual (innovation) obtained by fitting the model in (13.1) to the data in Figure 13.3 and the correlation coefficients are listed in Table 13.5.

The power spectrum $P(f)$ and power contribution of the individual variates are calculated by using only diagonal elements of Σ, despite the fact that the correlation coefficient of HT and PT is fairly large approximately 0.4. Results are sketched in Figure 13.4(a)–(e).

A peak of the power is observed in the vicinity of 0.8Hz of HA and HT. Contrarily to this, ST and PT have power in the low frequency. The power spectrum of RO appears to have the characteristics of the both. A stand out feature with respect to the power contribution is that the contribution from ST and PT to the other variables

Table 13.4 Functions of ARdock

Functions	Commands	Contents	Comments
Model fitting	FPEC	FPEC according to Akaike	Control-type AR model
	SYST	Revised version of MULMAR	For System analysis
Analysis of model	SPEC	Power spectrum	Power contribution also computed
	PBP	Power building profile	
	INPR	Impulse response	
	STPR	Step response	
	FRQR	Frequency response	
Simulation	SIML	Generation of data	White noise or residual of real data inputted
Modification of model	CUT	Circuit cut	Interactive processing
	OPTC	Optimum control design	

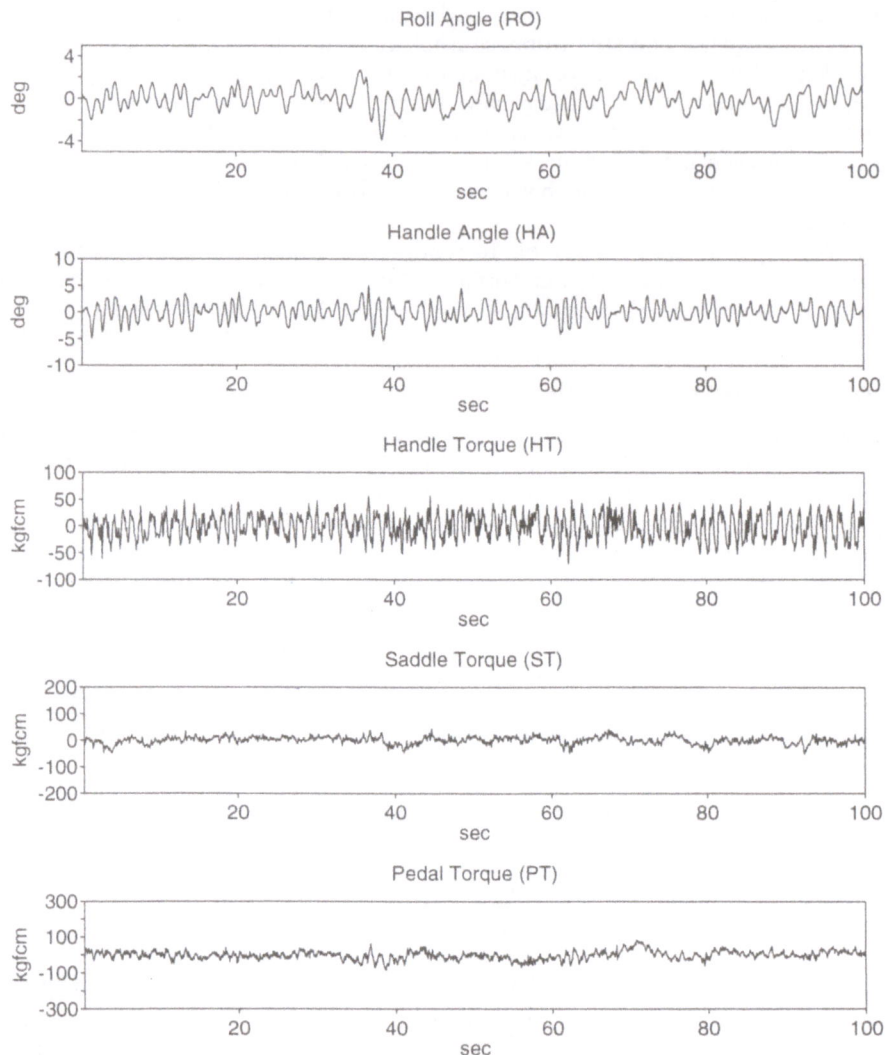

Figure 13.3 Mr. T's running record (20km/h, on the road)

is small. Further details with more of the power spectrum and power contribution is discussed in the following.

[Route Connection Degree and Route Characteristics]   The route connection degree table indicates the output of SYST. The tabulated values are the increase of AIC caused by fixing the elements of the coefficient matrix of AR, which is related to the route, as 0. This implies the deterioration of the model fit caused by the restriction

Power Spectrum   Power Contribution

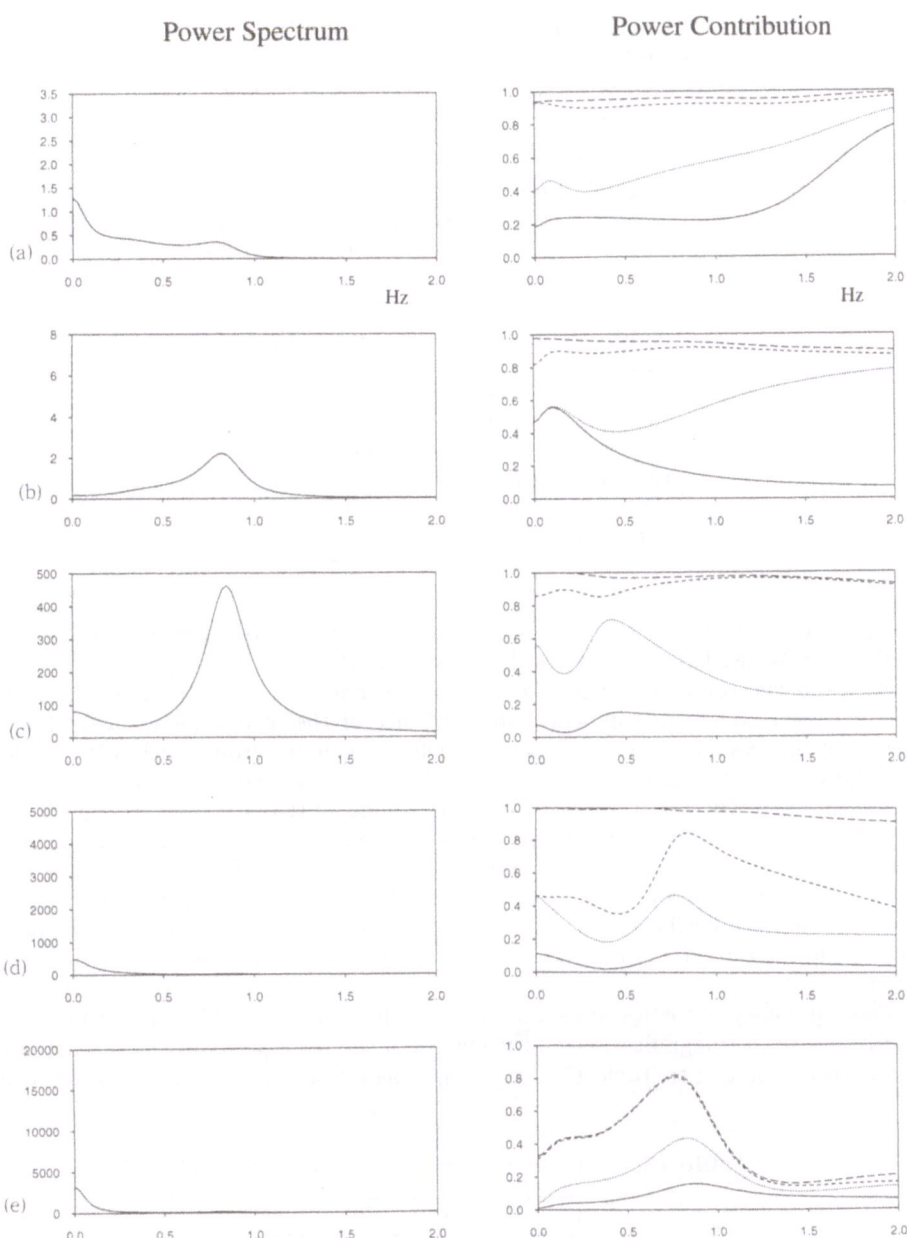

Figure 13.4 Power spectrum and power contribution (a) RO, (b) HA, (c) HT, (d) ST, (e) PT

Table 13.5 Variance/covariance of the residue (the delta on the upper side of the right) and the correlation co-efficient (the lower side of the left)

	RO	HA	HT	ST	PT
RO	0.01	0.00	0.12	−0.01	0.11
HA	0.106	0.04	0.37	0.24	0.42
HT	0.165	0.209	79.47	8.61	31.70
ST	−0.041	0.292	0.238	16.54	2.37
PT	0.153	0.234	0.394	0.065	81.26

Table 13.6 Route connection degree

	RO	HA	HT	ST	PT
RO	–	346.2	13.5	39.1	
HA	247.4	–	309.0	66.3	63.4
HT	223.9	215.9	–		58.1
ST	2.3	301.8	262.1	–	161.2
PT	129.9	71.1	86.3	206.5	–

of the AR coefficient, and the greater the said value is, the greater the role of the AR coefficient possessed is. Thus it can be considered that important information flows through the said route. In Table 13.6, the route connection degree obtained from the data is shown. The table indicates the influence of the individual variables in the upper row exercised on the variables in the left end column, from which it is revealed that the value of AIC is raised by 346.2 on the assumption that the 1-row/2-column component A_{12m} $(m = 1, 2, \ldots, M)$ that is equivalent to HA→RO = 0. It is perceived that this route plays a very important role in the system. It is natural for the angle of the handlebar to exercise strong influence on the inclination of the bicycle. The part of PT→RO, ST→HT are vacant in the table: the said parts cause the decreasing of AIC by fixing the coefficient as 0. We may say that it will be a useless attempt to try to estimate those parameters from given set of data. Thus this route is viewed as being cut.

The route characteristics concerning RO are shown in Table 13.7. The route characteristics are the magnifications of the change of the total power of RO tabulated in a style corresponding to Table 13.6, when the coefficient of the route judged as "con-

Table 13.7 Route characteristics (concerning RO)

	RO	HA	HT	ST	PT
RO	–	∞	0.8	1.0	1.0
HA	7.4	–	2.2	1.2	1.1
HT	5.6	∞	–	1.0	1.0
ST	1.0	1.0	1.2	–	1.0
PT	1.2	1.2	1.0	1.0	–

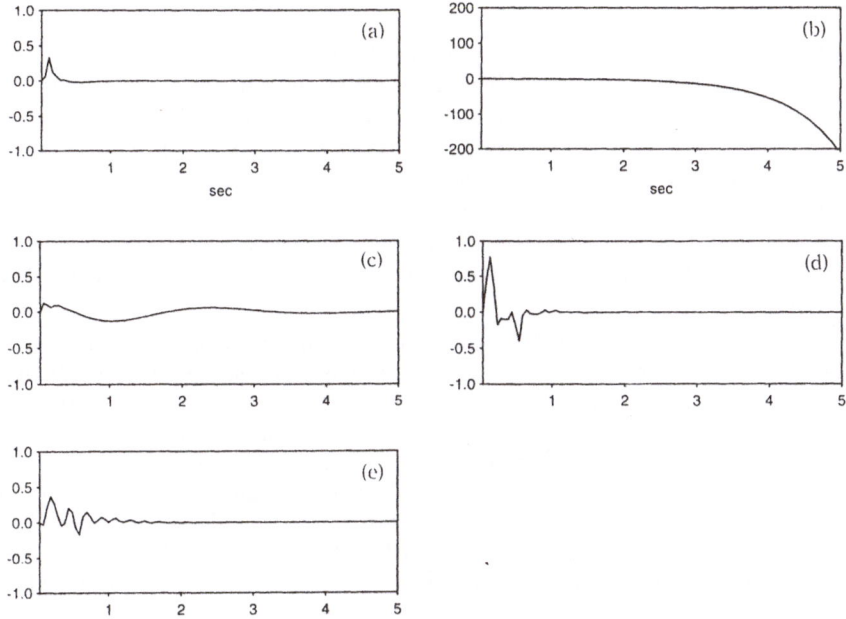

Figure 13.5 (a) RO⟹HA (open loop), (b) HA⟹RO (open loop), (c) HT⟹HA (bicycle→cuts man), (d) RO⟹HT (man→cuts bicycle), (e) RO⟹ST (man→cuts bicycle)

nected" in Table 13.6 is changed into 0. To investigate the details of the influence of the route cut, change of the power spectrum should be inspected. However only the change of total power is summarized in the table as the outline. In such case as the route characteristics close to 1 notwithstanding that the route connection degree is high, the total power of RO varies hardly, but the shape of RO power spectrum varies greatly by cutting its route. The symbol ∞ indicates that RO will become non-stationary by cutting the portion here and will tend to infinity in power.

In glancing at these tables, it is clear that among influences of a human to a bicycle in this system the greatest is HT→HA. It is also clear that RO, HA→HT have played the most significant role as a route of the information from a bicycle to a human, consequently the control via HT, that is to say, the control by handlebar operation is the greatest control circuit.

[Impulse Response] Graphs of the impulse response are sketched in Figure 13.5(a)–(e). To take account of the difference of the variation amplitude between the input and the output, "normalization" is executed in a manner of multiplying the standard deviation of the variation of the input value and furthermore dividing by the standard deviation of the variation of the output value using the numerical values in Table 13.3.

From these results, it can be construed that:

1) Open-loop; When the impulse RO to the left direction is inputted, the handlebar turns to the left (Figure 13.5(a)).

2) Open-loop; When the impulse HA to the left direction is inputted, RO is inclined to the right (Figure 13.5(b)).

3) In the case that the routes from bicycle to human are cut; When such impulse HT as of rotating the handlebar to the left comes in, the handlebar is rotated first to the left and then to the right (Figure 13.5(c)).

4) In the case that the routes from human to bicycle are cut; When the impulse RO of left direction comes in, handlebar torque is applied left and then right (Figure 13.5(d)).

5) In the case that the routes from human to bicycle are cut; When the impulse RO of left direction comes in, saddle torque is applied to the left at first (Figure 13.5(e)).

Here, 1) and 2) concern with the mutual relationships among the individual parts of the bicycle, and the bicycle's mechanical dynamic characteristics as the structural elements of the human/bicycle system are made known. 3) is about the response of the bicycle subsystem complying with the control input. 4) and 5) are related to the response of the "human-subsystem" in the human/bicycle system. The impulse response of the open loop of RO→HA expresses the dynamic stability (resulted from the steering system alignment referred earlier) of the handlebar angle itself where stability is accomplished even if the feedback is deprived of, and the open-loop impulse response of HA→RO reveals that the bicycle will be capsized if the feedback is devoided.

[**Power Building Profile**]   The power building profile is defined as an expected value of the variance of the individual components of $x_1, x_2, \ldots$, when $x_0, x_{-1}, \ldots, x_{1-M}$ are all put as 0 in the AR model Equation (13.1). The power building profile of the k-th variable is expressed by a symbol of $P_{kk}^B(t)$. The power spectrum $P_{kk}(f)$ has a meaning but only when

$$P_{kk}^B(\infty) = \lim_{t \to \infty} P_{kk}^B(t) < \infty$$

Thus the following is established.

$$P_{kk}^B(\infty) = \int P_{kk}(f) \, df = \mathrm{Var}\{x_k\}$$

If the power building profile of all the variables is confined in a finite value, then the whole of the system will be stable.

The power building profile (the curve marked with B in Figure 13.6) of RO in case the transmission control onto the bicycle is cut does not diverge to the infinity. This is contradictory to the reality and is revealing the limitation of the model in question. However when the matter is compared with the profile of no route-cut (the curve marked with A in Figure 13.6), it is conceived that the variation is rapidly increased. In observing the power spectrum (Figure 13.7(a)) of RO at that time, it is revealed that the variation of the frequency that was utterly invisible in Figure 13.4(a) is great. This has been suppressed by the rider's control. In observing the frequency response function of the rider's response, it might be expected that something discernible will be obtained.

When only the transmission from HT to the bicycle is left, the growth curve of RO is changed into the curve marked with C in Figure 13.6 and the power spectrum is changed into the one sketched in Figure 13.7(b). Thus it is shown that the control via

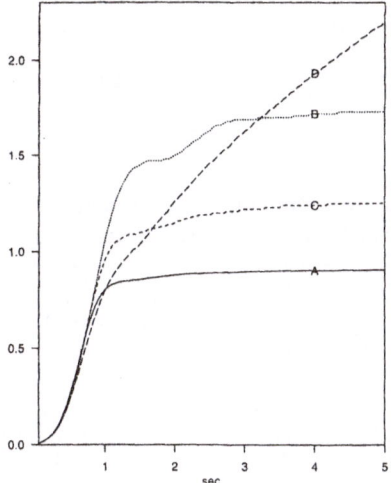

Figure 13.6 Power building profile of RO

Power Spectrum   Power Contribution

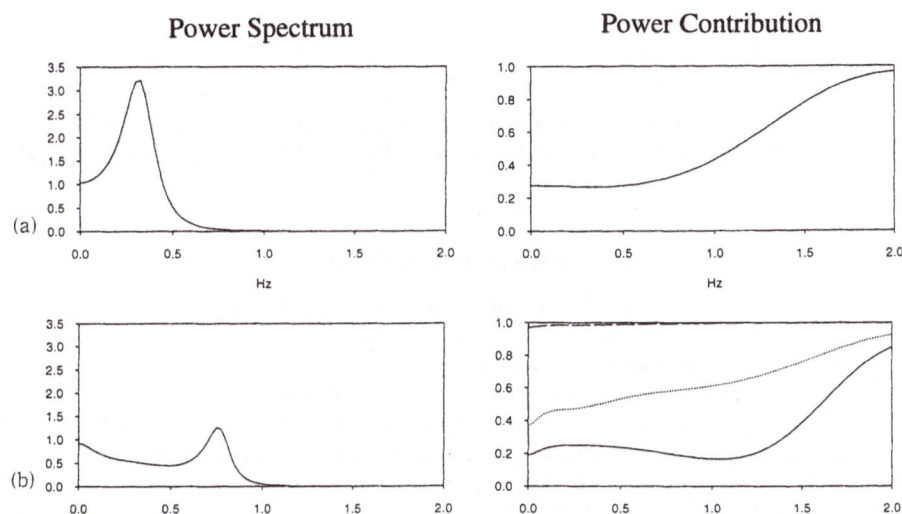

Figure 13.7 (a) Power spectrum and power contribution of RO when control is devoid of, (b) Power spectrum and power contribution of RO under the control by HT

HT suppresses the variation of RO. Contrarily to the above, the power of RO diverges (the curve marked by D in Figure 13.6) when the route from ST and PT to RO and HA is left. It is shown that none of effective control is possible only with ST and PT.

The results of the above are summarized as follows:

1) The bicycle is the system possessing such the dynamic response characteristics

Table 13.8 Standard deviation (in case of being hands-free)

RO	HA	HT	ST	PT
1.5deg	3.1deg	0.76kgf·cm	85kgf·cm	160kgf·cm

Table 13.9 Variance/covariance matrix (the delta zone on the upper side of the right) of the residue and correlation coefficient (on the lower side of the left)

	RO	HA	HT	ST	PT
RO	0.01	−0.01	0.0	−0.41	0.54
HA	−0.238	0.05	0.0	0.79	0.15
HT	0.0	0.0	0.48	−0.26	1.49
ST	−0.449	0.390	−0.041	83.44	−71.43
PT	0.203	0.025	0.079	−0.290	729.5

as having the inclination to stable intrinsically (the steering system alignment).

2) The operation for the purpose of accomplishing stability is made mainly by handlebar operations.

3) The operation referred to above is supplemented by the control with leaning rider's body.

Proviso; Item (2) and (3) are the conclusion deduced from abovementioned data. In the case of different rider, control with leaning rider's body may be stronger than that with handlebar.

13.5 Analysis of the Hands-Free Steering

Records concerning the hands-free running experiment are depicted in Figure 13.8. The running was made in a room. The scales on the ordinate are the same as those in Figure 13.3. As can be expected as a matter of course, it is evident that the variation width is as a whole has been made greater except that HT is very small (Table 13.8, Figure 13.9(a)–(e)). It is impressive that the power spectra of HA, ST, and PT resemble each other in shape, but none of definite explanation has been made by the study at the stage of preparing this paper as to why such resemblance is brought about. It is imagined, however, that it is related to the fact that just single control steering is made exclusively based on the leaning body in case of hands-free riding, while two types of control operations, i.e. handlebar steering and leaning body are available in case of ordinary running.

Variance/covariance matrix and correlation coefficient of the residual obtained by applying the model to the data in question are listed in Table 13.9, whereas a route connection degree and the route characteristics are listed in Tables 13.10 and 13.11, respectively.

In case of hands-free riding, no control by the route of HT→HA is, as a matter of course, available. From these Figures and Tables it can be explicitly construed that control by leaning body is being made.

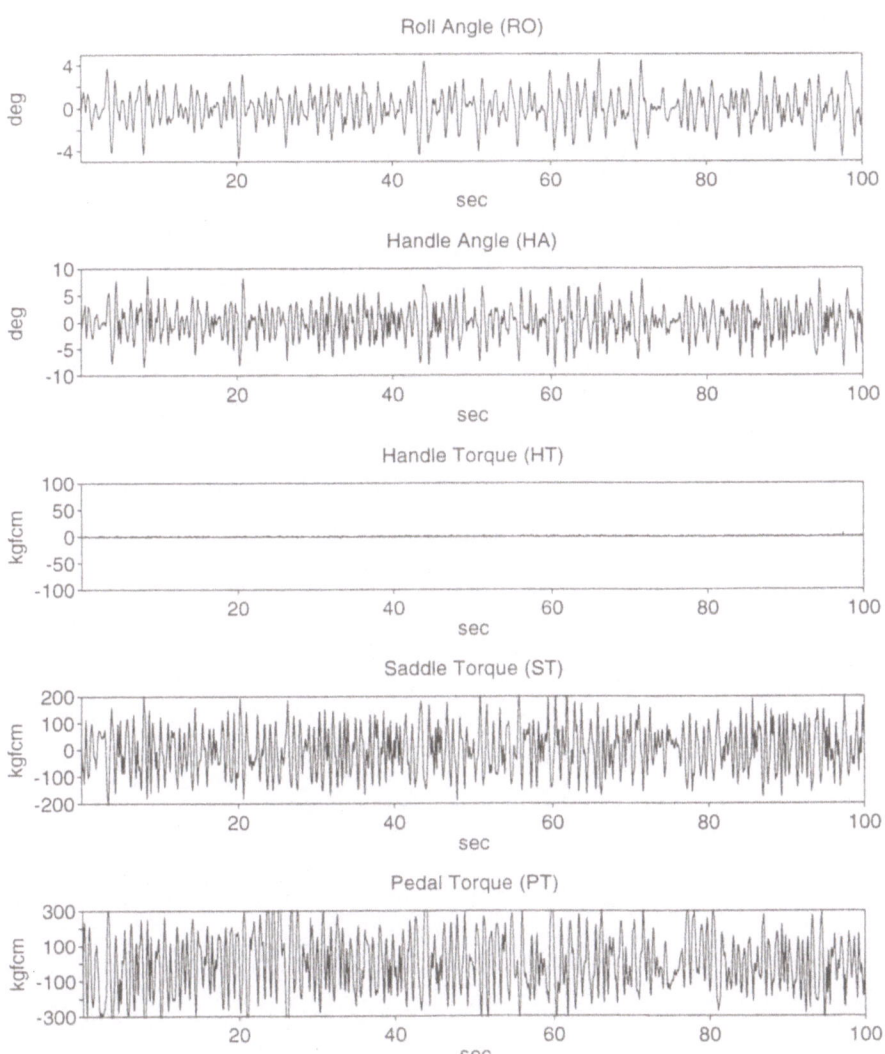

Figure 13.8 Mr. T's running records (20km/h, indoor, hands-free)

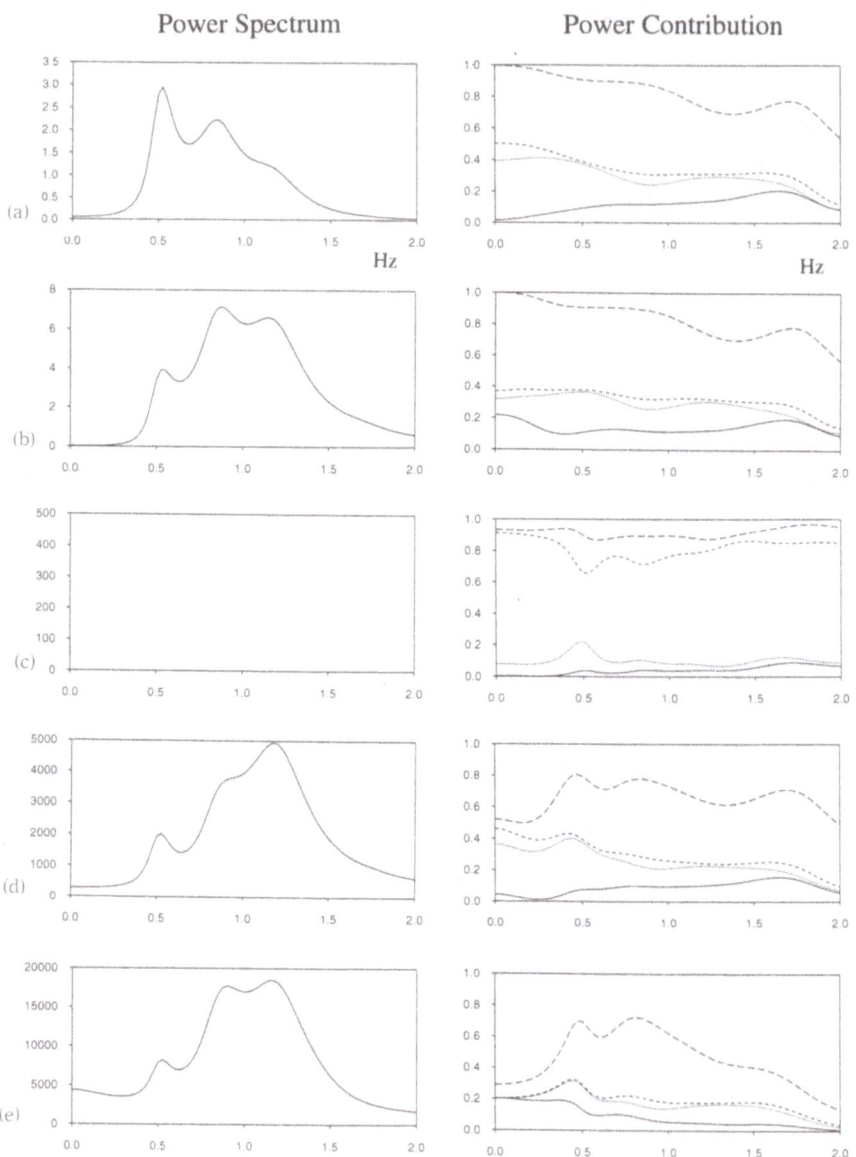

Figure 13.9 Power spectrum and power contribution (a) RO, (b) HA, (c) HT, (d) ST, (e) PT

Table 13.10 Route connection degree (in case of being hands-free)

	RO	HA	HT	ST	PT
RO	–	473.6	1.0	397.7	441.6
HA	279.7	–	37.8	773.0	638.8
HT	27.7	55.3	–	50.2	
ST	745.7	440.3	3.2	–	939.6
PT	205.3	90.1	4.7	337.5	–

Table 13.11 Route characteristics concerning RO (in case of being hands-free)

	RO	HA	HT	ST	PT
RO	–	∞	1.0	∞	0.5
HA	∞	–	1.0	∞	1.8
HT	1.2	1.4	–	0.9	1.0
ST	∞	∞	1.0	–	∞
PT	∞	18.1	1.0	∞	–

Comparing the impulse response of hands-free steering (Figure 13.10(a)-(d)) with that of ordinary running (Figure 13.5), it is comprehended very well that the pattern of the "bicycle$\Longrightarrow$rider" is different contrarily to the fact that a qualitative style of the "bicycle$\Longrightarrow$bicycle" is similar to the case of the ordinary running. Glancing at RO$\Longrightarrow$ST, when the bicycle's body is inclined to the left by the leftward impulse, the action of applying ST to the right and then applying to the left, which has never been seen in case of the ordinary running using a handlebar, can be seen. This reveals the control action to tilt the upper body of the rider with the intention to rotate rapidly the handlebar to the left (utilization of the natural handlebar rotation mechanism by steering system alignment).

13.6 The Optimum Control

An attempt is made with designing the optimum control using a model fitted to the data in Figure 13.3. If an equation of the system is written denoting the value of RO, HA at the time t by y_t and the value of HT, ST, PT by w_t, then the following is obtained.

$$\left[\begin{array}{c} y_t \\ w_t \end{array} \right] = \sum_{m=1}^{M} \left[\begin{array}{cc} A_m^{yy} & A_m^{yw} \\ A_m^{wy} & A_m^{ww} \end{array} \right] \left[\begin{array}{c} y_{t-m} \\ w_{t-m} \end{array} \right] + \left[\begin{array}{c} \zeta_t \\ \eta_t \end{array} \right] \tag{13.2}$$

From the above equation, the influence of $w_{t-l}, w_{t-2}, \ldots$ exercised on y_t can be examined. Utilizing the result from the above, w_t can be determined so that the undermentioned equation that is defined by the weight $q_1, q_2, \ldots$ and $r_1, r_2, \ldots$ would be minimum.

$$J = \sum_{k}^{K_y} q_k E\{y_{kt}^2\} + \sum_{k}^{K_w} r_k E\{w_{k(t-1)}^2\} \tag{13.3}$$

Table 13.12   Standard weighted coefficient

RO	HA	HT	ST	PT
1.0	0.4	0.002	0.005	0.002

Table 13.13   Weighted coefficent II

RO	HA	HT	ST	PT
1.0	0.4	100.0	0.001	0.001

Here, y_{kt} and $w_{k(t-1)}$ are respectively the k-th components of y_t and $w_{(t-1)}$, whereas K_y and K_w are the dimensions of y_t and $w_{(t-1)}$. In calculation, it is found that the behavior of the whole system under this control is described in a style of the equation of the form

$$\begin{bmatrix} y_t \\ w_t \end{bmatrix} = \sum_{m=1}^{M} \begin{bmatrix} A_m^{yy} & A_m^{yw} \\ \tilde{A}_m^{wy} & \tilde{A}_m^{ww} \end{bmatrix} \begin{bmatrix} y_{t-m} \\ w_{t-m} \end{bmatrix} + \begin{bmatrix} \zeta_t \\ G_1\zeta_t \end{bmatrix} \tag{13.4}$$

G_1 and the equation shown below are the matrixes determined by depending on (A_m^{yy}, A_m^{yw}) $(m = 1, \ldots, M)$ and $q_1, q_2, \ldots, q_{K_y}, r_1, r_2 \ldots, q_{K_w}$.

$$(\tilde{A}_m^{wy}, \tilde{A}_m^{ww}) \quad (m = 1, \ldots, M)$$

In viewing the impulse response (Figure 13.11(a)–(b)) of "the optimum control I" designed using the numerical values (Table 13.12) close to a reciprocal of the square of the standard deviation of the individual variables listed in Table 13.3, it is clear that the behavior almost similar to the measured values in Figures 13.5(d) and 13.5(e) is seen.

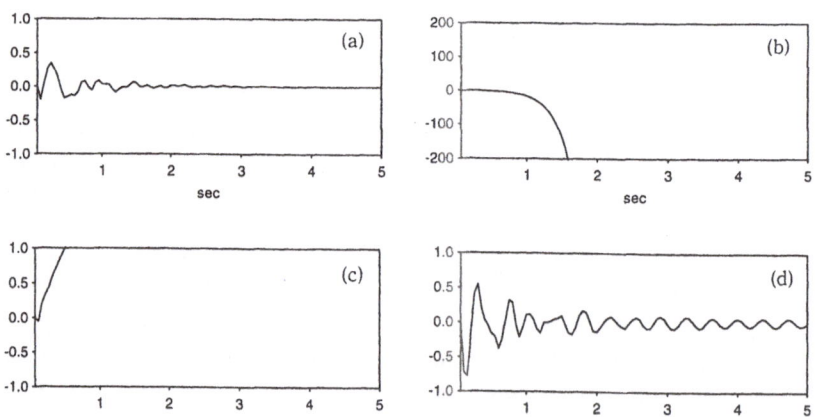

Figure 13.10   Impulse response (a) RO⟹HA (open loop), (b) HA⟹RO (open loop), (c) ST⟹RO (bicycle→cuts man), (d) RO⟹ST (man→cuts bicycle)

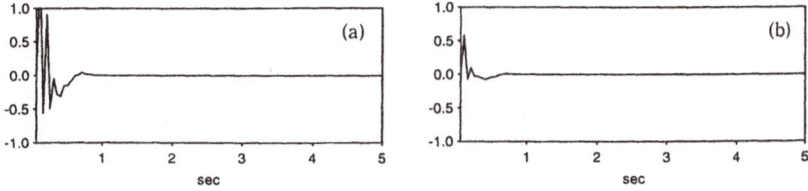

Figure 13.11   Impulse response (a) RO⟹HT, (b) RO⟹ST of "the optimum control I"

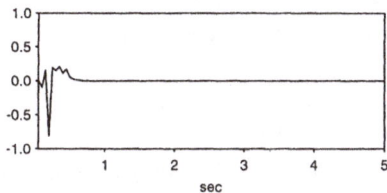

Figure 13.12   Impulse response RO⟹ST of "the optimum control II"

Contrarily to the above, it is explained that the response (RO ⟹ ST, Figure 13.12) of "the optimum control II" by the weighted coefficient II with which the penalty complying with the variation of HT as listed in Table 13.13 resembles to the value obtained by the analysis of the measured values of the hands-free steering (Figure 13.10(d)). Thus the action exhibiting great ST to the right when bicycle's body is inclined to the left can be seen.

As shown by this simple example, impulse responses varying according to how the weighted coefficient is taken in case of designing the optimum control is obtained. On the supposition that the human composing part of the system is doing the optimum control in some sense, it might be possible to some extent to infer the human's intension with a method to estimate parameters of the control design in question. This will be taken up as the subject for investigation in future.

Note 1: Part of the study reported here was supported by the specific grant-in-aid for "Development of the Data Analysis Supporting System" and the Grant-in-Aid for International Scientific Research (University to University Cooperative Research), for which the authors of this paper wish to express their sincere gratitude. The authors are also appreciative of the fact that the computation was conducted by the Institute of Statistical Mathematics' Stochastic Calculation System. In connection with ARdock, its detailed explanation and the source list will later be publicized separately.

References

Akaike, H. and Nakagawa, T. (1972), *Statistical Analysis and Control of Dynamic System*, Saiensu-sha.

Aoki, A. (1979), "Experimental study on motorcycle steering performance," *SAE*

Paper, 790265.

Ishiguro, M. (1989), "System analysis with multivariate AR model," *Operations Research*, Vol. 34, No. 10, 547–554 (in Japanese).

Ishiguro, M. (1994), "System analysis and seasonal adjustment through model fitting," *Proceedings of The First US/Japan Conference on the Frontiers of Statistical, Modeling: An Informational Approach*, Bozdogan, H. (ed.), Kluwer, Netherland, 79–91.

Ozaki, T. (1988), *Time Series Analysis*, University of the Air Press, Tokyo (in Japanese).

Oya, T., Ishiguro, M., Ogino, H. and Hirayama, K. (1991), "Identification of a stability control systems of the running two-wheeled vehicle," *JSME (C)*, Vol. 57, No. 535, 848–853.

Iguchi, M. (1962), "Dynamics of running two-wheeled vehicle," *Science of Machine*, Vol. 14, No. 7, 890–894; No. 8, 1009–1017.

Hattori, Y. *et al.* (1979), "Dimensions of bicycle effecting steering characteristics," *Technical Center Report (Japan Bicycle Production Institute)*, No. 5, 15–24.

Kageyama, I. and Kogo, A. (1984), "Human factors in the steering system of two-wheeled vehicles," *JSME, (C)*, Vol. 50, No. 458, 129–136.

Nagai, M. (1986), "Directional/lateral control of two-wheeled vehicles at low speeds," *Transactions of the SAE of Japan*, No. 32, 113–118.

Nishimi, T. *et al.* (1985), "Analysis of straight running stability of motorcycles," *The Tenth International Technical Conference on Experiimental Safety Vehicles*, July, 1080–1112.

Prem, H. and Good, M. C. (1983), "A rider-lean steering mechanism for motorcycle control," *8th IAVSD Symposium, Dynamics of Vehicle on Road and Tracks*, 422–435.

Sharp, R. S. (1971), "The stability and control of motorcycles," *JMES*, Vol. 13, No. 5, 316–329.

Tsukada, Y. and Oya, T. (1981), "A study on dimension of stability factor of bicycle," *Design and Drafting*, Vol. 16, No. 88, 24–31.

Yokomori, M *et al.* (1992), "Rider's operation of a motorcycle running straight at low speed," *JSME International Journal, Series III*, Vol. 35, No. 4, 553–559.

Zellner, J. W. & Weir, D. H. (1979), "Moped directional dynamics and handling qualities," *SAE Paper*, 790260.

Chapter 14

Vibration Data Analysis of Automobiles

Shinzi Yamakawa
Kogakuin University
1-24-2 Nishi-Shinjyuku, Shinjyuku-ku, Tokyo 160-0023, Japan
yamakawa@cc.kogakuin.ac.jp

14.1 Preface

Roads have generally random surface undulations, which cause the vertical vibration of automobiles. Those road inputs act on their bodies, suspensions and axles and then have severe effects on the durability of the components and the riding comforts. The concept of the random data process for automobiles has been applied for a long time. The stress occurrence analysis for durability study was previously carried out with scales on pen-oscillograms. The analysis is now automatically executed with advances of electronic devices. Spectral analysis for study on frequency characteristics of vibration and stress has been popular since 1960's and now becomes one of routine works at engineering divisions with the spread of FFT equipment. ISO/DIS 8608 Mechanical Vibration-Road Surface Profiles-Reporting Measured Data which is the accumulation of measured road surface undulations for the standard property presentation was published. We can evaluate wear of the component materials and the riding comfort through the DIS data. Most of time series data are currently processed through FFT methods for ease of computation. It is still the problem whether the instantaneous values of road surface undulation during running can be obtained with enough accuracy. The frequency response analyses among accelerations and stresses of components during running are usually carried out in practice. The power contribution method is also used in case of feedback system. When the system has human factors, auto-regressive models for short time data are often in consideration of the performance change with time. Sometimes time domain models are also adopted for the time-variant system besides in the case of human factor. Studies on the decision of continuity of data characteristics and the identification of nonlinear vibration systems and other examples are presented in the following sections.

14.2   Road Surface Input-Wear of Component Material and Riding Comfort

Because the frequency component of the response induced by road surface input is mostly less than 50 Hz, usual sampling interval for the spectral analysis is 0.01 s. Data of 1000 to 2000 points or 10 to 20 seconds are usually processed. The maximum lag number for the correlation function is 100 and the Hanning window (see Appendix in this chapter) is used to smooth the spectrum. Averaging of 20 points for the FFT frequency method is suitable. In most cases, adequate result would be attained in practice. It is recommended to use the auto-regressive method together with traditional methods, if the spectral properties are unknown empirically or if the length of the data is insufficient. The data should be preprocessed after removal of the mean value and trend and the test of stationarity and normality.

We could calculate occurrences of extreme values through power spectra and the equation indicated by Rice (1944) and then estimate damage of the materials, if the data are stationary and normal. However, we cannot put excessive expectation for estimation of absolute life of the components. We have many barriers such as life estimation for random stresses, scatter of material fatigue and linearity of component stress to the road input, if the power spectrum of the road surface is readily known. On the other hand, we might estimate the effect of a slight change in the power spectrum to the life of the component that have been damaged through an endurance test (Yamakawa 1973). Figure 14.1 shows an example of stress power spectrum with some peaks. It is assumed that each peak can be reduced by half through individual means. How much life extension can we expect by the those means? Table 14.1 shows an estimation result, where σ^2 is the variance, $(\sigma/\sigma_0)^2$ is the ratio of the variance to the experimental value, and the Fatigue Guide Number is a relative fatigue damage index calculated through accumulated fatigue damage (so called Miner's) law and

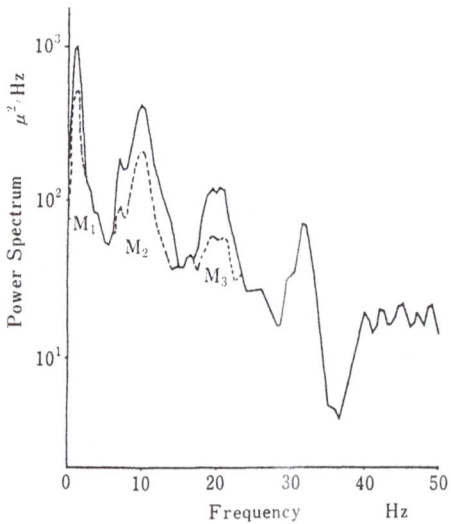

Figure 14.1   Reduction of Peak Value

Table 14.1   Reduction of Fatigue Guide Number

	Original	M_1	M_2	M_3
Content of Modification	—	Reduction of 0-2.5Hz Freq. Component by Half	Reduction of 6-14Hz Freq. Component by Half	Reduction of 18-23.5Hz Freq. Component by Half
σ^2	9127	7718	7510	8624
$(\sigma/\sigma_0)^2$	1.00	0.85	0.82	0.94
F.G.N	9.50	6.19	5.62	7.92
$(F.G.N.)_0/(F.G.N.)$	1.00	1.54	1.69	1.14

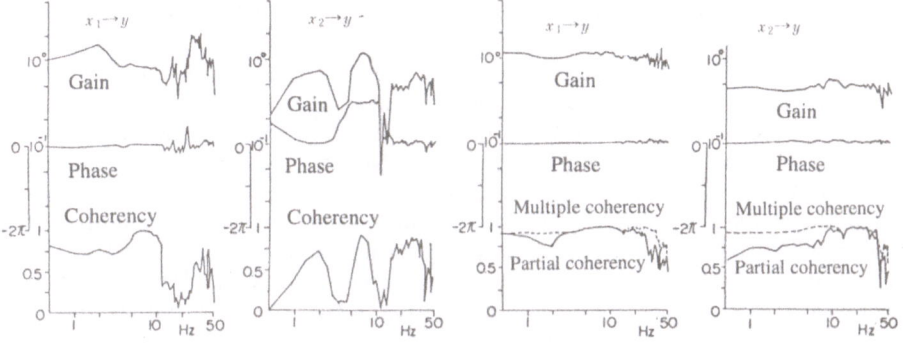

Figure 14.2   Example of Result for Multiple Input System Left Side-Estimated Result as Single Input System Right Side-Estimated Result as Double Input System

the peak occurrence estimation. The expected value for life extension is given at the bottom row as the inverse ratio to the standard value.

The output power spectrum is given by the product of input spectrum and the squared value of the frequency response function in the linear system. Therefore we can estimate some properties through it. In the case of riding comfort, sufficient result would be practically produced by equivalently linearized characteristics, even if human body responses are nonlinear. The method to obtain a frequency response through the cross-spectrum between the input and output records are often used in development stages of automobiles. In most cases many inputs should be considered.

A sample data which might draw a wrong conclusion by treating inherent multiple input data as single input data are shown. Figure 14.2 shows an example of the spectral analysis for a multiple input system. The system is assumed to have 2 inputs and an output: all of them are stresses in 3 steering linkages of an automobile. Because there are at least 3 inputs in the steering system, namely road inputs from the left and the right wheels and the driver's input from the steering handle, it is the problem to process as a single input system. The left part of Figure 14.2 shows

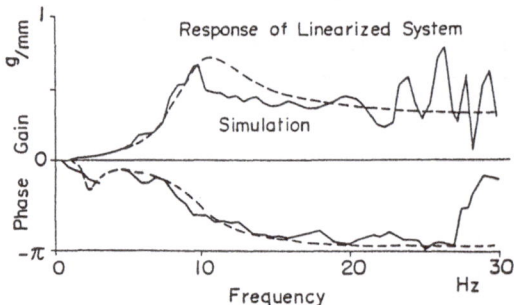

Figure 14.3   Response of Equivalently Linearized System

the result of single input system x_1 and x_2, respectively. The ordinary coherence function that indicate the coherence between the input and the output is partly large enough and further the gain and the phase of the frequency response function vary significantly. However, the right part of Figure 14.2 shows that the links transmit only proportional forces. Sometimes estimation of a frequency response function through the cross-spectrum produces the result that corresponds to equivalently linearizing of a nonlinear system and achieves an essential aim. Figure 14.3 shows a comparison between the frequency response function obtained from numerical simulation and that of equivalently linearized nonlinear system with a saturation element (Yamakawa 1971). They correspond well in the frequency range less than 20 Hz. In this case, the equivalent gain is calculated through iteration of the variance from the modified power spectrum.

14.3   Separation of Correlated Power Components in a Multiple Input System by Means of Power Contributions

Analysis of multiple input system with plural correlated random inputs is an important task in the most development processes of automobiles. For example, those analyses are necessary to make the following question clear: What kind of a part or which direction has a serious effect on the vibration and stress of the components of interest and what kind of guidance for the countermeasures can we get through the result? Power contribution analysis in the time series data analysis program package TIMSAC gives us a direct answer to look for the input that is closely relevant to the output in those cases. It is also an effective method in the case that the system has feedback loops and the signal flows are considered to be complicated (Akaike and Nakagawa 1972). Arrangements of exhaust gas pipes of most heavy duty trucks with V-engines are much complicated. When insufficient durability of the trial parts is pointed out and countermeasures such as shape changes of the pipes or their mountings are requested, determination of the object by intuition is difficult (Yamakawa 1983). Figure 14.4 shows the stress power spectrum of the exhaust gas pipe resulted from trailed test on a rough road. Evident peaks are recognized in the frequency ranges about 5, 6, 17.5, 20.5 Hz. To study the relevancy of those peaks to the vibration of the exhaust system parts many acceleration pickups were attached on the pipes

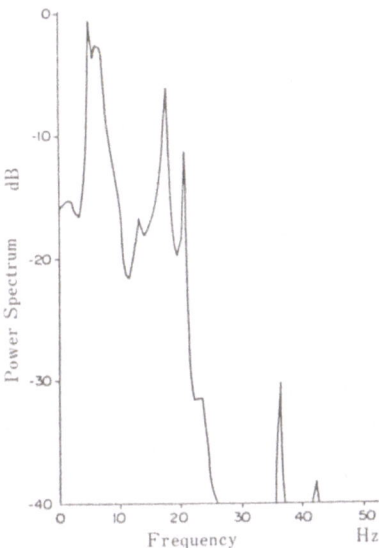

Figure 14.4   Power Spectrum of Exhaust Pipe Stress

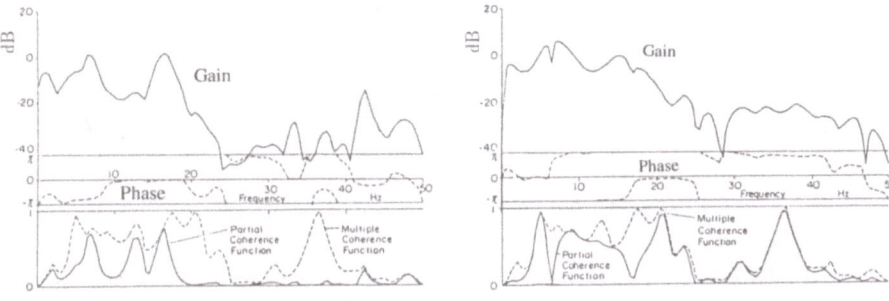

Figure 14.5   Response to
Vertical Acceleration

Figure 14.6   Response to
Horizontal Acceleration

and acceleration and stress were measured simultaneously. The data were processed according to traditional frequency response and coherence analysis method first. The exhaust system with a muffler is attached to the frame through their several mountings. Figures 14.5 and 14.6 show the result of frequency response and coherence function analysis, where vertical and horizontal acceleration at a fore part of the pipe are set to the 2 inputs and the stress of interest is the output. The peaks of stress at 5 Hz and 20.5 Hz are seemed to be relevant to the horizontal acceleration and the peak at 6 Hz to the vertical acceleration according to the coherence function.; and the peak at 17.5 Hz is much affected by the vertical acceleration. It would be difficult to decide the object of change in the pipe mountings according to those results only. Though the input and the output are set to be as above, the relation between acceleration and stress is not always one-sided. Therefore, the power contribution might give a more effective information.

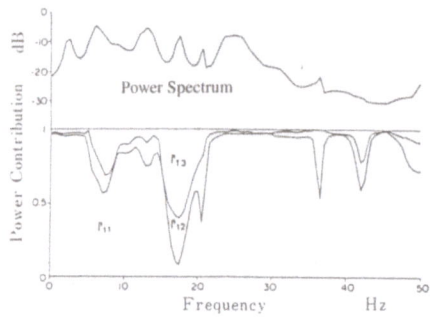

Figure 14.7 Power Contribution of Vertical Acceleration

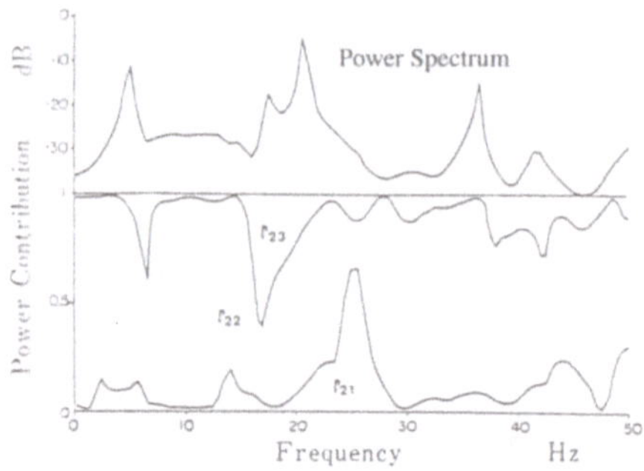

Figure 14.8 Power Contribution of Horizontal Acceleration

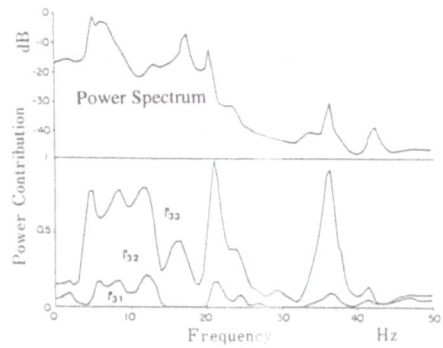

Figure 14.9 Power Contribution of Stress

Figures 14.7 to 14.9 show power spectra and power contributions of the vertical and horizontal accelerations and stress respectively. The power contribution r_{ij} is the ratio of the part caused by the inherent component of x_j to the power spectrum of x_i and so indicated that sum of lengths of individual components is to be unity in the ordinate. The covariance matrix of the normalized inherent component noises is shown in Table 14.2. Because the values of diagonal elements are relatively large, the inherent noise components are seemed to be thoroughly separated. Coherence function and power contribution r_{3j} at the frequency of 4 peaks of the stress power spectrum are shown in Table 14.3. Because the power contribution of the component inherent in the stress at the 17.5 Hz peak is large, existence of another quality that

Table 14.2 Covariance Matrix of Inherent Noise

1	−0.0122	−0.0193
−0.0122	1	−0.1887
−0.0193	−0.187	1

Table 14.3 Coherence Function and Power Contribution

	Hz	5	6	17.5	20.5
	$\gamma^2_{31.2}$	.305	.500	.326	.025
Coherence Function	$\gamma^2_{32.1}$	.951	.289	.213	.901
	γ^2_0	.954	.745	.972	.986
	r_{31}	.080	.180	.008	.105
Power Contribution	r_{32}	.694	.411	.315	.667
	r_{33}	.226	.409	.667	.228

has a serious effect on the stress would be suggested. Thus dominance of the effect of the horizontal vibration is made clear not through coherence function analysis but through power contribution analysis and improvement of the durability is expected by the change of the exhaust pipe mounting characteristics.

14.4   Decision of Continuity of Data Properties

It is a question whether some part of the road surface is regarded to be the same as the adjacent parts in the analysis of automobile rough road test. Decision of the processing unit length is an alternative task. Though it is preferable to set a long time section as long as the statistical properties continue because of accuracy, a forced linkage of sections might an erroneous result. Concept of the local stationarity (Ozaki and Tong 1975) is applied to decide the continuity of data properties here. In this case a globally nonstationary process is regarded as a chain of various locally stationary processes, using the auto-regressive model technique. The range where a certain model gives minimum AIC is regarded as stationary in this method. Whole data are divided into blocks with provided length and then adjacent two blocks are decided whether they should be linked. We decide the optimum auto-regressive model of the beginning block with N_0 data according to the information criteria AIC first. Then we decide the optimum AR model of N_1 data including N_2 new data. In addition, we calculate the model of N_2 data alone in the same way and compare between AIC of the extended model with $N_1 = N_0 + N_2$ data with that of the separate model with N_0 data and N_2 data.

$$\mathrm{AIC}_2 = N_0 \log V_{e0} + N_2 \log V_{e2} + 2(p_0 + p_2 + 4)$$

$$\mathrm{AIC}_1 = N_1 \log V_{e1} + 2(p_1 + 2)$$

$$\mathrm{DAIC} = \mathrm{AIC}_2 - \mathrm{AIC}_1 ,$$

where V_{e0}, V_{e1} and V_{e2} are estimates of variance and p_{e0}, p_{e1} and p_{e2} are number of model order respectively. Then we choose a new model according to the sign of DAIC.

$$\mathrm{DAIC} \geq 0 : adopt\ an\ extended\ model .$$

$$\mathrm{DAIC} < 0 : adopt\ a\ separation\ model .$$

The next step is to set the newly accepted model with N_1 data or N_2 data to N_0 data and repeat the same algorithm to the last block. Table 14.4 is an example

Table 14.4   DAIC and Number of Model Order

Data	Road	Number of data	Number of block of data	DAIC 1–2	DAIC 2–3	Model Order M1	Model Order M2	Model Order M3
1	A	2880	960	35	43	34	36	34
2	A	1440	480	27	22	9	12	12
3	A	1440	480	11	10	10	25	27
4	A	1440	480	15	29	12	16	16
5	A	1440	480	46	14	14	19	19
6	B+A	1440	480	31	6	13	21	46
7	B+A	1440	480	45	−79	15	35	39
8	B/2+A	960	480	–	−66	–	27	39
9	B+A	1440	480	8	−12	15	23	25

of automotive rough road data (Yamakawa 1985). In the case of No.1 to No.5, three blocks are linked and the single model is accepted, where the road A is a special paved track for endurance test (so called Belgian block road) and the statistical properties are considered to be identical over the whole length. No.6 to No.9 are the test data on a paved test track: a third after part of the track has the same properties as the road A and the two thirds fore part of the track has a quite different properties from the after part. No.6 data shows that the value of AIC keeps a slight plus sign between the blocks 2 and 3 too and the change is not detected, but the changes of the generation mechanisms in No.7 and No.9 are detected. A separation model is adopted in the case of No.8 that is the two thirds after parts of No.8. To study sensitivity of this method modified No.2 data were tested where the amplitude is decreased gradually by half at the end. The results show that the change is detected between blocks 2 and 3. The running data on a unpaved road are shown in Figure 14.10 and divided into 4 blocks. The result shows the fore two blocks are separated and the after two blocks are linked. Figure 14.11 shows the power spectra of the individual block data. The shapes of the power spectrum, which are the positions of peaks and relative magnitude, are nearly identical. However the total level of the spectrum of data (1) and that of data (2) in the frequency range less than 7 Hz are low. This result shows that the square mean property of the whole block data would affect the decision of linkage or separation between data blocks in this case. This data process method through application of locally stationary process is a forward step of nonstationary data processing. The author expects a further advance in practical application including the improvement.

14.5   Identification of Nonlinear Vibration System Through Bispectral Analysis

The tires of automobiles frequently leave the road surface during running on a rough road. That phenomenon is apt to occur especially under the empty load condition of trucks, because the spring constant of the tire corresponds to the full load condition. The wave-form of the acceleration of the axle is markedly asymmetrical at that time. Automotive suspension system model in dynamics is shown by the left part of Figure 14.12, whereas that of the empty truck on the rough road is nonlinear system with

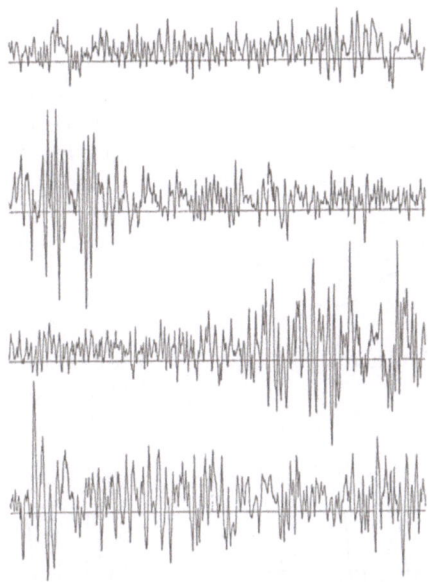

Figure 14.10   Time Series Record on Unpaved Road

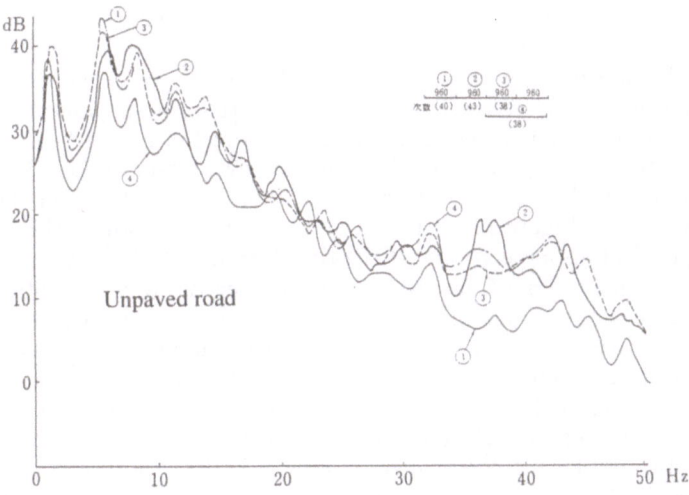

Figure 14.11   Power Spectra for Block Data

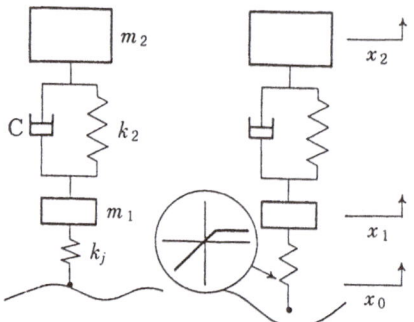

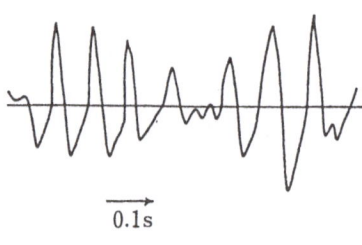

Figure 14.12 Mathematical
Model

Figure 14.13 Wave-form of
Experimental Data

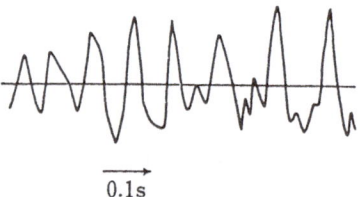

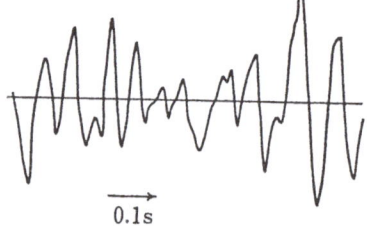

Figure 14.14 Wave-form of
Simulation Output

Figure 14.15 Wave-form of
Gaussian Signal

asymmetrical restoring force characteristics of the tire shown in the right part of the
figure (Yamakawa 1975). To confirm whether the actual running automobiles are under these conditions, numerical simulation for the model in Figure 14.12 are executed
and the wave-form is compared with that obtained by actual running. The wave-form
by actual running is shown in Figure 14.13, and that obtained from simulation in Figure 14.14. Moreover, Figure 14.15 shows the Gaussian wave-form of the linear system
with the equivalent spring constant of the tire. The direct objective of this study is
to generate an input signal from the axle for an electric-hydraulic fatigue tester in
the development stage of automobiles. We can obtain acceleration data records of
the axle on the road easily, whereas it is difficult to obtain the exact instantaneous
value of road surface undulation as the correspondent input. Therefore, identification
of nonlinear characteristics is tried through comparison between the bispectrum of
actual running data of the axle and that of simulation data. As for computation of
bispectrum, two methods are considered: the double Fourier transform of the third
order autocorrelation function and the triple product of the direct Fourier transform.
The former method was adopted. The third order correlation function $C(\tau_1, \tau_2)$ and
bispectrum $B(f_1, f_2)$ of x_i are given as follows;

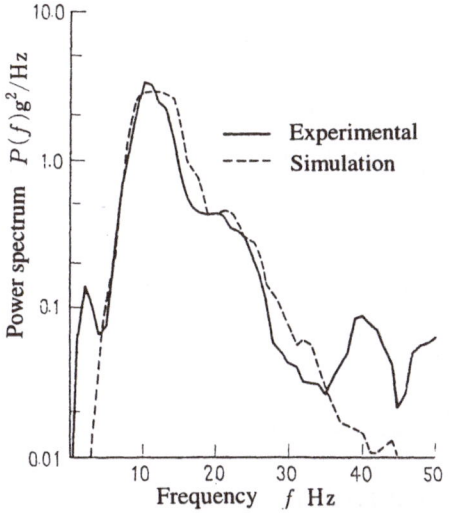

Figure 14.16 Power Spectrum

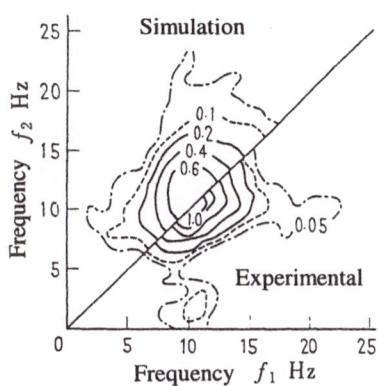

Figure 14.17 Bispectrum

$$C(\tau_1, \tau_2) = \frac{1}{N} \sum_{t=t_0}^{N} x_t x_{t+\tau_1} x_{t+\tau_2}$$

$$B(f_1, f_2) = \sum_{\tau_1=1-N}^{N-1} \sum_{\tau_2=1-N}^{N-1} C(\tau_1, \tau_2) \exp\{-i2\pi(f_1\tau_1 + f_2\tau_2)\},$$

where $t_0 = \max(1, \tau_1, \tau_2)$. Instead of the crude estimates of the bispectra obtained through the above equations, good estimates can be obtained by averaging in three directions. Figure 14.16 shows the comparison of the power spectrum of the experimental value and that of the simulation result, where the sampling interval is 0.01 second, the number of data is 1440, the maximum lag number is 50 and the Hamming window is applied in this case. The spectral values of concern in the neighborhood of 10 Hz, the fundamental frequency and 20 Hz, the second harmonic frequency are nearly equal. Because the value of the spectrum has symmetrical properties with six times reputation, both the experimental value and the simulation result are shown respectively within 45 degrees of range in Figure 14.17. When the absolute value of bispectrum is large at some point in a contour map, there is a question whether the real dependence of phase relations among the related frequency component exists. The estimate of bispectrum may include large errors caused by large powers of components. Then the normalized value, that is the ratio of the estimated power of the phase-dependent components to the power of the total components is adopted as a measure that indicates the mutual dependence of the components as follows.

$$S(f_1, f_2) = k \frac{B(f_1, f_2)}{\{p(f_1)p(f_2)p(f_1 + f_2)\}^{\frac{1}{2}}}$$

If the constant k is fixed to unity, usual normalized bispectrum is obtained. It is known that squared standard error of the bispectrum is given as follows (Akaike 1964)

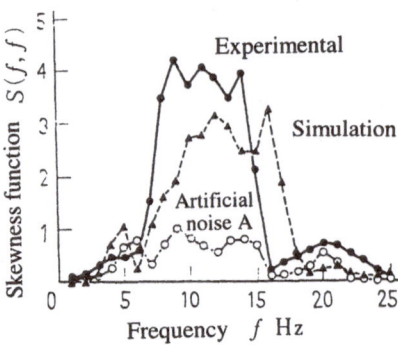

Figure 14.18 Skewness Function

If the constant k is fixed to unity, usual normalized bispectrum is obtained. It is known that squared standard error of the bispectrum is given as follows (Akaike 1964)

$$E|B(f_1, f_2)|^2 \doteq \frac{h^2}{T} \{p(f_1)p(f_2)p(f_1 + f_2)\}^{\frac{1}{2}},$$

where T is the data length (second) and h is the maximum lag of the correlation function (second). If $k = T^{\frac{1}{2}}/h$ (the unit is $s^{-1/2}$), then the normalizing means dividing by the standard error and the normalized value indicate directly the meaning in comparison with the error. Now $S(f_1, f_2)$ is tentatively termed the skewness function. Figure 14.18 shows the absolute value of the skewness function of the experimental and simulation data. The absolute values of the skewness function on the 45 degrees line which indicate the relation between the fundamental and the second order components are shown in Figure 14.19. Noise A in the figure is the Gaussian noise whose power spectrum is similar to that of the experimental data. Figure 14.20 shows the real and imaginary parts of the skewness function on the 45 degree line. The absolute values of the experimental data and the simulation result are nearly equal and the ratios of the real part to the imaginary part, which indicate the phase relation, are slightly different. If the one-sided property of the suspension damper is considered in the dynamic model, good accordance of the skewness function (bispectrum) is obtained. If we can get instantaneous values of the input and the output simultaneously to identify a nonlinear vibration system, it might be better to use the concept of the cross bispectrum which measures the effect of the product of the inputs to the output.

14.6   Continuous Measurement of Time-variant Spectrum

It is said that thick "rumbling" noise gives us unpleasant feeling as to the tone of automotive noise during acceleration. Sound pressure components of $n/2$th order of engine revolution which is treated as canceled in a elemental textbook of machine dynamics still remain because of scatter among respective cylinder outputs and the structural asymmetry in the real engines. Then those components cause the unpleasant rumbling noise; these phenomena are called the beat. Figure 14.21 shows an

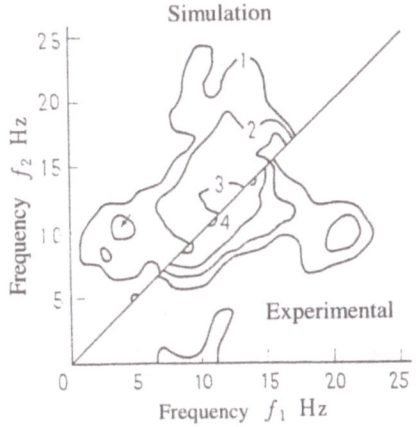

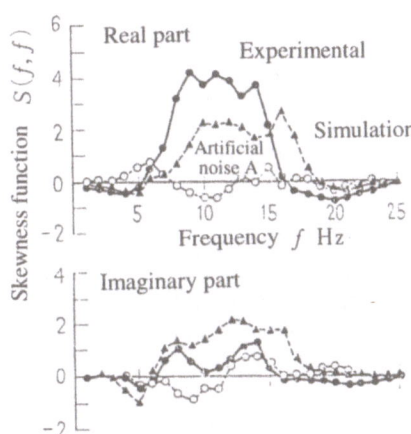

Figure 14.19 Skewness Function on 45 degrees Line

Figure 14.20 Skewness Function on 45 Degrees Line (Real Part and Imaginary Part)

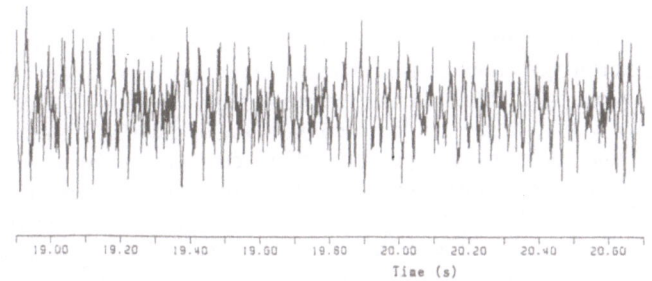

Figure 14.21 Noise Record During Acceleration

example of a noise record during acceleration which is sampled at 0.4 ms and involves the from low to high a wide frequency range. Those components vary every moment gradually with engine speed including their amplitudes during acceleration.

As for those kinds of data analyses, expressions through averaged FFT spectra of short time noise data are used. We can recognize existence of $n/2$th order of sound pressure components besides the remarkable second order component in Figure 14.22 which shows the crude power spectrum through FFT (periodogram). However, it is difficult to find out which components is superior besides the second. Figure 14.23 shows the averaged result three times with the neighboring squared spectra in the ratio of 1:2:1. The more the number of averaging times increase, the clearer the magnitude of components are made and the wider the width of the peaks grows. The wave-form in this case should be recognized as the syntheses of the very narrow range random noises with the gradually changing frequency range. The time domain model technique is suitable for sharp spectral peaks. The optimum order model for the process is decided through the Yule-Walker method and Akaike's AIC or FPE

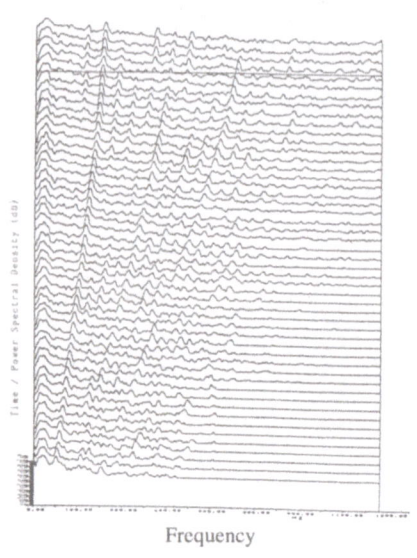

Figure 14.22 Result Through FFT Method (Crude)

Figure 14.23 Result Through FFT Method (3 Times Averaging)

(Akaike 1970, 1973). Figure 14.24 shows a result. The optimum order for this kind of data would generally get much larger. The maximum number of order is limited to 50, because the estimates of the variance of errors σ^2 by the Yule-Walker method might have a negative value caused by the accumulation of calculating error for the large order model. The equations are directly estimated to minimize the squared sum of forward and backward estimation errors using the partial autocorrelation coefficient in the so called Burg's method. This method is advantageous to the spectrum with sharp peaks. The method using the correlation coefficient between forward and backward estimated error series through the algorithm of the computational program developed by the Institute of Statistical Mathematics is adopted instead of the Burg's own method. Computation is carried out using the coefficients in the last step according to the following recurrence formulas (Kitagawa 1986).

$$f_{0n} = b_{0n} = y_n \qquad (n = 1, \dots, N)$$

$$a_{kk} = -\frac{\sum_{n=k+1}^{N} f_{k-1,n} b_{k-1,n-k}}{\left(\sum_{n=k+1}^{N} f_{k-1,n}^2 \sum_{n=1}^{N-k} b_{k-1,n}^2\right)^{\frac{1}{2}}}$$

$$f_{kn} = f_{k-1,n} + a_{kk} b_{k-1,n-k} \qquad (n = k+1, \dots, N)$$

$$b_{kn} = b_{k-1,n} + a_{kk} f_{k-1,n+k} \qquad (n = 1, \dots, N-k)$$

Among the above equations, the equation to calculate a_{kk} is different from the Burg's own equation. Figure 14.25 shows an example of analysis result through this

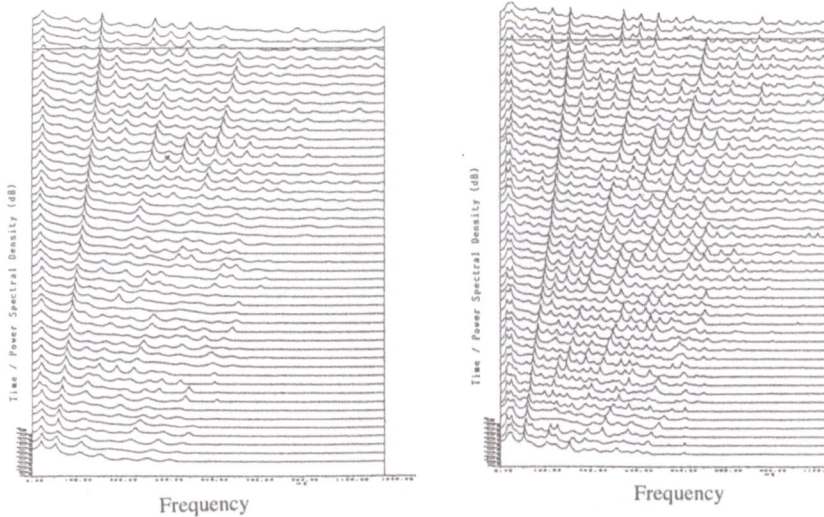

Figure 14.24 Result Through Yule-Walker's Method

Figure 14.25 Result through Burg's Method

algorithm (Yamakawa 1990). In the figure peaks of some of the $n/2$th order are seen very clearly. It is found that concerning this automobile the 3rd order and the 3.5th order components are considerably large at a certain speed range and the 1.5th to 2.5th order components at the other speed range. Because the scale of the ordinate in the figure is logarithmic, we have to study whether the amplitude of each component is so close to that of the other that the beat phenomena occur. Besides, tens Hz of very low frequency component in the figure have been beforehand removed through the numerical filter in the traditional analysis. Another example of analysis on the other automobile is shown in Figure 14.26. In this figure, $n/2$th components appear more conspicuously.

14.7 Afterword

The number of reports of the time series data analysis is not so many among the study on automobiles recently. This fact would be based on the result that the techniques of the time series data analysis have been established in the usual works. Among those studies analyses through the AR method regarding driving stability (Soma *et al.* 1993; Noguchi *et al.* 1994) attracted our attention, and besides an analysis of stationary random vibration with bumps to the suspension stopper (Aisaka *et al.* 1993) was published. As the result that time series data analysis techniques have spread and put down roots, the applicability to the automotive field would gradually spread hereafter too.

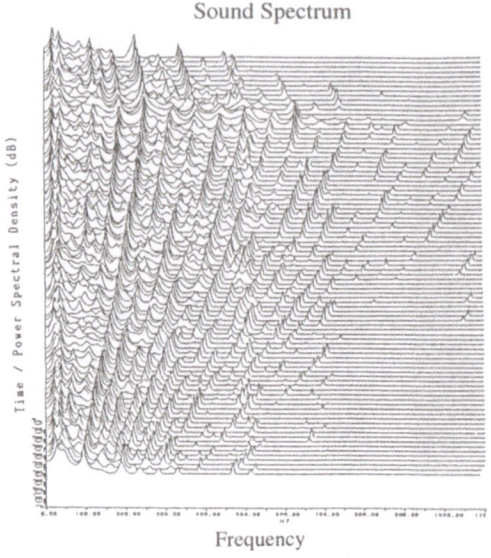

Sound Spectrum

Figure 14.26   Result Through Burg's Method

Appendix: Smoothing Through the Hanning Window

We smooth the crude spectrum p_j obtained through the Fourier transform of the correlation function using the coefficients w_i $(i = 0, \pm 1, \ldots, \pm m)$ those are called the spectral window according to

$$\hat{p}_j = \sum_{i=-m}^{m} w_i p_{j-i} \quad (j = 0, 1, \ldots, k),$$

where $p_{-j}=p_j$, $p_{k+i}=p_{k-i}$. The Hanning window is the spectral window, in which m=1, w_0=0.5, w_1=w_{-1}=0.25.

References

Aisaka, M. *et al.* (1993), "Study on stationary random oscillations of engine mouting system with vibration stopper," *SAEJ Annual Meeting Preprint*, 993, 113–116 (in Japanese).

Akaike, H. (1970), "On a semi-automatic power spectrum estimation procedure," *Proceedings of 3rd Hawaii International Conference on System Sciences*, 974–977.

Akaike, H. (1973), "Information theory and an extension of the maximum likelihood principle", 2nd International Symposium in Information Theory (Petrov, B. N.

and Csaki, F. eds., Akademiai Kiado, Budapest, 267–281. (Reproduced in Break-throughs in Statistics, Vol. 1, S. Kotz and N. L. Johnson, eds., Springer Verlag, New York, 1992)

Akaike, H. (1964), "Introduction to probability and Fourier transform," Short Course Textbook, Institute of Statistical Mathematics (in Japanese).

Akaike, H. and Nakagawa T. (1972), "Statistical analysis and control of dynamic system," Science Inc. (in Japanese (English translation is published from Kluwer Academic Publishers in 1988.))

Kitagawa, G. (1986), "Trend of spectral analysis," *Measurement and Control*, Vol. 25, No. 12, 1074–1081 (in Japanese).

Noguchi, Y. *et al.* (1993), "Study of a method for evaluation handling of bus by measuring drivers psychological reaction," *AVEC '94*, 67–72.

Ozaki, T. and Tong, H. (1975), "On the fitting of non-stationary autoregressive models analysis," *Proceedings of 8th Hawaii International Conference on System Science*, 224–226.

Rice S. O. (1944, 1945),
Mathematical analysis of random noise," *Bell Technical Journal*, 23-3, 282, and 24-1, 46–156.

Soma H. *et al.* (1993), "Identification of vehicle dynamics under lateral wind distri-bution using autoregressive model," *SAE Paper* 931894, IPC-7.

Yamakawa S. (1971), "Nonlinear vibrations of axles of automobiles moving over rough roads," *Proceedings of US-Japan Seminar on Stochastic Methods in Dynamical Problem*," 6.1, 1–20.

Yamakawa, S. (1973), "Statistical process of random vibration," 374th Short Course Textbook, *JSME* (in Japanese).

Yamakawa, S. (1976), "Investigation of peculiarity in some wave-forms through bis-pectral analysis," Bulletin of the JSME, Vol. 19, No. 127, 29–36.

Yamakawa, S. (1983), "Separation of correlated power components in a multiple input system by means of power contribution analysis," *Bulletin of the JSME*, Vol. 49, No. 449, 2061–2067 (in Japanese).

Yamakawa, S. (1985), "Vibration test and data analysis," 594th Short Course Text-book, *JSME* (in Japanese).

Yamakawa, S. (1990), "Analysis of automotive acceleration noise data," *SAEJ Annual Meeting Preprint*, 902, 2, 233–236.

The author would like to acknowledge the guidance and the assistance of the former director-general H. Akaike, Prof. G. Kitagawa and Prof. Y. Tamura of the Institute of Statistical Mathematics.

Chapter 15

Auto-regressive Spectral Analysis of RR-Interval Time Series in Healthy Fetus and Newborn Infants

——Continuity of autonomic nervous system function from prenatal to postnatal life——

Teruyuki Ogawa
Professor Emiritus of Pediatics
Oita Medical University, School of Medicine
Midorigaoka 5–13–7, Oita 870–1172 Japan
http://www.lukaster@oec-net.or.jp

15.1   Introduction

A fetus's heart rate variation is regulated by the heart rate control center existing in the brainstem, and its rhythmic movement is a sharp information reflecting the ever-changing autonomic nervous activity state (for example, antagonistic regulation of vagus nerves and sympathetic nerves). However, no report describing the continuity of the sway of the heart rate variation ranging from a fetus to a newborn infant has been available until now. With such a situation in mind, analysis is made in this chapter with the maturity process of the autonomic nervous activity ranging from a human fetus to a newborn infant by applying an auto-regressive model. Thus, studies have been carried out with the purpose of estimating the physiological significance lurking in the background of the above process.

15.2   Subjects and Methods

15.2.1   Subjects

41 healthy fetuses who are in an age range from 26 to 39 or 40 gestational weeks and 41 healthy newborn infants including those babies and young children who are in infancy period were taken up as subjects. Among these fetal subjects, observation was made by equipping the Hewlett-Packard 8031A Doppler ultrasonic fetal heart rate instrument onto the mother's abdominal wall. The fetal ECG were confirmed

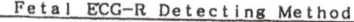

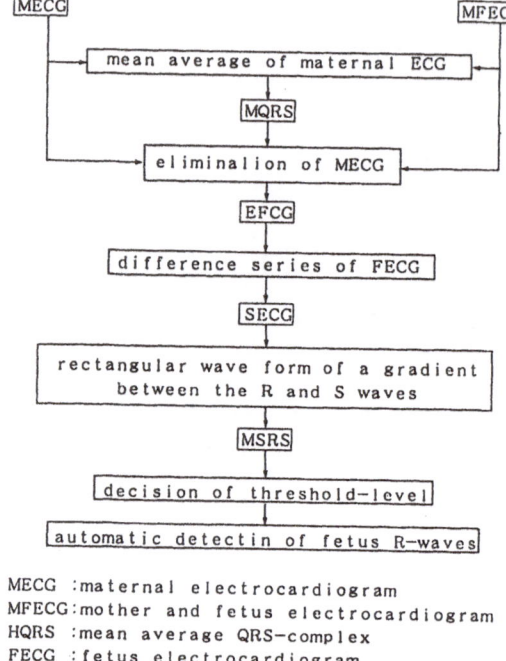

MECG :maternal electrocardiogram
MFECG:mother and fetus electrocardiogram
HQRS :mean average QRS-complex
FECG :fetus electrocardiogram
SECG :difference series of fetus ECG
MSRS :rectangular wave of a gradient
          between the R and S waves

Figure 15.1   Flowchart of the fetus's electrocardiograph extraction method.

by means of fetal echoes after the observation of the fetuses that are in Pretchtl's state 1F (a state equivalent to a newborn infant's quiet sleep state). After that, electrocardiogram R wave was extracted with the software developed by us, and 3 or 4 sections were selected per fetus with the 250-beat RR-interval time series as one section. Thus 128 sections in all were taken up as subjects for analysis.

Also with the newborn infants, the RR-interval time series is obtained at the quiet sleep stage with the aid of polygraphy records from the 41 cases of the healthy newborn infants (AFD infants: Apgar Score more than 8 points after 1 minute) who are in an age range from immediately after birth to 7 days. Thus 238 sections in all were dealt with as subjects by selecting 3 to 7 sections from each newborn infant.

15.2.2   Data Acquisition

Figure 15.1 illustrates the detection procedure of the R wave of a fetus electrocardiogram. At first, by feeling the mother's abdominal wall, electrodes are equipped onto it with the infant's head as reference (the electrodes placed within the electric field of the fetus electrocardiogram), the infant's back as non-references, and a portion where no infant feels the mother's abdominal wall as ground electrodes. After that, recording was made by introducing maternal and fetal electrocardiogram (MFECG) as synthesis of the electrocardiogram of the mother and the fetus. Also at the same

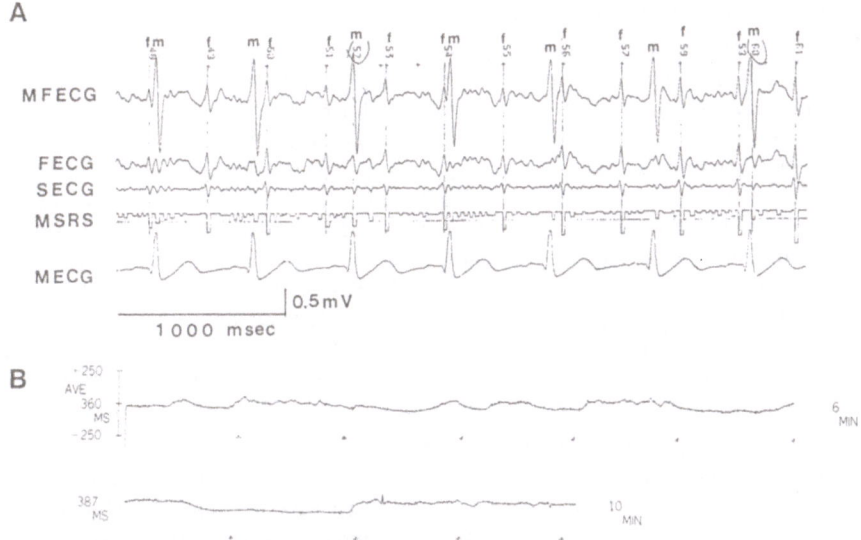

Figure 15.2 Fetus R wave detections by the "Fetus ECG-Monitor" (Toitsu,Co LTD)" (A) By eliminating the mother ECG wave from the mother and fetus electrocardiograph (MFECG), the fetus electrocardiographic R wave was visually detected. (B) RR interval Time series (16 minutes) of the fetus.

time, recording was made with the maternal electrocardiogram (MECG) exclusively for the mother's body from the mother's breast. In the MFECG, high-amplitude maternal electrocardiogram, electromyogram, noise, etc. are mingled. For this reason, it is not easy to detect low-amplitude fetus R wave. With such a situation in mind, an automatic detecting method was developed as shown in the flowchart illustrated in Figure 15.1 in order to find the R-wave.

1) Let the AC components and adaptive noise cancellation be eliminated from MFECG via a hum filter and high pass filter (more than 53Hz). By simultaneously AD-converting the analog data of MFECG and MECG at a sampling spacing 1 msec, let recording be accomplished.

2) Secondly, by obtaining, a mean value after 30-time addition of the data of around 200 msec of the mother's R wave (MR point) on MFECG with the aid of R wave of MECG used as trigger, let an average cardiogram (MQRS) be constructed to eliminate fetus cardiogram or noise.

3) Furthermore by allowing MQRS to be subtracted after synchronization of MFECG with MR point with the aid of the R wave of MECG used as a trigger, let the maternal electrocardiogram be eliminated to obtain a fetal electrocardiogram (FECG). (Figure 15.2(A)).

4) At this stage, let attention be paid to a gradient between the R and S waves to recognize a fetal QRS waveform on FECG. After that, let a rectangular waveform series (MSRS) be obtained as depicted in Figure 15.2(A) using the difference series (SECG) of FECG. Finally let a fetal RR spacing time series be secured

under visual recognition of the fetal R wave.

By sketching the time series (Figure 15.2(B)) for 10 minutes at the fetal RR-inteval time series obtained by this procedure, the 250 beats that could be believed to be stationary are determined as one section. Furthermore, by taking out 3 sections from each subject, auto-regressive spectral analysis was carried out.

With the newborn infants, the first polygraph record was taken for more than 2 hours within one hour to 22 hours (7.7±6.4 hour in average) after birth. The same process was executed successively for 7 days. The record of the electrocardiogram was accomplished at the quiet sleep stage using a newborn infant's heart-beat and respiration monitor called "Life Scope 6" (Nihon Koden K.K.'s OEC-6301:UHF). Thus the RR-interval time series was obtained with the R wave as a pulse trigger using a pulse-interval measurement instrument. After that, auto-regressive spectral analysis was applied to the 250 data of the RR-interval time series using a mini-computer. Later, the power spectrum was calculated and its characteristics was analyzed.

15.2.3   Auto-regressive spectral Analysis

To measure the sway of heart rates (beat-to-beat), one beat has been determined as a spacing unit as illustrated in Figure 15.3. In this occasion, let the time series $y_1, y_2, \ldots, y_k, \ldots, y_N$ of the deviation from the average of the RR interval time (unit msec) be an objective for observation. As deduced from Figure 15.4, the mean RR-interval time expresses approximately normal distribution in most of the cases. Let an auto-regressive model

$$y_k = a_1 y_{k-1} + a_2 y_{k-2} + \cdots + a_M y_{k-M} + n_k , \qquad (15.1)$$

be considered for the time series y_k, where n_k is normally distributed with the mean zero and the constant variance, and is independent of y_{k-j}. The characteristics of the sway activity phenomenon are determined by the M AR coefficients. M is called order of the auto-regressive model.

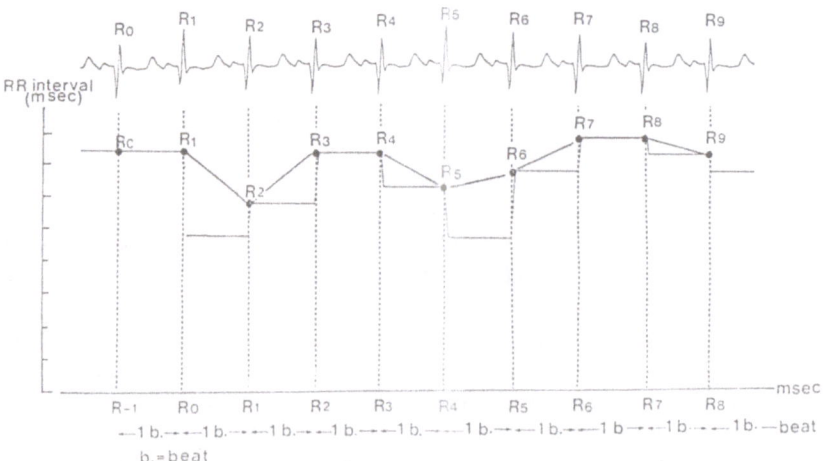

Figure 15.3   AD conversion of the heart rate variation. Taking $R_0 - R_1, R_1 - R_2, \ldots, R_8 - R_9$ spacings on an abscissa, the time series was obtained on the assumption that the RR interval time of the individuals is 1 beat (heart rate).

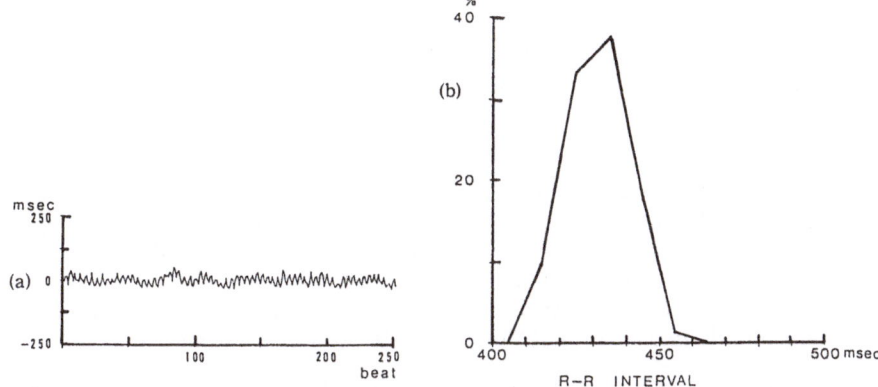

Figure 15.4 RR spacing time series of a 36-week premature infant after impregnation and the series's histogram (a) RR spacing time series (250 beat) (b) histogram (class width 10 msec): Normal distribution was shown at RR spacing time 431.75 msec in average.

Though it is necessary to determine the value of the order M, it can be practically estimated by obtaining the minimum value of the final prediction error FPE of Akaike (1969, 1972, and 1974) or of the information criterion AIC (15.3) such that

$$\text{FPE}(M) = \frac{N+m+1}{N-m-1}\sigma_n^2, \tag{15.2}$$

and

$$\text{AIC}(M) = N\log\sigma_n^2 + 2(M+1) \tag{15.3}$$

respectively, where N is the lenght of the time series y_k in (15.1).

The power spectrum density $p_M(f)$ is obtained by

$$p_M(f) = \frac{\sigma^2}{\left|\sum_{m=0}^{M} a_m e^{-i2\pi mf}\right|^2}, \quad (0 \leq |f| \leq 1/2) \tag{15.4}$$

where σ^2 is the variance of n_k, and $a_0 = -1$.

By substituting the backward shift operator (Box and Jenkins,1970), say B, such that

$$B^m y_t = y_t - m, \quad B^m n_t = y_t - m \tag{15.5}$$
$$|B| \leq 1, (1,2,3,\cdots).$$

into expression (15.1), it is written as

$$g(B)n_k = y_k. \tag{15.6}$$

Where $g(B)$ is the *transfer function* of RR-interval time series given as the inverse of the characteristic function $a(B)$ of y_t in (15.1) such that

$$a(B) = 1 - \sum_{m=1}^{M} a_m B^m, \tag{15.7}$$

that is,

$$g(B) = 1/a(B) \tag{15.8}$$

The characteristic equation $a(B) = 0$ of order M with respect to B can be factorized as follows

$$a(B) = (1 - S_1 B) \cdots (1 - S_M B) = 0 \tag{15.9}$$

Where $\{S_m^{-1}\}$, $(m = 1, 2, \cdots, M)$ is the root of the characteristic equation $a(B) = 0$ and practically roots can be assumed to be simple. When one of the roots is imaginary, its complex conjugate roots are W, say $\{ \eta_w^{-1}, \eta_w^{*-1} \}$, $(w = 1, 2, \cdots, W)$ and real roots are V, say $\{ \phi_v^{-1} \}$, $(v = 1, 2, \cdots V)$.

Then, the above equation can be rewritten by products of the first and second order characteristic functions, $a_{1v}(B)$ and $a_{2w}(B)$, respectively, such that

$$a(B) = \{\Pi_{v=1}^{V} a_{1v}(B)\} \cdot \{\Pi_{w=1}^{W} a_{2w}(B)\}, \quad (V + 2W = M) \tag{15.10}$$

where

$$
\begin{align}
a_{1v}(B) &= 1 - \phi_v B \tag{15.11} \\
a_{2w}(B) &= 1 - \phi_{1w} B - \phi_{2w} B^2 \tag{15.12} \\
\phi_{1w} &= \eta_w + \eta_w^*, \tag{15.13} \\
\phi_{2w} &= -\eta_w, \eta_w^*, \tag{15.14}
\end{align}
$$

$$(\eta_w, \eta_w^* : \text{complex conjugate}).$$

The value characteristic function of the first and second order processes $a_{1v}(B)$ and $a_{2w}(B)$ are easy to obtain by letting $M = 1$ and 2 in equation (15.7), respectively.

Consequently, the RR time series y_t having AR-process of order M can be decomposed into the first and second order process of V and W in number.

Furthermore this expression is factorized (Sato, Ono et al. 1977, Sato 1988) as the first and second order transfer function in equation (15.14) and (15.15), respectively (see Appendix).

$$g_{1v}(B) = \frac{1}{1 - a_{1v} B} \tag{15.15}$$

$$g_{2w}(B) = \frac{1}{1 - a_{21w} B - a_{22w} B^2} \tag{15.16}$$

From the secondary order element, damping frequency (FD) is obtained. Although the signals fed from the cardioregulatory center are believed to be cyclic, this cyclic activity is modified by a variety of stimuli produced from the external and internal environments. Accordingly, sway is generated with the cyclic heart-beat activity, and the auto-covariance function or auto-correlation function displaying an average time pattern of the sway are directed to exhibit damping oscillation. The frequency of the damping oscillation hereby referred to is damping frequency (FD).

The impulse response of the filter corresponding to the individual elements depicts a damped oscillating curve depending on time as shown in Figure 15.5. The frequency characteristics of the first order element shows the maximum value at 0Hz when $a_{1v} > 0$, and a damping exponential curve lowering gradually accompanied by the increase of the frequency is depicted. It is noted with the impulse response of this element that the width of the crest is narrowed (widened) in the inverse proportion of the time constant TC (the time of the response lapsed to be reduced to the value of $1/e$

(= 0.368, approximately 37%) as the maximum value starting from the original point). The impulse response of the secondary order element depicts damping oscillation, and its frequency is the damping frequency (FD). The time constant of the envelope of this oscillation is the damping time (DT).

The auto-covariance series of the auto-regressive process of M-th order can be expressed using the roots of the characteristic equation (refer to equation (15.17) in the Appendix of this chapter). From this, it is understood that approximation is made by the sum of the first or secondary auto-regressive process corresponding to the individual roots (Zetterberg 1969, 1977), in case that the crests of the power spectrum vector density corresponding to the individual roots are sufficiently separated. Sato (1977) calls them elementary waves, because he believes them to be the waves corresponding to the individual elements. By using (15.14) and (15.15), he obtains the power of the individual element waves (the element wave power). This approximation method will be used in this chapter as well.

15.3   Results

15.3.1   AR-analysis of Fetal RR-interval Time Series

Figure 15.5 shows an exemplification of the auto-regressive analysis of a fetus in 34-gestational week. On the uppermost row, the RR-interval time series is shown. A mean value of the RR interval time series at 250 beats is 422.4 msec. In Figure 15.5 (a) in the 2nd row, the power spectrum is shown. The total power (TP) as variance of the time series is 32.0 log msec^2. On the right of it, Akaike information criterion equivalent to FPE is shown. Since AIC gets its minimum at order 6, the characteristic function of the denominator of the transfer function $G(B)$ is an expression of degree 6 of B. Accordingly, an element of degree 2 displaying 3 damping oscillation curves is obtained. The damping frequency FD of the element of degree 2 showed 0.01, 0.27 and 0.43 cycle/beat. The left column in Figure 15.5 (b) shows an auto-correlation function and the duration time of the oscillation can individually be expressed with $1/e$ of the envelope of the damping oscillation as illustrated in the figure. The waves of 0.01, 0.27 and 0.43 cycle/beat show individually the durational properties of 4.96, 3.92 and 2.23 beats.

15.3.2   Development of the RR-interval Time

As Figure 15.6 illustrates, the RR-interval time has significantly increased ($P <$ 0.001) with the development of either the fetus and the newborn infant.

15.3.3   Damping Frequency Histogram

When a histogram was constructed with the damping frequency obtained by element decomposition at class spacing cycle/beat (c/b), 3 groups have been obtained as sketched in Figure 15.7. They have been classified as less than 0.15 c/b (LF), 0.15 to 0.27c/b (MF) and 0.28 to 0.45 c/b (HF) for fetus (right) and newborn infant (left). The MF group, which is distinctly seen with the newborn infant, is not necessarily explicit with the fetus, and therefore was excluded from this study. Thus, investigation was conducted with various types of characteristics of the LF and HF groups.

15.3.4   Development of the Total Power

The total power (Figure 15.8) started to take high values since the 32nd gestational week in case of the fetus. From the 36th gestational week on, distinct development was

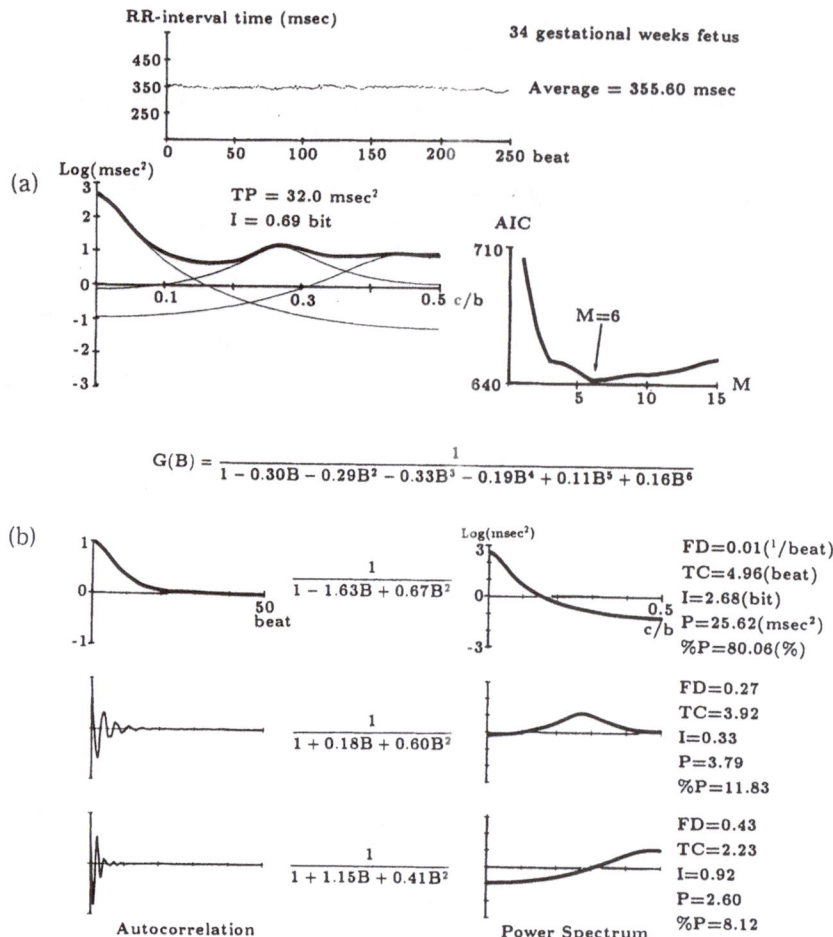

$$G(B) = \frac{1}{1 - 0.30B - 0.29B^2 - 0.33B^3 - 0.19B^4 + 0.11B^5 + 0.16B^6}$$

Figure 15.5 RR interval time-series auto-regressive analysis of a 39-gestational-week fetus (a) 6-degree auto-regressive spectrum. TP = power (msec^2), I = information activity amount (bit) (b) Auto-correlational diagram (left column) of the component wave activity, transfer function (center), and the elemental wave power spectrum FD = damping frequency (cycle/beat), TC = time constant (beat), DT = damping time (beat), P = power of the component wave (log msec^2) The parentheses indicate the percentage of the elemental wave power accounted for in the entirety.

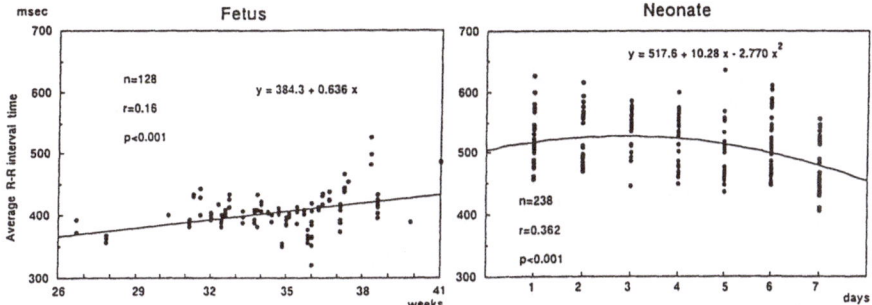

Figure 15.6 Developmental change of the RR interval times

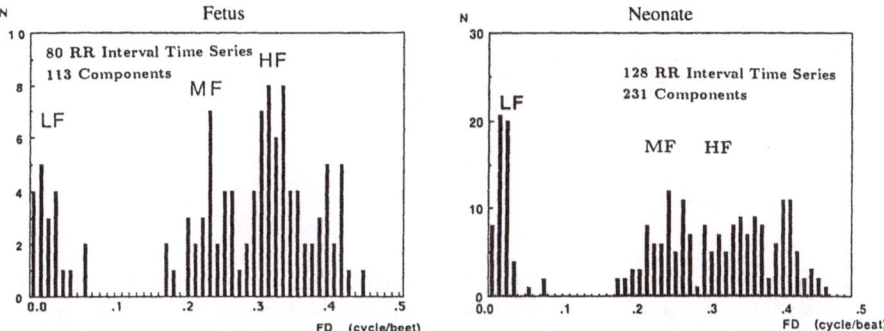

Figure 15.7 Damping frequency histogram with fetuses and newborn infants. The histograms of the elemental wave FD constructed at the class width 0.01 cycle/beat have been classified into the 3 groups, i.e. 0–0.15 c/b(LF), 0.15–0.27 c/B (MF), and 0.28–0.45 c/b (HF). Fetuses (left), newborn infants(right)

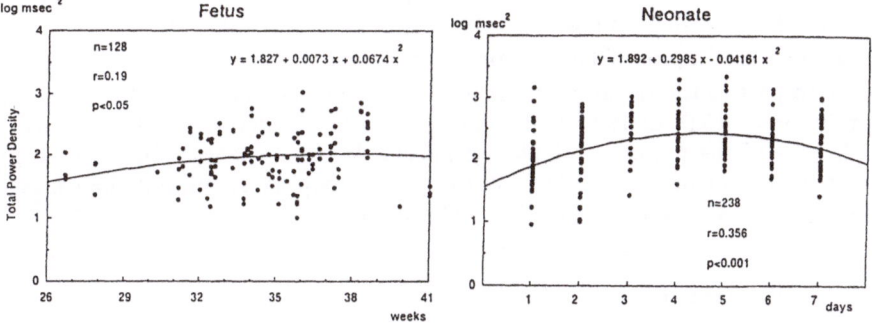

Figure 15.8 Continuity of the developmental change of the total power with fetuses and neonates. Left side: fetuses, Right side neonates. Ordinate: power density (log msec^2). Abscissa: Fetus indicates the weeks during the impregnation, whereas neonate indicate the days after birth.

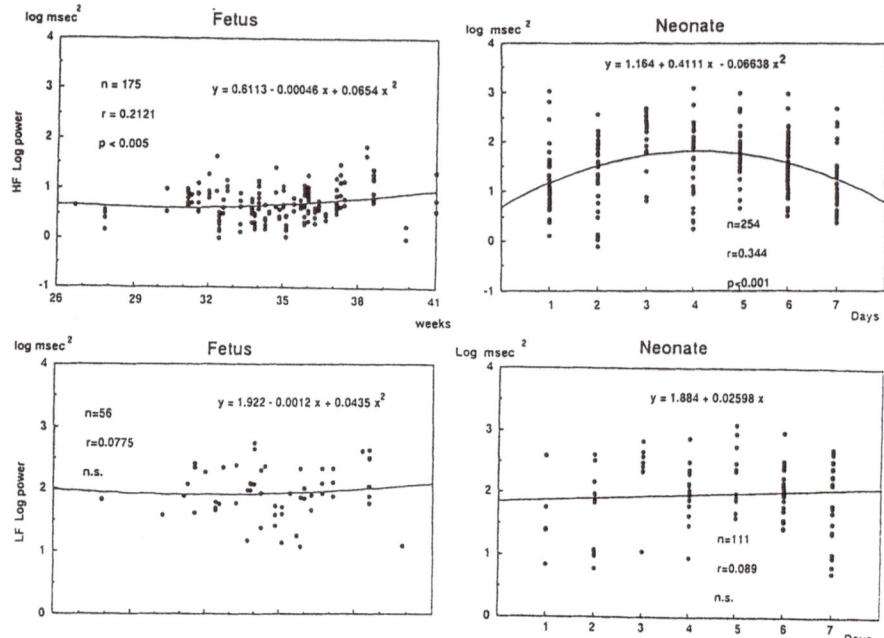

Figure 15.9 Continuity of the developmental change of the elemental wave power with the fetuses and neonates. The left side of the individuals indicates the fetuses' developmental change, whereas the right side indicates the neonates'. Meanwhile the upper row indicates the power of the HF group, whereas the lower row indicates the developmental change of the power of the LF group. Ordinate: power density (log msec^2). Abscissa: the fetus indicates the weeks during impregnation, whereas the neonate indicates the days after birth.

observed. In the case of neonates, the development was followed by other significant development exhibiting the maximum value at 4 to 5 days after birth.

15.3.5   Development of the Component Wave Power

With respect to the HF group, significant development can be seen for the fetus as shown in Figure 15.9. Especially after the 38th week, the development was a distinct one. In case of the newborn infants, significant development was observed reaching its maximum value on the 4th day . With the LF group, no significant change was to be seen either with the fetuses or the newborn infants.

15.4   Discussion

According to the reports of Ibarra-Polo (1972) and Yagi (1986), rapid elevation of the basal fetal heart rate is noted between the 5th week and the 9th week and the rate is mildly lowered until the 30th week after pregnancy to be finally stabilized.

Estimation has been made in sheep fetus with the fetal dynamic analysis of autonomic nervous system function using a method that provides medicines or autonomic nerve cut. However from the estimation of the study on fetal lambs by Dalton and Dawes (1983), it is known that reactions of parasympathetic nervous system are noted at the period equivalent to the 15th gestational week of human fetus. Reviewing the development of the autonomic nervous system controlling the fetus heart rate through the extensive reports published by many researchers, it is clear that the heart rate rhythm begins to have regularity at around the 5th week of pregnancy to be followed by lowering of the basal heart rate accompanied with the development of the atrisventicular node at the 9th week of pregnancy (Hirose *et al.* 1987). Furthermore, completion of the stimulus-conductive system can be seen at the 16th week of pregnancy, and the autonomic nervous node is generated around 20th week. Thus it is widely known that heart-rate controlling functions are gradually expressed from the postganglionic fibers. In this period, the parasympathetic nervous system accomplishes predominant development for the sympathetic nervous system (Woods 1977). Finally, around the 28th gestational week, the networks of the autonomic nervous system are extended throughout the entirety of the heart and fine heart-rate control becomes possible.

Cerutti *et al.* (1989) seek elemental components by applying auto-regressive analysis onto RR-interval time series by taking up adults as objectives as we do in this paper. However they carry out their study by classifying the components into 2 groups with 0.05 to 0.15 Hz as a lower frequency band (LF) and with 0.20 to 0.40 Hz as a higher frequency band (HF). In addition, judging from the fact that HF is synchronous with respiration and that the respiration ratio exercises influence on the electrocardiograph, it is considered that HF is respiratory sinus arrhythmia (RSA). It is also considered that the power of the spectrum at that portion expresses quantitatively the vagus nerve control. On the other hand, LF is a wave slower than HF and its component is a sympathetic nerve control activity strongly subjected to the variation of the blood pressure. Thus the LF power is raised due to blood pressure raise.

According to researches (Cerutti 1989; Lindercrantz 1993: Pagani 1986), it is understood that the power spectrum-composing element waves of the RR-interval time series might become useful evaluation of the sympathetic/parasympathetic nervous system balance. It is widely known that the fetus well endures the stress at a hypoxic state owing to the compensatory mechanism, but activation of the sympathetic/parasympathetic nervous system is said to play an important role in maintaining homeostasis during hypoxia. Therefore at an acute hypoxic state, the sympathetic nervous system is strained and followed by emission of cathecholamine (Stange 1977; Widmark 1989). Then the blood pressure is elevated allowing further increase in the sympathetic nervous functions. In this occasion, it is verified from an experiment on amniotic fetus that the power of LF is increased (Cerutti 1989; Dalton 1983; Stange 1977).

On the other hand, the power of HF increases due to the strain of the parasympathetic nerve activity. Accordingly it is noted that the heart rate increases with application of venous injection of noradrenaline onto the amniotic fetus. A report has already been released by Lilja (1986) in connection with the above matter, and it is also in use as differential index of the sympathetic/parasympathetic nervous system (Yeh 1973). By applying venous injection of noradrenaline onto an amniotic fetus (117 to 135 days) and by executing AR-analysis onto the RR interval time series, Lindecrantz (1993) proves that LF distinctly increases. From these reports and the investigation of the author, it can be said that LF exhibits the function-promoting state of the sympathetic nervous system, whereas HF displays the correspondent one of the parasympathetic nervous system. Furtheremore, significant development was seen with the fetus at the 35th or the 36th gestational week, especially with the fetus of more than 38 gestational weeks. From this it can be considered that the function promotion of the parasympathetic nervous system is seen at this period. It can also be assumed that existence of the fetus's respiration preparatory state for the aftermath of the birth has been started.

Heart-rate (HR) variability data can basically be considered as being driven from an irregularly sampled process. When AR-analysis is applied directly to the HR or RR-interval series, as sequentially measured in time, it yields a spectrum that is measured in cycle per beat (instead of cycle persecond or hertz). The author made the assumption that the sampling interval is 1 heart rate, i,e, 1 beat, and therefore the maximum frequency obtained by the power vector has become 0.5 cycle/beat. To express this in hertz, it is necessary to make adjustments in accordance with the average RR-interval time per each person. The fetal heart rate is stable with a narrow bandwidth in State 1F. Therefore, when the frequency band of the fetal heart rate is adjuted with the average interval time and expressed in hertz, the detailed information concerning the sway between RR-interval time could be lost. That is why in this chapter cycle/beat was used rather than hertz.

Also with the fetus (or the preterm infant) homeostasis is maintained as a result of complicated responses brought about by many of the parameters such as heart rate, respiration, blood pressure, body temperature and so on. When homeostasis collapses, various kinds of disease occur. Besides, the author has already released a report concerning the individual types of the response among the heart rate, respiration, and blood pressure by applying a multi-dimensional auto-regressive model (Ogawa 1990, 1991; Ogawa 1993; Wada and Ogawa, 1996). The authors are looking forward to make more detailed investigations in the future.

15.5   Conclusions

With the purpose to have knowledge of the continuity of heart-rate autonomic nervous control activity from fetus to the newborn infant period, 41 healthy fetuses ranging from the 26th to the 40th gestational week together with healthy newborn infants (38 to 40 conceptional weeks) were taken as subjects in this investigation. Then, their RR-interval time series were recorded at the quiet sleep stage (Prechtl's state 1F with the fetus). Furthermore by obtaining the power of the auto-regressive spectrum and various characteristics of the compositional elements (damping frequency, damping time), investigation was given to the developmental change of the individual parameters accompanied with the number of the gestational weeks. As a result, significant increase was noticed exclusively with the power in the frequency band of 0.28 to 0.45 cycle/beat ranging from fetuses to newborn infants. Especially with the periods between the 35th or 36th gestational week and the 4th day after birth, drastic increase was to be seen. This fact suggests that promotion of parasympathetic nervous function is seen with the fetus of the 35th to the 36th gestational week, and it can be assumed that existence of the fetus's respiratory preparation state for the newborn period is shown.

Appendix

Calculating the elemental wave power (by Ono 1976)

Assuming that M roots (characteristic roots) of the equation of degree M

$$1 - a_1 B - a_2 B^2 - \cdots - a_M B^M = 0 \tag{15.17}$$

concerning the B obtained by putting the characteristic function $a(B)$ defined in relation to equation (15.4) as 0 are s_j^{-1}, $(j = 1, 2, \ldots, M)$, the auto-covariance r_k of the auto-regressive process y_t of order M is given by

$$r_k = H_1 s_1^k + H_2 s_2^k + \cdots + H_M s_M^k, \tag{15.18}$$

where $H_1, H_2, \cdots, H_M$ are the constants determined by the characteristic roots. Supposing hereby that the p-th real root is s_p^{-1} and the classes of the q-th conjugate complex roots are s_q^{-1}, $\bar{s}_q^{-1}$, respectively, it can be expressed as

$$r_k = \sum_{p=1}^{V} H_p s_p^k + \sum_{q=1}^{W} (H_q s_q^k + \bar{H}_q \bar{s}_q^k). \tag{15.19}$$

Therefore

$$r_0 = \sum_{p=1}^{V} H_P + \sum_{q=1}^{W} H_{q1} \tag{15.20}$$

and $H_{q1} = H_q + \bar{H}_q$, where $^-$ indicates the conjugate complex. By putting as

$$A(B) = M - (M-1)a_1 B - (M-2)a_2 B^2 - \cdots - a_{M-1} B^{M-1}, \tag{15.21}$$

the above equations are given by

$$H_m = \sigma_n^2 / A(s_m) A(s_m^{-1}), \tag{15.22}$$

where σ_n^2 is the variance of the innovation n_k of the auto-regressive process of order M. H_p, H_{q1} give the power of the individual elemental waves. For interpretation of the elemental wave power, it is necessary that the conditions for approximation described in this chapter be established.

References

Akaike, H. (1969), "Fitting autoregressive models for prediction," *Annals of the Institute of Statistical Mathematics*, Vol. 21, 143-247.

Akaike, H. (1974), " A new look at the statistical model identification," *IEEE Transactions on Automatic Control*, Vol. AC-19, 716–723.

Akaike, H. and Nakagawa, T. (1988), *Statistical Analysis and Control of Dynamic Systems*, Kluwer Academic Publishers, Dordrecht.

Box, P. G. and Jenkins, G. M. (1970), *Time series analysis; forecasting and control,* Holden-Day, San Francisco, USA.

Cerutti, S. et al. (1989), "Compressed spectral arrays for the analysis of 24-hr heart rate variavility signal: enhancement of parameters and data reduction," *Computers and Biomedical Research,* Vol. 22, 424–441.

Dalton, K. J., Dawes, G. S., Patrik, J. E. (1983), "The autonomic nervous system and fetal heart variability," *American Journal of Obstetrics and Gynecology,* Vol. 146, 45.

Hirose, K. Shimokawa, H. Koyanagi, K. (1987), "Development of baseline heart-rate in human fetus," *Syusanki Igaku,* Vol. 17, 671–674. (in Japanese)

Ibarra-Polo, A. A., Grriloff, E., Gometz-Rogers, C. (1972), "Fetal heart rate throughout pregnancy," *American Journal of Obstetrics and Gynecology,* Vol. 113, 814.

Lindecrantz, Ketal (1993), "Power spectrum analysis of the fetal heart rate during noradrenaline infusion and acute hypoxemia in the chronic fetal lamb preparation," *International Journal of Biomedical Computation,* Vol. 33, 199–207.

Ogawa, T., Sonoda, H., Sawaguchi, H. et al. (1990), "Interaction between respiration and heart rate in preterm infants and developmental changes," *The Autonomic Nervous System,* Vol. 27, 612–619.

Ogawa T, Sonoda H (1991), "Relationship between RR interval, blood-pressure and respiratory fluctuations in a preterm infants." *The Autonomic Nervous System,* Vol. 28, 59–65.

Ogawa, T., Kojo, M., Fukushima, N. et al. (1993), "Cardio-respiratory control in an infant with Ondine's curse: a multivariate autoregressive modelling approach," *Journal of Autonomous Nerve System,* Vol. 42, 41–52.

Ono, K. (1976), "On a minicomputer system for autoregressive analysis of biological sways," *Bulletin of Neuroinformation Laboratory Nagasaki University,* Vol. 3, 19–27.

Pagani, M. et al. (1986), "Power spectral analysis a beat-to-beat heart and blood pressure variability as a possible marker of sympaths-vagal interaction in man and conscious dog," *Circulation Research,* Vol. 59, 178.

Sato, K., Ono, K. et al. (1977), "Component activities in the autoregressive activity of physiological systems," *International Journal of Neuroscience,* 7, 239–249.

Sato, K., Ono, K., Fukata, K. (1978), "Multidimensional autoregressive activity models in physiological systems," *Internat, Symp. med. Inform. System. MEDIS,* Osaka, Japan.

Stange, L. et al. (1977), "Quantification of fetal heart rate variability in relation to oxygen in the sheep fetus," *Acta Obstet. Gynecol. Scand,* Vol. 56, 205–209.

Wada, M. Ogawa, T. Sonoda, H. (1996), "Development of relative power contribution ratio of the EEG in normal children : a multivariate autoregressive modeling approach,*Electroenceph. and Clin. Neurophysiol,* Vol. 98, 69–75.

Widmark, C. et al. (1989), "Electrocardiographic waveform changes and catecholamine responses during acute hypoxia in the immature and mature fetal lamb," *American Journal of Obstetrics and Gynecology,* Vol. 160, 1245–1250.

Woods, J. R. Jr. et al. (1977), "Autonomic control of cardiovascular functions during neonatal development and in adult sheep," *Circulation Research*, Vol. 40, 401–407.

Yatsuki, K. et al. (1986), "Embryonal heart rate before abortion," *Proceedings of the Japan Society of Ultrasonics in Medicene*, Vol. 48, 155.

Yeh, S. Y. et al. (1973), "Quantification of the fetal heart beat to beat interval differences," *Obstetrics Gynecology*, Vol. 41, 355–363.

Zetterberg, L. H. (1969), "Estimation of parameters for a linear difference equation with application to EEG analysis," *Mathematical Biosciences*, 5, 227–275.

Zetterberg, L. H. (1977), "Means and methods for processing of physiological signals with emphasis on EEG analysis," *Advances in Biology and Medical Physics*, J. H. Laurence, et al., eds. Academic Press, New York, 41–91.

Chapter 16

Information Processing Mechanisms in the Mammalian Brain: Analysis of Spatio-temporal Neural Response in the Auditory Cortex

Kohyu Fukunishi[1]
Advanced Research laboratory, Hitachi, Ltd.
Hatoyama, Saitama 350-0395 Japan

16.1 Introduction

The mammalian brain can be regarded as a huge and complicated dynamical information processing system composed of single units called neurons or nerve cells. The mechanisms of brain function have, traditionally, been elucidated with the aid of single microelectrodes to measure the responses in single neurons. This approach, however, seems to be insufficient for identifying the complex dynamical system as the brain. An optical recording method, on the other hand, has made possible real-time multipoint measurement of the evoked neural activities distributed in the brain. This new recording method can be used to explore new mechanisms responsible for the dynamical neural processing activities of the brain. Such neural activities always exhibit nonlinear and nonstationary characteristics, and so straight forward application of any system identification theory to the neural system is inappropriate. On the other hand, many industrial dynamical systems, which involve nonstationary and nonlinear dynamical phenomena, exquisitely are modeled and controlled by using the extensive linear theory regarding to system identification and control. From this fact, there is a possibility that a linear identification theory such as time series analysis could be used in exploring the functioning of a nonlinear and nonstationary brain.

In this chapter, I present the results obtained using an optical recording technique developed at ARL, Hitachi, Ltd., to investigate the spatio-temporal pattern of sound-evoked neural responses in the guinea pig's auditory cortex. I also presents the pattern time analysis of the spatio-temporal neural responses using a multivariate autoregressive (MAR) model.

---

[1]Present address: Department of Electronic Engineering, Graduate School of Engineering, Osaka University, Yamada-Oka 2-1, Suita 565-0871, Japan, E-mail: fukunisi@ele.eng.osaka-u.ac.jp

16.2   Instrumentation of and Information Processing in the Brain

In the sensory cortices of the human brain more than 10 billion neurons process information carried by auditory and visual signals. Each neuron receives synaptic input from as many as ten thousand other neurons and provides output to several thousand neurons. The neural activity generates electric impulses, rapid changes in transmembrane electric potential, that move along neural fibers at the speeds exceeding 100 m/s. A brain is thus a signal processing system having cerebral neural pathways, having as its fundamental elements neurons that are interconnected through synapses and having a huge number of neural networks combined by synapses.

Mammalian visual information processing is said to be understood relatively well. Visual information from the retina is projected to the primary visual cortex (V1) via the lateral genuculate nucleus of the thalamus. Although the retina and the lateral genuculate nucleus contain neurons, neural cells, that are responsive to small spots of light, the V1 cells respond only to a visual stimulus with line properties, such as a line or a bar (Hubel and Wiesel 1965). These V1 cells, which are called simple cells, provide input to complex cells, which in general are particularly responsive to a visual stimulus moving across a receptive field. It has been thought that these cells in turn provide input to super complex cells that are responsive to more sophisticated visual stimulus. A three-dimensional scene may be decomposed into various fragments of shapes, patterns, and other kinds of visual information (such as color) which together become a visual stimulus specific to one or more of these super complex cells. These specific cells have been sought along visual processing pathway in the cortex of the monkey. As a results, ultimate super complex cells that respond to faces, the so-called grandmother cells, have been founded in a specific area of the visual cortex.

A hierarchical neural structure for processing auditory information in the auditory cortex, like the one for processing visual information in the visual cortex, is not known clearly. The tonotopical organization which shows regular arrangement of specific neural cells by sound frequency is a significant feature in the auditory processing of the mammal brain. Frequency as a tonal component of sound and vocal calls is a predominate element in sound and vocal information. Therefore, tonotopic cells might play a fundamental role in the neural processing of auditory information processing of vocal sound in the auditory cortex. The functional role of the tonotopic cells in the auditory cortex, however, are still unclear.

The specific cells in the auditory cortex which respond to a vocal call, like as ultimate super cells in the visual cortex that are called grandmother cells, have been named pontiff cells. Finding such cells has been one of the challenging research themes in auditory neuroscience. Unfortunately, however, even the existence of cells in the mammal auditory cortex that are specific cells to a vocal call remains suspect after years of study. The characteristic difference between the visual signal and the auditory signal apparently that the auditory information is temporal information, and this difference is the reason that grandmother cells have been found in the visual cortex but pontiff cells have not been found in the auditory cortex. This hypothesis might be true to some extend, but not definite one.

Although the signals of neural activities recorded from electrodes from brain are noisy and dynamic, the noise and the dynamic characteristics of the neural activities have long been thought to be meaningless with regard to the neural processing and neural coding in the brain, since 60's. The schema of neural coding based on this conviction, called mean rate coding or rate coding, is one in which the mean value of the response signal is a measure to evaluate the cell's characteristics to a stimulus. The finding of cell's characteristics as simple cells and complex cells are based on the hypothesis of this mean rate schema.

If the brain is a dynamical processing system, this schema might be inadequate and the results inferred from the findings obtained while using it might be unreliable. In fact, the correlative analysis of the output of two micro-electrodes at different sites in the visual cortex of a mammal brain has recently revealed a feature of the brain's dynamical processing related to a visual stimulus. Specifically, the correlation between the oscillatory activities of two neurons in the visual cortex varies according to changing of the visual stimulus. It can be said that dynamics of nervous cell has become more important than before for considering the cell's characteristics.

Significant aspects of dynamical neural behavior have been revealed much more clearly by multipoint measurements of neural activities in the cortical field. One such method, the optical recording with a voltage-sensitive dye, which is a new method measuring patterns of neural activity (De Weer and Salzberg 1986), has revealed various spatio-temporal features of the neural responses to sound stimuli in the auditory cortex of a mammal (Fukunishi *et al.* 1992). Therefore, discussing the dynamics of the brain might be interesting controversy and one of the hot spots in the study of neuroscience involving theoretical dispute as well.

16.3   Optical Multipoint Observation in the Mammalian Auditory Cortex

An optical recording method can be used to measure the neural activity at multiple points in an animal brain at the same time. The principle of the method is based on detecting the change of optical signal, the fluorescence change, using an optical sensor converted from the potential change of a cell's membrane by the effect of voltage-sensitive dye. A distribution pattern of the optical signal changes is revealed by arranging multiple optical sensors. An optical recording system, with a data acquisition system, developed by the author's group and used for measuring the neural activity evoked in the auditory cortex of guinea pigs is described below.

A 128-channel optical measurement system used for the measuring the sound induced responses of the mammalian auditory cortex is illustrated in Figure 16.1. Before stimulus induced responses were measured, a small part of the skull over the auditory cortex was surgically removed under anesthesia so that the cortical area could be stained with a voltage-sensitively dye. The membrane potential change associated with the neural response induced by an auditory stimulus was transduced by the dye into a fluorescence change and was detected by photo sensors. A 12-by-12 array of photodiode optical sensors was installed in a microscope system. The spatial resolution of the system corresponds to 130 - 220 mm on the surface of the cortical field. Since the potential change induced by the auditory stimulus itself and the signal converting efficiency of the dye were very small, the signal-to-noise ratio, corresponding to the ratio of the fluorescence change to the background fluorescence level, was at most 0.1

The actual response data was obtained by subtracting the nonstimulus activities from the stimulus responses. These two measurements were synchronized with the animal's pulse or respiration. Furthermore, the data gathered in response to continuous auditory stimuli were accumulated in order to improve the signal-to-noise ratio. From the next section on are shown the experiment results in which the sound stimulus was either clickwith 0.1 ms pulse duration (Figure 16.2) or tone bursts with 10 ms rise and fall times and a 30 ms continuation time (Figure 16.5). The animal was in a sound-shield room and the sound stimulus with delivered by a speaker or earphone having wide-range frequency characteristics.

16.4   Spatio-Temporal Neural Activity Observation

How auditory information like that in a vocal call is coded in the auditory cortex of an animal's brain is not known. The sound vibration carrying the auditory information, is transmitted to cochlea, where it causes the basilar membrane of the cochlea to vibrate. Different sound frequencies affect different portions of the basilar membrane. The organ of corti along the basilar membrane in the cochlea contains hair cells and the different frequencies of the membrane vibration stimulate individual hair cells at different points along the cochlea. Vibrations of the hair cells corresponding to individual frequencies are transformed into electrical signals in the auditory nerve and are transmitted to the auditory cortex via many auditory nuclei in the brain. The auditory cortex (primary auditory cortex) contains tonotopic maps of frequency spectrum which show a regular band-like arrangement (found by microelectrode experiments) of neurons encoded by frequency along the rostacaudal axis. The tonotopic map is a representation of the cochlea functional structure of frequency tuning. Physiological

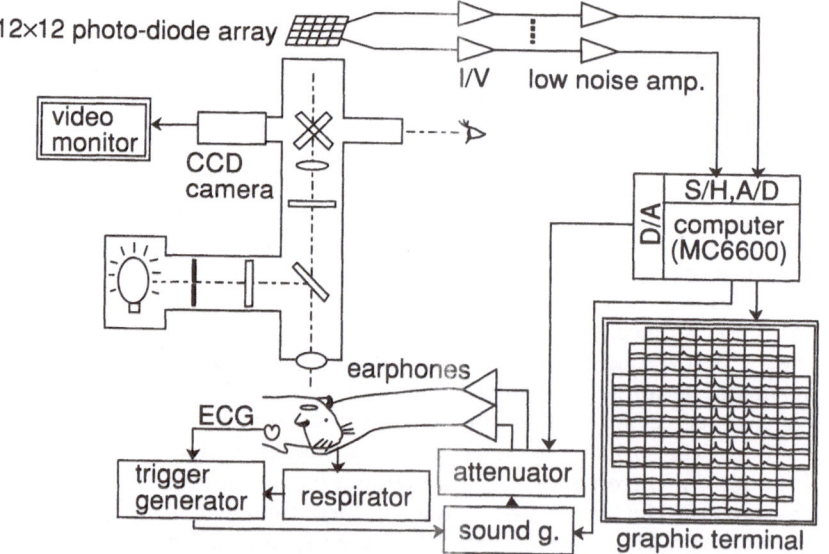

Figure 16.1   Optical recording system for spatio-temporal neural activity measurement of animals' auditory cortex

experiments using micro-electrodes have revealed a tonotopic map in the primary auditory cortex of many mammalian species, and we thought it interesting to visualize the tonotopic map of guinea pig by using the optical pattern measurement instead of by successive point measurements with microelectrodes.

The time courses of the click-induced response in the primary auditory cortex of the guinea pig by optical measurement are depicted in Figure 16.2 (Fukunishi *et al.* 1992). One handled times of trials were accumulated for this visualization. This response region is equivalent to approximately 1/4 of the whole primary auditory cortex. One pixel is equivalent to a 130×130-mm cortical area. The spatial-temporal characteristics of this click-induced neural activity distributed throughout the auditory cortex are shown in Figure 16.3. The response amplitude is given by the size of the small black circles on the individual pixels (measured points on the cortex). It is noticed that the strong response area (large black circle) shift from rostradorsal to caudraventral in the cortical field. Thus, the synchronized multipoints measurement of the neural activity visualized by optical recording provides a glimpse of the dynamic neural activity not evident in the results provided by a set of single unit recordings with micro-electrodes.

The click-induced neural responses in a broader area of the primary auditory cortex can be shown in a 3-dimensional patter (Figure 16.4). The observed area here is equivalent to approximately 2/3 of the primary auditory cortex (approximately 3×3-mm) and the spatial resolution was 220 mm on the surface of the cortical field. It is shown that the high response part (the peaks of the mountain in the figure) shifts from the rostradorsal to the caudal and to the rostraventral, like as in a shape of a crescent, at the latency from 18 ms to 33 ms after the stimulus. A complex sound like as click, in which various sound waves with various frequencies are present at the

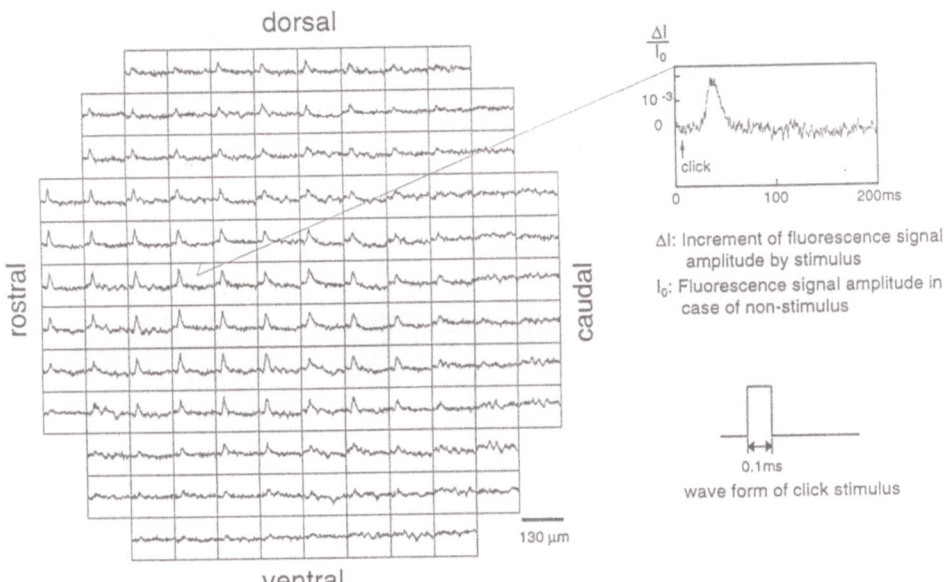

Figure 16.2 Distribution of time series data of click stimulus evoked response in auditory cortex by optical recording

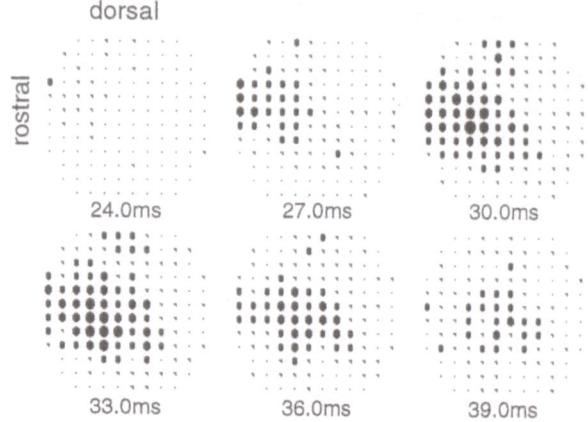

Figure 16.3   Temporal shift of response amplitude pattern evoked by click stimulus in primary auditory cortex (corresponding to Figure 16.2)

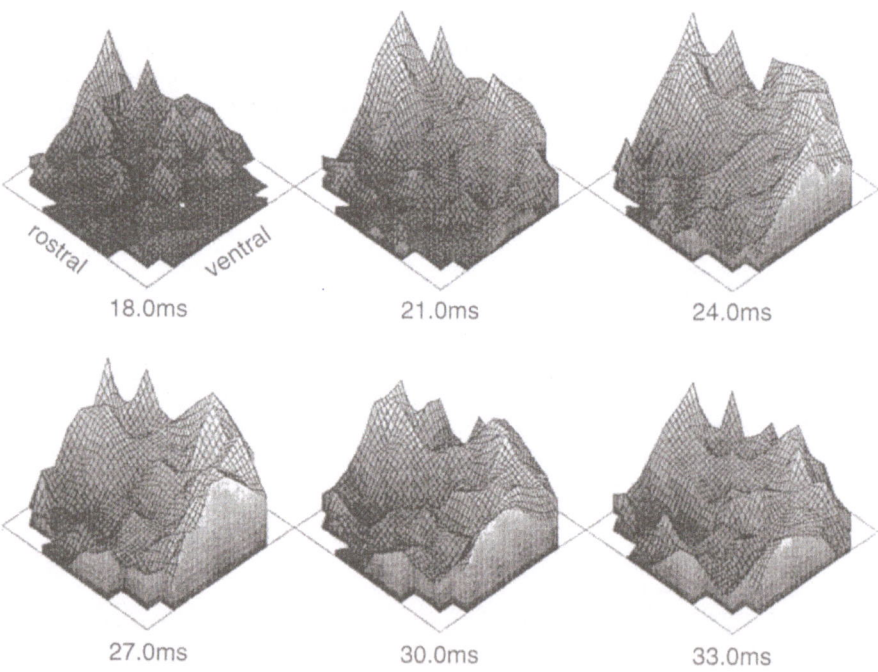

Figure 16.4   3-dimensional expression of click evoked response shift in primary auditory cortex AI as latency (observed area: about 2/3 of AI, 3.0 mm×3.0 mm, spatial resolution:0.22 mm×0.22 mm)

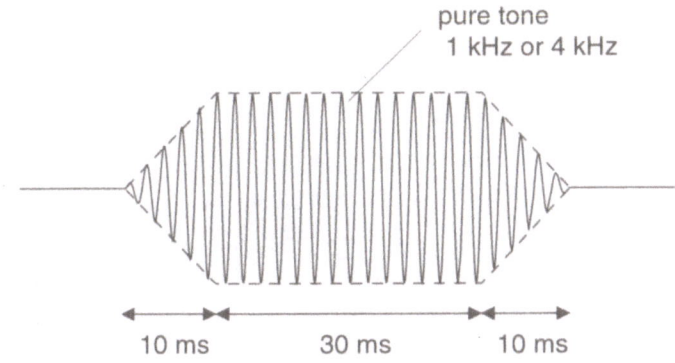

Figure 16.5 Sound stimulus wave form of tone burst

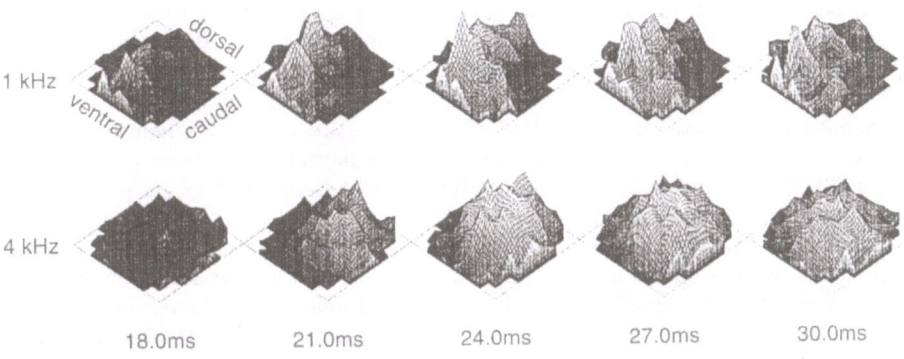

Figure 16.6 Evoked response pattern in primary auditory cortex (about 2/3 of AI) for 1 and 4 kHz burst stimulus (the same measured part of the same animal)– Visualization of tonopicity–

same time, might be processed by assembling temporally the neural cells encoded by frequencies in the different parts of the primary auditory cortex. It is suggested that such a complex sound as a click can not be encoded by a specific neuron or a specific neural population.

The tonotopic map, tonotopic patterns, observed in the auditory cortex are not exactly same according to the experiments by successive unit recording or optical recording. The results of our optical imaging indicate that the tone-induced response regions in the auditory cortex have wide island shapes instead of the band shapes obtained by the microelectrode recording and overlap each other at their peripheral part. The specific neurons, neural population or neural assemblies, that respond to 1-kHz and 4-kHz tone bursts, for example, can be shown individually at the rostral area (1-kHz) and the rather caudal (4-kHz) area in the observed primary auditory cortical field (approximately 3×3-mm) as given in Figure 16.5. The optically imaged topographic pattern of response to tone bursts has cast new doubt on the existence of

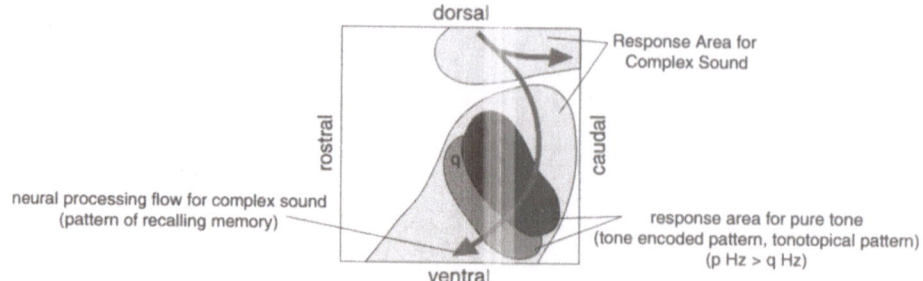

dorsal

Response Area for
Complex Sound

rostral

caudal

neural processing flow for complex sound
(pattern of recalling memory)

response area for pure tone
(tone encoded pattern, tonotopical pattern)
(p Hz > q Hz)

ventral

Figure 16.7 Concept of complex sound information processing in a guinea pig's primary auditory cortex (imaging from experimental results)

the one-to-one correspondence between a frequency and a neuron that was indicated by experiments with microelectrode. The optical experimental results with click-stimulus and tone-stimulus revealed different neural behaviors as spatially dynamic in case of a click and spatially stable or static in case of a tone.

The dynamical neural processing of a complex sound in the auditory cortex can be symbolized as illustrated in Figure 16.7. This illustration images the neural processing for a complex sound in the auditory cortex where neural populations corresponding to individual frequencies are assembled dynamically from the high-frequency to the low-frequency. The dynamical neural behavior may indicate that the frequency information encoded by the neural assembly (memory) in the auditory cortex is recalled successively when a sound having frequency components in a wide bandwidth (such as a click is heard (Fukunishi *et al.* 1992). The real meaning of the dynamical neural behavior in the auditory cortex that can be visualized by the optical recording has not yet been explained.

16.5   Functional Modules in the Auditory Cortex

Neurons are strongly bound in the direction perpendicular to the surface of the cortex and the visual cortex is known to contain neurons organized in columns. With the anatomical neural structure built up by such neurons in correspondence with the individual functions of the simple cells and or complex cells, functional modules by the neural organization are formed.

In the auditory cortex, however, neither column structure nor functional modules have been found except in one species of bat. If modularity of the neural activity can be found in the cortical field, then it is expected that comprehension of the meaning possessed by the optical spatio-temporal data of the neuron population activity is promoted. The pattern time series data of the click-induced responses revealed by the optical imaging shown in Figure 16.2 were used to calculate correlation functions between the responses at different cortical positions. As already noted with respect to Figure 16.3, the neural activity shifted from rostradorsal to caudaventral as response time after the sound stimulus. Illustrated in Figure 16.8 are the cross correlation functions among the time series data at the cortical positions (the eight points on the cross section of the measured pixel in Figure 16.8) on the straight line along the

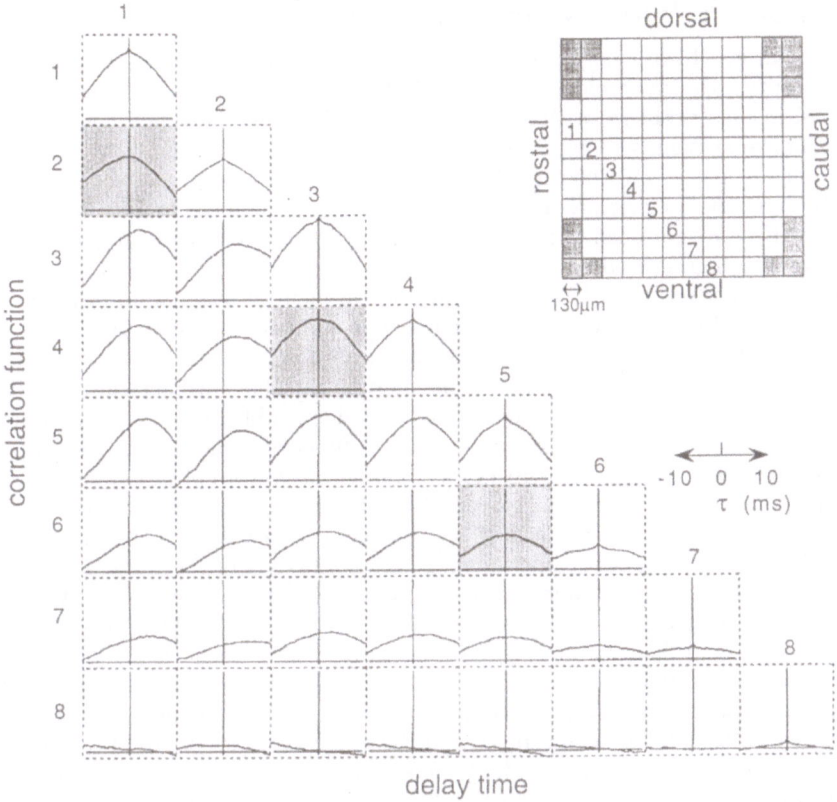

Figure 16.8 Cross correlation function of click evoked responses at eight pixels in primary auditory cortex (analysis of data in Figure 16.2)

direction of the shift of the response.

The values on the diagonal line give the autocorrelation functions at the eight points. In parallel with the diagonal line, the crosscorrelation functions corresponding to the cortical distances among the measurement pixels are illustrated (Fukunishi et al. 1993a). When the crosscorrelation function between the pixels j and i that are adjacent to each other is expressed at position $[i, j]$, the crosscorrelation functions at positions $[2, 1]$, $[4, 3]$, and $[6, 5]$ take a functional pattern symmetric to the time delay axis. However the crosscorrelation functions at positions $[3, 2]$ and $[5, 4]$ take an asymmetric pattern. This result implies that the cortical area corresponding respectively to the pixels 1 and 2, 3 and 4, and 5 and 6 are individuals of a module, and the neural activity conducts parallel processing with the individuals. On the contrary, in the area corresponding to the pixels 2 and 3, and 4 and 5, are not a module, and it is believed that the neurons became active serially to the pixels. A single pixel in this measurement is equivalent to a square 130μm on a side, and the results described above suggest that one module of the auditory cortex has a diameter of about 300μm. The auditory cortex thus seems to have a structure made up of functional modules.

16.6   Pattern Time Series Analysis

A problem in which time-variant states of a dynamical system are measured and used to identify the dynamical structure of that system is in general defined as a dynamical inverse problem of the system. And to identify the functional structure of a dynamical system, we might need to solve such an dynamical inverse problem. Optical imaging has already shown us clearly that the brain is a dynamical information processing system. Solving the inverse problem of the dynamical brain will therefore a new and effective approach to explain the brain's information processing mechanism with dynamical neural bindings.

It is impossible to uniquely solve the inverse problem of a dynamical system like that seen in a brain, one that is composed of a huge number of neurons, whose activity has nonlinear characteristics. Despite that, the result described in the previous section implies that the brain's cortex has functional module and that we can treat the state of the brain system in term of the activity of the neural assemblies or neural populations. Meanwhile, as the neural activity from a cortical functional module is equivalent to the total neural activities of the neurons in a module, the output of a module is subject to an output of a linear dynamical system. Such being the case, the auditory cortical system can be expressed by the neural activity of the cortical module corresponding to the adjusted pixel size of the optical imaging, and its dynamical characteristics can be identified inversely by the analysis of the multipoint optical measurements.

The author adapted a multivariate autoregressive model (MAR) that had been used to identify and diagnose the dynamical characteristics in nuclear plants (Fukunishi 1977). This method is applied to identify the cortical dynamic system. The multi-pixels of the optical recording of the sound-stimulus-evoked neural responses in the auditory cortical field can be considered the multi-variables or the state variables of a multivariate autoregressive model (Akaike and Nakagawa 1972). Since the data obtained by optical recording in a single trial (a one-time stimulus) experiment were noisy (with optical, electrical, and biological artifacts), the averaged data obtained in appropriate trial times are used to data analysis. The multivariate autoregressive model, which was essentially developed for the time series analysis of stationary linear system, is applied to analysis the multiple transient responses of a brain system complying with the deterministic input in order to extract the linear corresponding relations among the responses. Actually, the temporal behavior of a spatially distributed state pattern, pattern time series, is analyzed by applying the model (Fukunishi et al. 1995). On the other hand, the results obtained from the stationary linear optical data are discussed in the final section for this chapter.

The n-dimensional vector X_t at the time t represents the neural activities at the n-points from multipoint optical measurement of the auditory cortex, and the multivariate autoregressive model is expressed by

$$X_t = \sum_{k=1}^{p} A_k X_{t-k} + E_t, \tag{16.1}$$

where A_k $(k = 1, \ldots, p)$ is a series of an $n \times n$ parameter matrix. E_t is an n-dimensional vector whose elements are a mean value 0 and variance-covariance matrix Σ_p, which is equivalent to the residue of the model. Here, the average of X_t is also designed to be 0. On the assumption that E_t becomes Gaussian m-dimensional white noise, the model order p is determined by the Akaike's information criterion AIC (Akaike 1974) with which the order is selected so as to meet an optimal balance in relation to the data length N.

$$\text{AIC}(p) = N \log|\Sigma_p| + 2n^2 p, \tag{16.2}$$

where $|\Sigma_p|$ represents a matrix of the variance-covariance matrix of the estimated E_t. The estimation of the parameter matrix A_k $(k = 1, \ldots, p)$ is obtained by solving a normal equation for the least-square estimation of equation (16.1).

The spectrum density function matrix of the vector X_t can be expressed by

$$S(f) = A(f)^{-1} \Sigma_p A^*(f)^{-1}, \tag{16.3}$$

where $A(f)$ is the Fourier transform of the matrix A_k $(k = 1, \ldots, p)$. The asterisk $*$ denotes a complex conjugate operator. If the residues becomes noncorrelative white noises and nondiagonal elements of the variance-covariance matrix Σ_p can be neglected, then it is deduced from equation (16.3) that the power spectrum density (PSD) of $X_{i,t}$, $i = 1, 2, \ldots, n$ (the i-th element of the vector X_t) is given by

$$S_{ii}(f) = \sum_{j=1}^{n} |\{A(f)^{-1}\}_{ij}|^2 \Sigma_{jj,p}, \qquad (i = 1, 2, \ldots, n) \tag{16.4}$$

where $\Sigma_{jj,p}$ is the j-th diagonal element of the variance-covariance matrix Σ_p and where $\{\cdot\}_{ij}$ is the (i, j) element of the $n \times n$ matrix $\{\cdot\}$. This implies that the power of the neural activities at the i-th cortical position (i-th state) is generated by the sum of the transfer effects from the source powers of neural activities at the n analyzed cortical positions (n states).

The strength of the functional neural binding between different cortical positions, the cortico-cortical binding, can be expressed quantitatively by using the power contribution ratio which mean a relative power of one state by the power generated at the other state. The ratio of the power contribution where the neural activity $X_{j,t}$ at the cortical position j affects, in the meaning of power, on the neural activity $X_{i,t}$ at the cortical position i can be defined by the relative power contribution by

$$\gamma_{ij}(f) = \frac{|\{A(f)^{-1}\}_{ij}|^2 \Sigma_{jj,p}}{S_{ii}(f)}. \tag{16.5}$$

We can use this expression not only to evaluate the correlative intensities of the neural activities between different cortical positions, but also to estimate the direction of the signal flows of the neural activity. Thus, using MAR model to analyze the pattern time series of the optical measurement data makes it possible to evaluate the cortco-cortical functional binding relation of the neural activities at different cortical positions. The MAR model, which was originally intended to be used with stationary time series data is expected - provided that the mutual relation between the time series is stationary even in case of an nonstationary time series data - to be effective in the analysis of the nonstationary data.

16.7   Neural Correlation of and Neural Binding

There are many recent articles discussing the relations between the temporal coding, the binding of the neural activities, and the cortical neural oscillation. This new trends in brain research can be based on an idea antagonistic to the single-neuron doctrine. That is, it is considered that synchronized neural oscillations may serve to bind the different neurons into a unique representation of a specific feature (Engel *et al.*). The synchronous neural oscillation is usually evaluated by the correlation of the neural activities measured by two electrodes at different points in the cortical field. Actually, the synchronized neural oscillations around 40 Hz in case of specific pattern stimulation are measured at the distributed neurons in the visual cortex of mammals as cat and monkey. The synchrony binding for feature detection in the visual cortex might be necessary for segregating the target object from the background scene. The meaning of the synchronized neural oscillatory can be explained by an analogous idea that the multiple visual targets fluctuating synchronously with a specific frequency can be easily recognized from many other objects moving randomly. Thus, dynamical analysis of the neurons behavior has been considered very important for understanding the brain's information processing mechanism.

There is, an inherent problem in the application of correlation analysis to multivariable systems that contain inner feedback signal flows (Fukunishi *et al.* 1995). Now we assume a neural system with three neurons and suppose that none of direct binding exists between the neurons A and B, but the strong bindings exist between the neurons C and A, and between the neurons C and B. Correlation analyses between the neural activities of each two neurons from the neurons A, B, C show usually a strong correlation between the neurons A and B by intermediate effects of the neuron C. That is, the correlation analysis might derive the result an imaginary binding between the neurons A and B.

The coherency function equivalent to the square of the correlation function coefficient of two variables can be calculated from the evoked response obtained by optical recording, and in this analysis we use the data measured when the response was evoked by a 4 kHz tone burst (Figure 16.6). The cortical responses area is represented by circles whose sizes mean the response intensities 20 ms after the tone stimulus. The evoked response at the adjacent five points $(5, 7), \ldots, (9, 7)$ in the rostacaudal axis were analyzed as illustrated in Figure 16.9 (a). The coherency functions between each pair of responses from the five positions are denoted in Figure 16.9 (b). The mutual coherency function were calculated by using the cross- and auto-power spectrum density functions defined by equation (16.4). As shown in the figure, the all coherency functions at frequencies lower than 60 Hz were close to 1.0. This suggests that the neutral activities at these five points in the auditory cortex are clearly correlated and that there is functional neural binding between each pair points of the analyzed cortical points.

16.8   Evaluation of Cortical Neural Binding

The neural oscillation which is observed in the impulsive response to a specific stimulus in the visual cortex also can exist in the impulse response in the auditory cortex and also in the response of the neural populations as measured by optical recording. The data obtained at the cortical field by optical recording are the total sum of the field potentials due to neural activities by the stimulus under the each sensor. Since cortical modularity is noted with the neural activity of the auditory cortex as discussed previously, the dynamic neural characteristics maintained in the evoked response data by means of the optical recording can include the dynamical characteristics of the impulse neural activities of an neuron. Assuming the dynamical

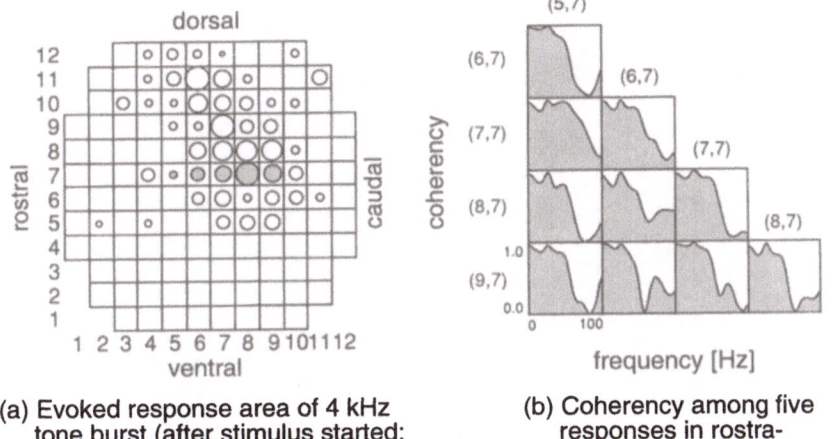

(a) Evoked response area of 4 kHz tone burst (after stimulus started: 20 ms, analyzed part: gray circle)

(b) Coherency among five responses in rostra-caudal cortical part

Figure 16.9 Cortico-cortical coherency of 4 kHz burst evoked responses (analysis of data in Figure 16.6)

feature of the neural assemblies like this way, the cortical distribution of the neural oscillatory sources and the functional binding structure with neural transmission flows is expected to be estimated by the pattern time series analysis.

Time series data of 4-kHz evoked responses at the five points (region) $(5, 7), \ldots, (9, 7)$ in the rostacaudal axis were selected in similar to the coherency analysis. The multivariate autoregressive model (MAR) shown in equation (16.1) was applied to the time series data of these five response variables. The response data used were accumulated by four trials of stimulus-response subtracted by control activity. Data length were 200 ms and sample rate was 0.2 ms. Figure 16.10 shows the PSD of these response variables and their relative power contribution. The order of the MAR model, p, obtained by AIC was 18, whereas the nondiagonal term of the covariance matrix Σ_p of the model's residue which is normalized by the diagonal elements was at most 0.18.

The PSD of each evoked responses at each of cortical positions exhibits a strong peak value in the vicinity of 30 Hz. This suggests that oscillatory neural behaviors exist in the neural population activities in the auditory cortex like as previously introduced discussion in visual cortex. The relative power contributions to each variable from other four variables and from the variable itself are illustrated like as a matrix form under each PSD in the same figure. The power contribution corresponding to the oscillatory frequency of the each PSD shows peak value at the same frequency as marked in gray in the same figure. The power contribution ratios from the variable at region (8, 7) to the other variables including the variable itself have reached 0.7–0.8. It is perceived that the neural populations that show oscillations with frequencies of about 30 Hz and that are found in the five regions $(5, 7), \ldots, (9, 7)$ at the rostacaudal axis of the primary auditory cortex have the same oscillatory source generated at the cortical region (8, 7). We can interpret the fact as a sign that the neural population activity in the region (8, 7) is predominant in the rostacaudal axis (known as the tonotopical axis) that there is unidirectional functional binding between the region

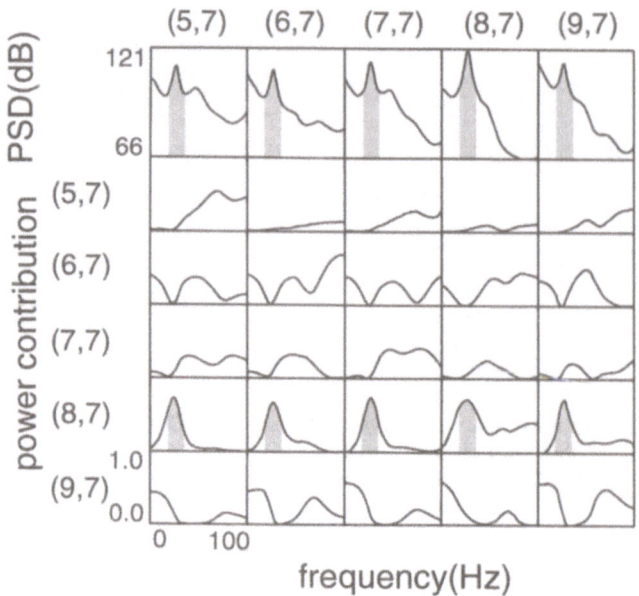

Figure 16.10 Pattern time series analysis by MAR model of 4 kHz burst evoked responses at five positions (Figure 16.9) of therostacaudal, and estimation of PSD and relative power contribution

(8, 7) and each region in the axis, and that nerve signals propagate from the region (8, 7) to the other regions.

Comparing the results obtained by the coherency analysis and those obtained by the pattern time series analysis using MAR modeling, we find that the strong coherency among the popuration activities at different regions that obtained in the analysis of the coherency does not necessarily mean the direct binding among the neuron populations at the regions as discussed by the relative power contribution of the oscillatory PSD. We can conclude from this analysis that the experimental approach for evaluating the binding structure based on analysis of the correlation between neural activitie is itself a problem (Fukunishi *et al.* 1995).

As shown in Figure 16.6 , neural activity area of a population of neurons responding to a tone stimulus - which means the distribution pattern of the specific neurons tuned to one frequency (a tonotopic pattern) - expands differently in the rostrocaudal and dorsoventral directions. The difference of the expansion pattern by the direction is due to the dynamical neural activities in dostventral than rostcaudal (Fukunishi *et al.* 1997) . Keeping in mind such tonotopical characteristics, the pattern time series analysis using MAR model are applied to evaluate the functional neural binding structure of the tonotopical pattern. The 4 kHz burst-evoked data used to the previous analysis were adopted again. Various data sets which were involved in the evoked data and, therefore, were a part of the tonotopical pattern were analyzed using a MAR model.

The results of a data set with the evoked responses at the five regions $(8, 5), \ldots, (8, 9)$ in the dorsoventral axis can be described as bellow. The PSDs of the evoked responses

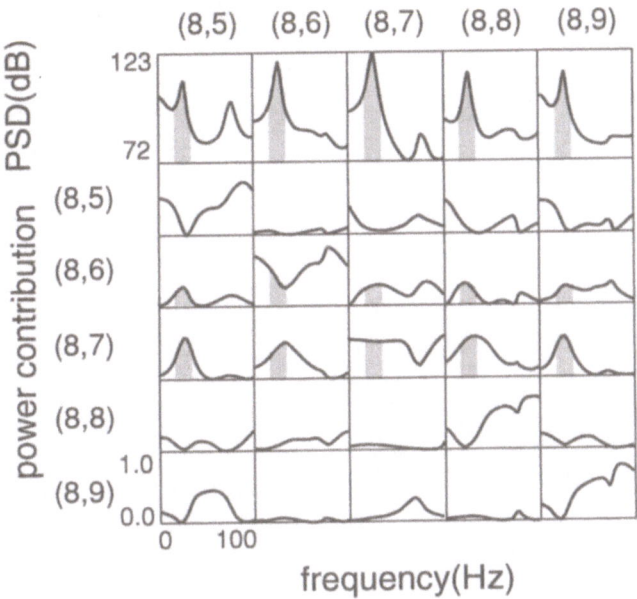

Figure 16.11 Pattern time series analysis by MAR model of 4 kHz burst evoked responses at five positions (Figure 16.9) of the dorsoventral, and estimation of PSD and relative power contribution

at the regions $(8, 5), \ldots, (8, 9)$ and the relative power contributions are illustrated in Figure 16.11 like as illustration of Figure 16.10. The neural activities at these regions show oscillatory at the frequency about 30 Hz similar to the previous results at the regions in the rostacaudal axis. The relative power contributions corresponding to the oscillatory frequency components at each region in the dorsoventral axis, however, show a dynamical feature different from the one evident in the result observed at the regions in the rostacaudal axis. The relative power contributions associated to the variable of each region from the variables of the positions $(8, 7)$ and $(8, 6)$ are significantly larger than those from the variables of other regions (as marked in gray in the Figure).

A large power contribution from a variable (A) to another variable (B) means not only a activity source existence in the variable A but also the activity transfer direction from the variable A to B in general. The previous results of the neural activity source at the regions along the rostacaudal axis revealed an existence of a single predominant activity source at the region $(8, 7)$ and neural signal flow from this region to the other regions. This results, however, suggest that the neural activities at the analyzed regions, $(8, 5), \ldots, (8, 9)$, are influenced by the neural activities at the regions, $(8, 7)$ and $(8, 6)$. Furthermore, the neural activities at those five regions are transferred through the unidirectional functional binding from the two activity source regions $(8, 7)$ and $(8, 6)$. The neural activities at the regions $(8, 7)$ and $(8, 6)$, themselves, can be mutually induced through bidirectional bindings and self-feedback bindings associated to the two regions. Thus, the functional neural binding structures

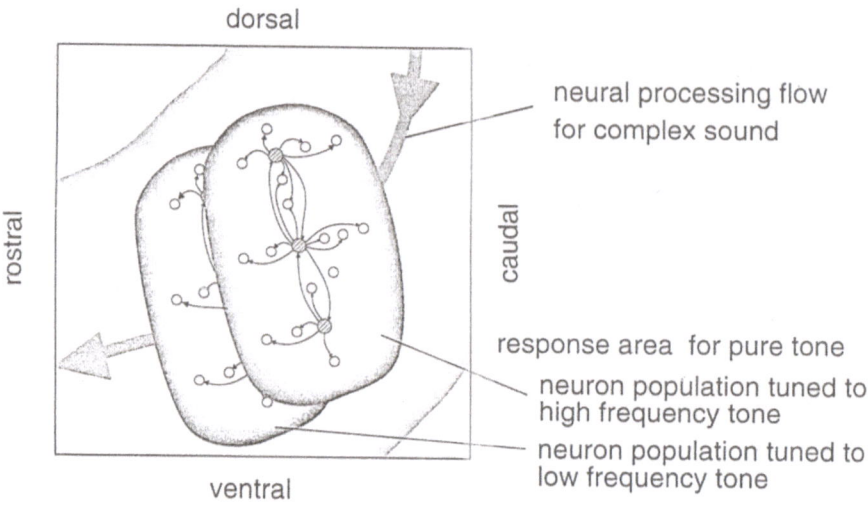

Figure 16.12   Concept of functional neural network for tonopiccal organization
of a guinea pig's auditory cortex (AI) (imaging from pattern time series analysis
of experimental data)

among the cortical regions in the tonotopical pattern to 4-kHz tone significantly differ
in the regions of the dorsoventral axis or the rostacaudal axis.

When the pattern time series analysis was applied to the same tone responses at
the other response variable (region) sets, the differences between the neural binding structures at the cortical regions in the dorsoventral and rostacaudal axes like as
discussed above were statistically significant. Similar relations were also seen in the
analysis of the tonotopical pattern to the 1-kHz stimulus (data shown in Figure 16.6).
(Fukunishi et al. 1993b). If these analytical results of tonotopical patterns are extensively interpreted, then the functional bindings that organize in a tonotopical pattern
in the primary auditory cortex can be imagined in a conceptual schematic diagram
as shown in Figure 16.12.

As seen above, pattern time series analysis, using the MAR model, of spatio-
temporal neural activities of the brain's auditory cortex that are measured by optical
recording provides us with a new way to evaluate the dynamic structure or the functional binding structure of the auditory cortex. The approach discussed here, however,
is preliminary and approximative. The dimension of the model corresponding to the
number of analyzed responses data (one data set) is restricted by the computational
procedure for estimating the parameter matrix A_k $(k = 1, \ldots, p)$ of the MAR model.
Therefore, the time series analyses to evaluate the functional binding in a data set is
obliged to repeated to evaluate them in the various data set of the tonotopical pattern.
A conclusion on the binding structure of a tonotopical pattern is extensively judged
by considering the evaluated results in various data set. Furthermore, the parameter
space of a MAR model estimated in a data set is not invariant to the parameter space
of the model estimated in other data set in general, it is necessary for strict analysis of

the pattern time series data to estimate one parameter space whose model can cover the whole area in a tonotopical pattern. A new method of the pattern time series analysis that can estimate the parameters of several hundred dimensions of a model is highly required to be developed.

16.9   Characteristics of Stationary Stochastic Response

Here, a recent study on stationary time series data of the tone evoked responses in the auditory cortex by optical recording is briefly introduced (Fukunishi *et al.* 1998). Spontaneous neural activity, random neural activity, that is autonomous neural behavior without any stimulus was not measured by using optical recording beforehand. Improving the sensitivity of our optical recording method, we have been able to instantaneously image the spontaneous response and the response evoked by a one-time sound stimulus; that is, without accumulating response data by repeating the stimulus. Thus, a spatio-temporal characteristic of stochastic neural activities in the spontaneous activity and the response to one-time sound stimulus can be discussed comparatively.

The pattern time series data of a one-time tone-evoked response and of a spontaneous activity in the primary auditory cortex of guinea pig were analyzed and compared using the MAR model. As the time series data of a one-time tone-evoked response also involves a deterministic response component that is usual nonstationary component and appear by accumulating repeatedly stimulus-responses data, this component was removed by subtracting the deterministic response obtained by the accumulation. Thus we can discuss two kinds of stationary stochastic neural activities, one with a sound stimulus and the other without a sound stimulus, and nonstationary stochastic activity which involves the deterministic evoked responses. By using the MAR model, we have found that the oscillatory stochastic responses and the deterministic responses, which are both components of one-time tone response, are generated by different neural mechanisms.

The oscillatory stochastic responses in the auditory cortex have also been shown to be generated by the the same mechanism generating the spontaneous (nonstimulus) activities. As example, the PSDs of these three response variables of the deterministic responses, the stochastic responses and the spontaneous responses regarding to the instantaneous neural activities at five positions (a)–(e) in the cortical field are shown in Figure 16.13. The PSD patterns of the deterministic responses (1) are different from the PSD patterns of the stochastic responses (2) and the spontaneous responses (3). Neural oscillations with frequencies about 30 Hz and about 50 Hz exist in the deterministic evoked response (1) with alpha wave range oscillations about 10 Hz. Both stochastic activities of the response (2) and the spontaneous (3) show monotonous power spectrum patterns decreasing with frequency and the alpha wave range oscillations. The oscillatory sources for both stochastic activities in the evoked responses and in the spontaneous activities were rather spatially stable in the cortex by the results of the relative power contributions. We can thus conclude that the stochastic responses evoked by tones are, like the spontaneous activities, unconstrained by the stimulus.

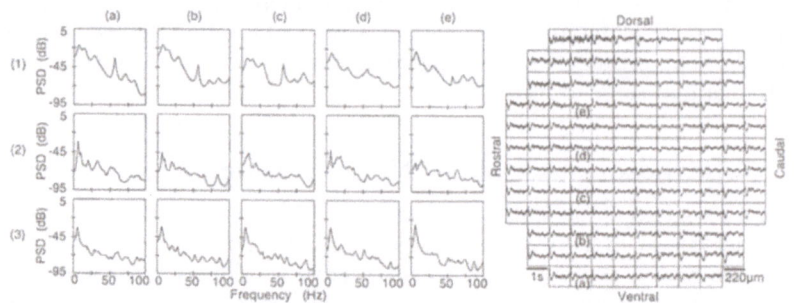

Figure 16.13 PSDs for deterministic evoked response to tone (1) , for stochastic evoked responses to tone (2) and for spontaneous activities (3) at five pixels in the dorsoventral in optically measured part (area:3.0mm×3.0 mm, resolution: 0.22 mm×0.22 mm) of a guinea pig's auditory cortex (AI)

16.10   Conclusions

The importance of the noise characteristics of neural activity has recently attracted a great deal of interest with regard to its possible relation to the neural coding of various kinds of information and to information processing in the brain. This new interest is a result of improvements in the techniques used to investigate distributed neural activities, such as optical recording and multiple-electrode recording. We discussed one way that the dynamical neural information processing mechanism in an animal brain might be revealed by applying a time series analysis to spatio-temporal response data measured by optical recording. The optical measurement method is still under development, an ideal recording system with better spatial resolution and time resolution will of course provide more precise information of the neural behavior. We need not only better measurement techniques, but also better methods for analyzing spatio-temporal pattern of neural activity.

References

Akaike, H. and Nakagawa, T. (1972), " Stochastic Analysis and Control of a Dynamical System," Science-sya (in Japanese)

Akaike, H. (1974), " A new look at the statistical model identification," *IEEE Transactions on Automatic Control*, AC-19, 716–723.

De Weer, P. and Salzberg, B. M., (eds.), (1986), "Optical Methods in Cell Physiology," *Wiley-Interscience*

Engel, A. K., Konig, P., Kreiter, A. K., Schillen, T. B. and Singer, W. (1992), "Temporal coding in the visual cortex: new vistas on integration in the nervous system," *Trends in Neuroscience* , Vol. 15, 218–226.

Hubel, D. H. and Wiesel, T. N. (1965) , "Binocular interaction in striate cortex of kitten reared with artificial squint," *J. Neurophysiology* , Vol. 28, 1041–1059.

Fukunishi, K. (1977), "Diagnostic analysis of a nuclear power plant using multivariate autoregressive processes," *Nuclear Science and Engineering*, Vol. 62, 215–225.

Fukunishi, K., Murai, N. and Uno, H. (1992), "Dynamical characteristics of the auditory cortex of Guinea pig observed with multichannel optical recording," *Biological Cybernetics*, Vol. 67, 501–509.

Fukunishi, K., Uno, H. and Murai, N.(1993a), "Spatio-temporal observation of guinea pig auditory cortex with optical recording, Japanese Journal of Physiology, Vol. 43, s 61–66.

Fukunishi, K., Murai, N. and Uno, H. (1993b), "Cortical neural networks revealed by spatio-temporal neural observation and analysis on Guinea pig auditory cortex," *Proceedings of 1993 International Joint Conference on Neural Networks*, IJCNN-93-Nagoya, 73–76.

Fukunishi, K. and Murai, N. (1995), "Temporal coding mechanism of Guinea pig auditory cortex as revealed by optical imaging and its pattern time series analysis," *Biological Cybernetics* , Vol. 72, 463–473.

Fukunishi, K., Tokioka, R., Miyashita, T. and Murai, N.(1997), "Species-specific vocalization in guinea pig auditory cortex observed by dye optical recording, Acoustic Signal Processing in the Central Auditory System: Syka, J. (ed.)," *Plenum Publishing Co.* , 443–449.

Fukunishi, K., Murai, N. and Tokioka, R.(1998), "On the stochastic neural characteristics of spontaneous activity and evoked response revealed by optical imaging and time series analysis in guinea pig auditory cortex. (to appear).

The author thank N. Murai for conducting his numerical calculations and he thanks the staff of Fukunishi Research Group of ARL, Hitachi, Ltd. for conducting the experiments.

Chapter 17

Time Series Analysis of Financial Asset Price Fluctuations

Hiroshi Tsuda
NLI Research Institute
1-1-1 Yurakucho, Chiyoda-ku, Tokyo 100 Japan
tsuda@nli-research.co.jp

17.1 Introduction

With the appreciation of stock prices in the late 1980s and subsequent crash in the 1990s, financial institutions and institutional investors have once again realized that asset management by holding large positions in volatile high-risk assets such as stocks carries the danger of incurring huge losses. This has focused attention on the need to improve risk management and asset management efficiency.

A promising solution to this problem is to use a scientific asset management approach based on quantitative analysis so that objective evaluations can be made using numbers. To make this scientific asset management approach work, it is important to detect regularities in the seemingly random price fluctuations of financial assets such as stocks and bonds, and to infer the underlying mechanism. Time series analysis of financial asset price fluctuations allows us to understand the underlying mechanism and provides clues for predicting future price movements.

17.2 Nonstationary Nature of Financial Asset Prices

Figure 17.1 shows the TOPIX index (TSE stock price index) during the period from January 1980 to December 1993. TOPIX is a price index of stocks in the first section of the Tokyo Stock Exchange, with a base value of 100 for January 4, 1968. The discernible trend of TOPIX clearly makes it unusable as a stationary time series. In general, price fluctuations of stocks and other financial assets are often nonstationary time series in which the auto-correlation does not approach zero quickly even with a large time lag. Thus a high degree of accuracy cannot be obtained by using a stationary time series model to predict prices. A stock price model that accurately predicts prices must use a nonstationary model that takes trends into account.

Methods used thus far to estimate trends involve observing fluctuation patterns in the data, hypothesizing simple functions or polynomial expressions that seem to fit, and identifying trends using regression analysis. However, such methods tend to be arbitrary.

Kitagawa and Gersch (1983, 1984) have proposed an approach to estimating trends that avoids this arbitrariness. This method is based on Akaike's (1980) Bayesian model and has the following features: (1) trends are treated as fluctuations that conform to stochastic processes; (2) the model's trend components are formulated based on a prior information using Bayesian methods; and (3) when estimating the model, not only trends but other fluctuating components such as cyclical and seasonal fluctuations can simultaneously be estimated. While the method resembles conventional methods in assuming the trend model beforehand, it differs significantly in several respects: it regards trends as stochastic processes, simultaneously determines other fluctuation elements such as AR series, seasonal fluctuations, and irregular fluctuations, and it makes objective evaluations using the Akaike information criterion (AIC).

17.2.1   Composition of a Nonstationary Model

We now describe the composition of our nonstationary model. We assume that the stock price model consists of weekly stock price P_t, trend variable Tr_t, cyclical variable V_t, and random fluctuation ε_t;

$$P_t = Tr_t + V_t + \varepsilon_t. \tag{17.1}$$

Here, trend variable Tr_t is assumed to be a smooth stochastic process and follows a second order stochastic difference equation model,

$$Tr_t = 2Tr_{t-1} - Tr_{t-2} + W_{1t}. \tag{17.2}$$

In addition, cyclical variable V_t is assumed to follow a k-dimension AR model,

$$V_t = \sum_{i=1}^{k} \alpha_i V_{t-i} + W_{2t}. \tag{17.3}$$

From equations (17.1)–(17.3), stock price P_t at time t can be expressed using state-space expression as

$$\begin{aligned} X_t &= FX_{t-1} + GW_t \\ P_t &= HX_t + \varepsilon_t. \end{aligned} \tag{17.4}$$

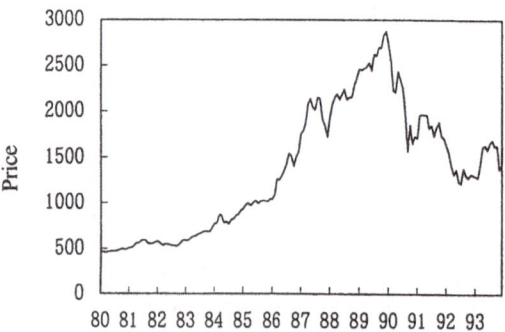

Figure 17.1   The TOPIX index

However, $W_t=(W_{1t}, W_{2t})^t$ and ε_t are stochastic variables with mean zero, unknown variance and covariance, and mutually independent normal distributions. The $*^t$ denotes a transposed matrix. Furthermore, X_t is a directly unobservable state vector that expresses the stock price fluctuation system. F, G and H have the following forms

$$F = \begin{bmatrix} F_1 & 0 \\ 0 & F_2 \end{bmatrix}, \quad G = \begin{bmatrix} G_1 & 0 \\ 0 & G_2 \end{bmatrix}$$

$$H = \begin{bmatrix} H_1 & H_2 \end{bmatrix}. \tag{17.5}$$

Here, F_1, G_1 and H_1 are the trend components, while F_2, G_2 and H_2 are coefficient matrices related to cyclical components.

Taking the example of a 2-dimensional AR model, the state vector is $X_t = [Tr_t, Tr_{t-1}, V_t, V_{t-1}]^t$, and the state-space expression of the stock price model is:

$$\begin{bmatrix} Tr_t \\ Tr_{t-1} \\ V_t \\ V_{t-1} \end{bmatrix} = \begin{bmatrix} 2 & -1 & 0 & 0 \\ 1 & 0 & 0 & 0 \\ 0 & 0 & \alpha_1 & \alpha_2 \\ 0 & 0 & 1 & 0 \end{bmatrix} \begin{bmatrix} Tr_{t-1} \\ Tr_{t-2} \\ V_{t-1} \\ V_{t-2} \end{bmatrix} + \begin{bmatrix} 1 & 0 \\ 0 & 0 \\ 0 & 1 \\ 0 & 0 \end{bmatrix} \begin{bmatrix} W_{1t} \\ W_{2t} \end{bmatrix}.$$

$$P_t = \begin{bmatrix} 1 & 0 & 1 & 0 \end{bmatrix} X_t + \varepsilon_t \tag{17.6}$$

Here, W_{1t}, W_{2t} and ε_t are all stochastic variables with an expected value of zero, unknown variances τ_1^2, τ_2^2 and σ^2, and mutually independent normal distributions

$$\begin{bmatrix} W_{1t} \\ W_{2t} \\ \varepsilon_t \end{bmatrix} \sim N \left(\begin{pmatrix} 0 \\ 0 \\ 0 \end{pmatrix}, \begin{pmatrix} \tau_1^2 & 0 & 0 \\ 0 & \tau_2^2 & 0 \\ 0 & 0 & \sigma^2 \end{pmatrix} \right). \tag{17.7}$$

Given the state space model in (17.4) and parameter $\theta=(\tau_1^2, \tau_2^2, \sigma^2, \alpha_1, \alpha_2)$, state variable X_t can be estimated using the Kalman filter. Given the system expressed in state space model (17.4), we can use estimated value $X_{t-1|t-1}$ in period $t-1$ and its variance-covariance matrix $S_{t-1|t-1}$ to obtain the predicted state value $X_{t|t-1}$ in period t and its variance-covariance matrix $S_{t|t-1}$ as follows

$$X_{t|t-1} = F X_{t-1|t-1}$$
$$S_{t|t-1} = F S_{t-1|t-1} F^t + GQG^t. \tag{17.8}$$

Here, Q is the variance-covariance matrix of system noise W_t.

Next, using the Kalman gain K in (17.9), we calculate state variable $X_{t|t}$ at period t and its variance-covariance matrix $S_{t|t}$ by filtering,

$$K = S_{t|t-1} H^t (H S_{t|t-1} H^t + R)^{-1}$$
$$X_{t|t} = X_{t|t-1} + K(P_t - H X_{t|t-1})$$
$$S_{t|t} = (I - KH) S_{t|t-1}. \tag{17.9}$$

Here, I is a unit matrix. Moreover, we set the components of the initial vector $X_{0|0}$ at zero, the initial variance-covariance matrix is $V_{0|0}$, and used a stationary distribution for the part corresponding to the autoregressive component, sufficiently

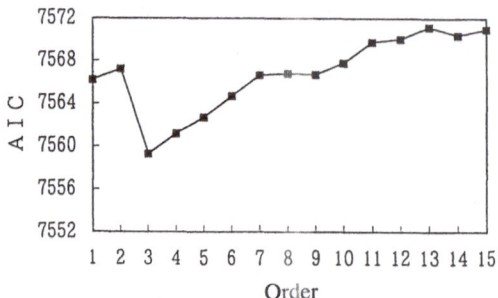

Figure 17.2   The relationships between orders of an AR model and AIC value

large values for the diagonal components corresponding to the trend component, and zero for all other components.

When we use the Kalman filter, the log likelihood function of the state space model is given as follows (Kitagawa 1993)

$$l(\theta) = -\frac{T}{2}\log 2\pi - \sum_{t=1}^{T}\log r_t - \frac{1}{2}\sum_{t=1}^{T}\frac{(P_t - HX_{t|t-1})^2}{r_t}. \qquad (17.10)$$

Here, $r_t = HS_{t|t-1}H^t + \sigma^2$. By maximizing the log likelihood in (17.10), we can obtain the maximum likelihood value $\hat{\theta}$ of parameter θ.

Furthermore, by minimizing the information criterion

$$\text{AIC} = -2l(\hat{\theta}) + 2(\text{dimension of } \theta), \qquad (17.11)$$

we can select the orders of the model.

17.2.2   Application of the Nonstationary Model to Stock Prices

We now apply the nonstationary Kitagawa-Gersch model described thus far to Toyota Motor's stock price and analyze its fluctuation characteristics.

We estimate the cyclical components with an AR model having from 1 to 15 orders. The relationships between the orders of an AR model and AIC value is the result showed in Figure 17.2. Figures 17.3 to 17.5 show the stock price and its trend line estimated using a $k=3$ model with minimal AIC value, and the AR and random fluctuation components over time. The stock price trend makes some adjustments but basically rises from the 1980s onward. Particularly during the so-called bubble economy years from 1986 to 1989, we see that the rate of increase is higher than in the early 1980s. However, from 1990, when the bubble burst, until the stock price bottomed out in 1992, the stock price trend is downward. In addition, for the cyclical component of the stock price, volatility increased as the stock price trended upward during the bubble era. The same is true of the random component.

Next, we look at the fundamentals (what can be explained by basic economic conditions) underlying the estimated stock price trend. Figure 17.6 shows the annual rate of change in the estimated trend as of June 31 of each year (this was the end of Toyota's fiscal year up to 1993), and annual rates of change for operating profit, ordinary profit, and term profit. The figure shows that changes in the stock price trend

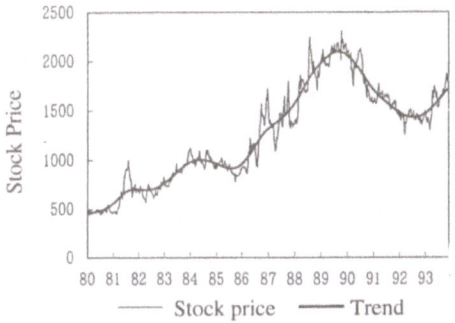

Figure 17.3 Stock price and its trend line

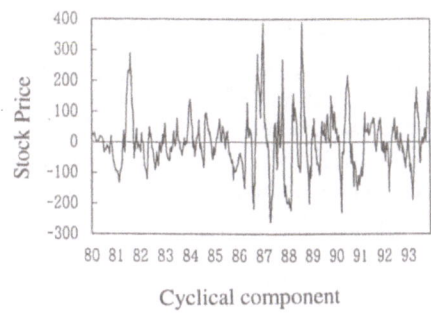

Figure 17.4 The cyclical components

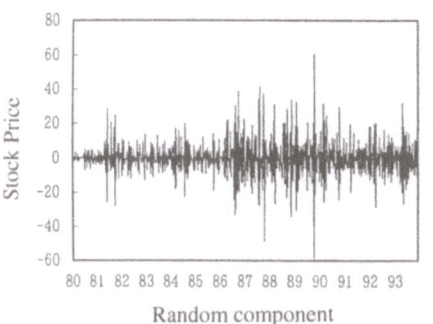

Figure 17.5 The random fluctuation components

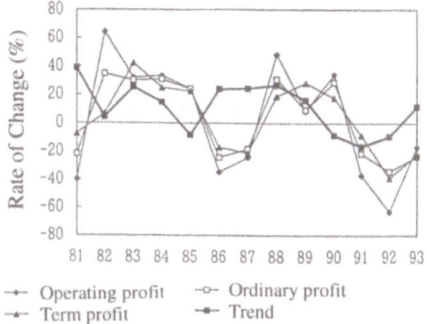

Figure 17.6 The random fluctuation components

lead changes in corporate profit by one year, and follow almost identical patterns. In other words, changes in the stock price trend are strongly linked to changes of corporate profit between the next fiscal year and the following fiscal year beginning July 1. This leading nature of stock prices with respect to corporate profit has also been clearly confirmed by Akaike (1954, p. 56) in his cross-sectional correlation analysis of stock prices and end-of-period profit rates for companies at each point in time.

Thus to predict the direction of Toyota's stock price trend in the new fiscal year, we need the predicted profit not only for the next fiscal year but for the following year as well. If corporate profit predictions can be obtained for both years, we can predict the change in the stock price trend. Information on the rate of change in the stock price trend provides investors whose asset management horizon is at least one year a valuable basis for deciding whether to invest in that stock, and how long to hold that position.

Furthermore, if we use the stock price model estimated here, we can separate the trend component and the AR component, which expresses cyclical fluctuations around the trend. Predictive information related to the AR component, which expresses cyclical fluctuations around the trend, provides critical information for short-term investment timing decisions. This application method is discussed again in Section 17.3.2.

17.3   Multivariate Analysis of the Time Series Model

Thus far we have used Toyota's stock price for a single variate time series analysis. But in reality, financial asset prices including stocks fluctuate under the influence of many factors either in the same period or with a lag, they fluctuate with cross-section correlations to these factors and time series correlation. Since a time series model such as a single variate autoregressive model do not explicitly consider factors other than its own past fluctuations, we cannot understand the relationship of other factors.

To expand the AR model so that it considers relationships with multiple factors, we use a vector autoregressive (VAR) model,

$$X_t = \sum_{l=1}^{M} A_l X_{t-l} + \varepsilon_t. \tag{17.12}$$

Here, A_l is an $m \times m$ matrix whose (i, j) components are $a_l(i, j)$, and can be called an autoregressive coefficient matrix. ε_t denotes m-dimensional white noise and satisfies the following conditions

$$E(\varepsilon_t) = \begin{bmatrix} 0 \\ \vdots \\ 0 \end{bmatrix}, \quad E(\varepsilon_t \varepsilon_t^t) = \begin{bmatrix} \sigma_{11} & \cdots & \sigma_{1m} \\ \vdots & \ddots & \vdots \\ \sigma_{m1} & \cdots & \sigma_{ll} \end{bmatrix} = U \tag{17.13}$$

$$\begin{aligned} E(\varepsilon_t \varepsilon_s^t) &= 0, \quad (t \neq s) \\ E(\varepsilon_t X_s^t) &= 0, \quad (t > s). \end{aligned} \tag{17.14}$$

Let $P(f)$ be the cross spectrum of the vector process X_t which is expressed by an $m \times m$ matrix

$$P(f) = \begin{bmatrix} p_{11}(f) & \cdots & p_{1m}(f) \\ \vdots & \ddots & \vdots \\ p_{m1}(f) & \cdots & p_{mm}(f) \end{bmatrix}. \tag{17.15}$$

If the time series is represented by the VAR model, the cross spectrum can be obtained by

$$P(f) = A(f)^{-1} W (A(f)^{-1})^*. \tag{17.16}$$

Here, A^* is the complex transpose of matrix A, and $A(f)$ is an $m \times m$ matrix whose (j, k) components are expressed by

$$A_{jk}(f) = \sum_{l=0}^{M} a_l(j, k) e^{-2\pi i l f}. \tag{17.17}$$

Furthermore, $a_0(j, j) = -1$ and $a_0(j, k) = 0$ $(j \neq k)$.

If all the components of white noise ε_t are uncorrelated, the power spectrum of component i (the i-th time series) can be expressed as

$$p_{ii}(f) = \sum_{j=1}^{m} |b_{ij}(f)|^2 \sigma_j^2. \tag{17.18}$$

Here, $\sigma_j^2 = \sigma_{jj}$, and $b_{ij}(f)$ are the (i, j) components of the inverse matrix of $A(f)$. The Akaike's relative power contribution ratio $r_{ij}(f)$ can be expressed by

$$r_{ij}(f) = \frac{|b_{ij}(f)|^2 \sigma_j^2}{p_{ii}(f)}. \tag{17.19}$$

This ratio expresses the proportion of the fluctuation of $X_t(i)$ at frequency f caused by $\varepsilon_t(j)$. The accumulated value of the ratio is called the cumulative relative power contribution ratio. Incidentally, $|b_{ij}(f)|^2 \sigma_j^2$ expresses the part of the power spectrum of component i caused by noise from component j, and is called the absolute power contribution ratio.

In the next section, we analyze the empirical results of the VAR model analysis of correlations among price fluctuations of several stocks.

17.3.1   VAR Model Analysis of Price Fluctuations of Selected Stocks

In the previous section, we used the Kitagawa-Gersch nonstationary model to separate Toyota's stock price fluctuations into a trend component, cyclical component moving around the trend, and random component, and examined the fundamental meaning of the trend component. Now we will use the same method to separate the trend component from the cyclical component for selected stocks, and then conduct a multivariate analysis of the cyclical component. We use the VAR model to see how each stock's cyclical component is influenced by price fluctuations of other stocks.

The analysis covers the period from the first week of January 1980 to the fourth week of August 1993, and uses end-of-week stock prices of four stocks-Nissan Motor Co., Toyota Motor Corp., Mazda Motor Corp., and Honda Motor Co. The stock prices and trend components of each stock are shown in Figure 17.7. The graphs show that stock price trends are similar for Nissan and Mazda on the one hand, and Toyota and Honda on the other.

Figure 17.8 shows the power spectra and absolute power contribution ratios obtained from a 4-variable VAR model of the cyclical components of the four stocks. A basic assumption in power contribution ratio theory is the non-correlation of the white noise component. In our case, since the maximum correlation ratio is 0.6 (for Toyota and Honda), this assumption does not hold. But we shall assume that the assumption is satisfied and proceed with the calculations below.

Figure 17.9 shows the cumulative relative power contribution ratios. In each graph, the four curves (in order from top to bottom) correspond to the contributions of Honda, Mazda, Toyota, and Nissan. For Toyota and Honda, their own stock price fluctuation components contribute greatly at any frequency, and receive little influence from other stocks. For Nissan, the power spectrum is at a maximum in the low frequency range (0-0.025, with a cycle of at least 40 weeks), where Honda exhibits a relative power contribution (approximately 10-40%). Likewise, when Mazda's power spectrum is maximized (0.042, approximately 24-week cycle), Nissan's relative power contribution is also maximized at 20%, and Honda's relative power contribution is approximately 5-20% in the low frequency range. These results suggest that the cyclical components of these stocks are influenced by the other stocks.

The above empirical analysis reveals new findings on the correlations among the cyclical components of the four automakers' stock prices. Further study is needed to determine what market structures are being expressed by these findings.

Empirical analysis using economic time series data such as stock prices must take into account the fact that results depend heavily on sampling intervals and periods

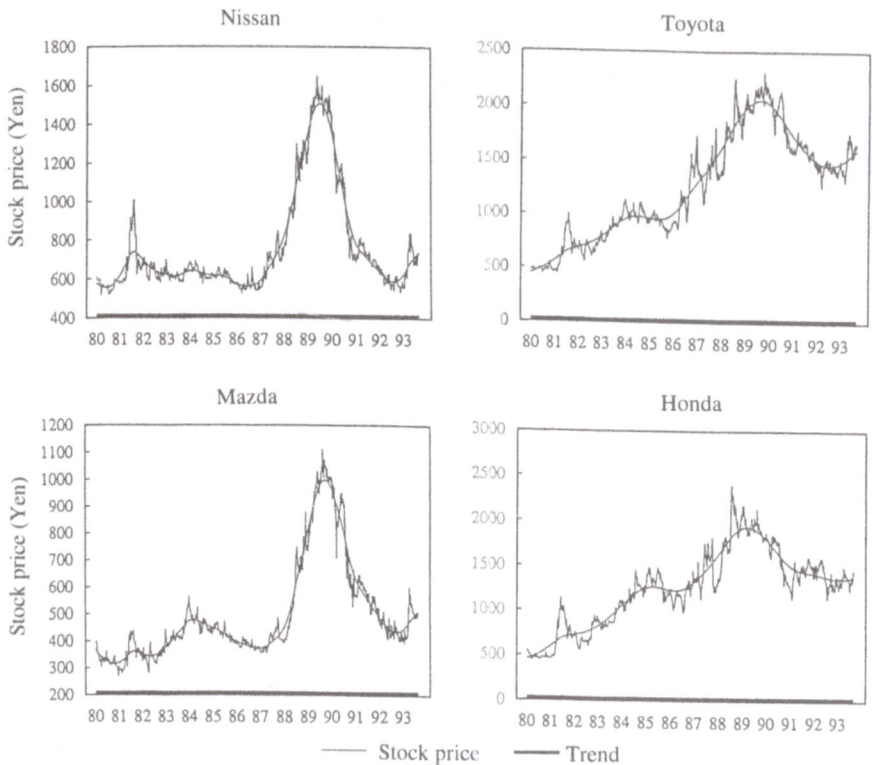

Figure 17.7   The stock prices and trend components of automakers

covered. Our results were obtained from weekly data during the period from the first week in January 1980 to the fourth week in August 1993. To ascertain the mutual interaction between stock price fluctuations of different stocks, we would need to redo the analysis using different sampling periods and time periods, and at the same time develop an explanation that conforms with market structures.

17.3.2   Analysis and Prediction of Industry Indexes

Generally in the stock market, stock groups that share certain features have similar observable stock price patterns. This phenomenon is caused by common factors that affect the stock group's price formation.

In this section, we will analyze how the prices of stock groups, especially their cyclical price fluctuations, mutually affect stock groups in other industries. Our unit of analysis is not individual stock prices but nine industry stock price indexes selected from the 33 industry classifications of the TSE: (1) construction; (2) steel products; (3) nonferrous metals; (4) machinery; (5) transport equipment; (6) precision instruments; (7) electric power and gas; (8) land transport; and (9) marine transport.

We chose to use industry indexes instead of individual stock prices because we assumed that if stock prices in one industry affect prices in another industry, then the influence would be easier to detect using industry indexes, which are an overall

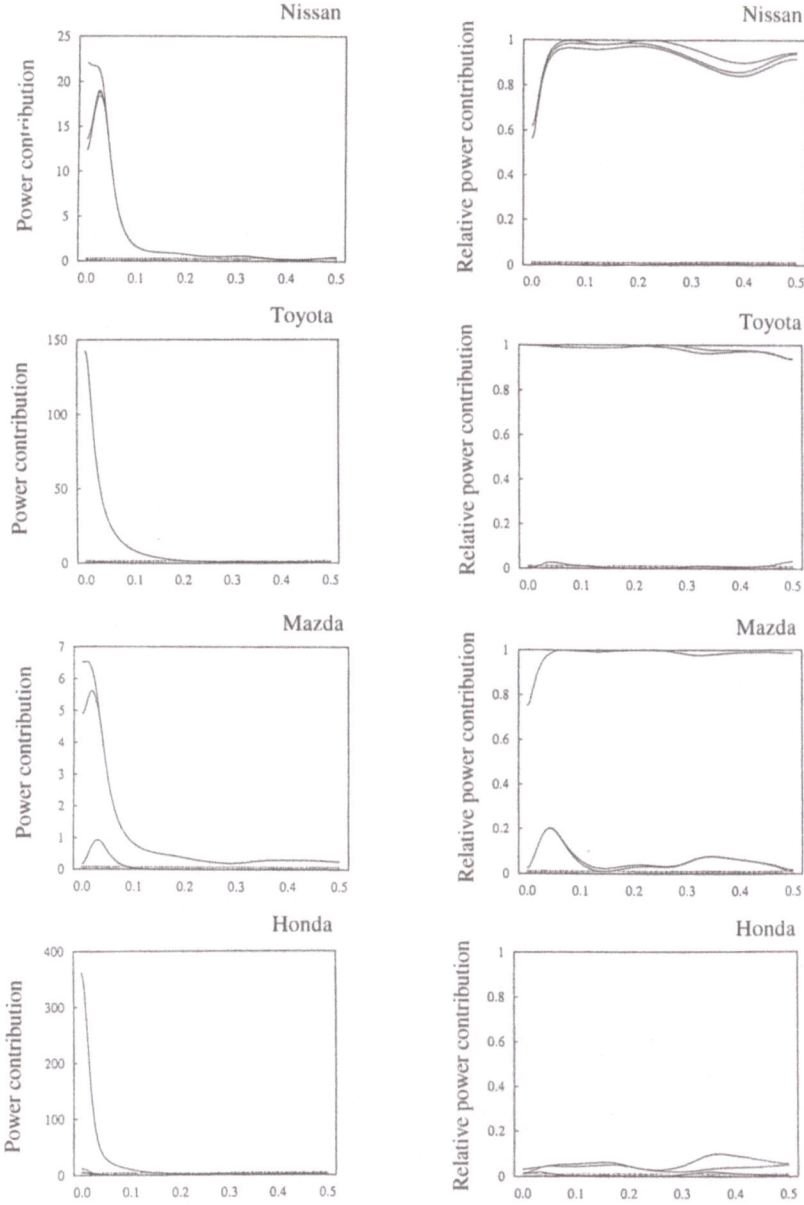

Figure 17.8   The power spectra

Figure 17.9   Cumulative relative power contribution

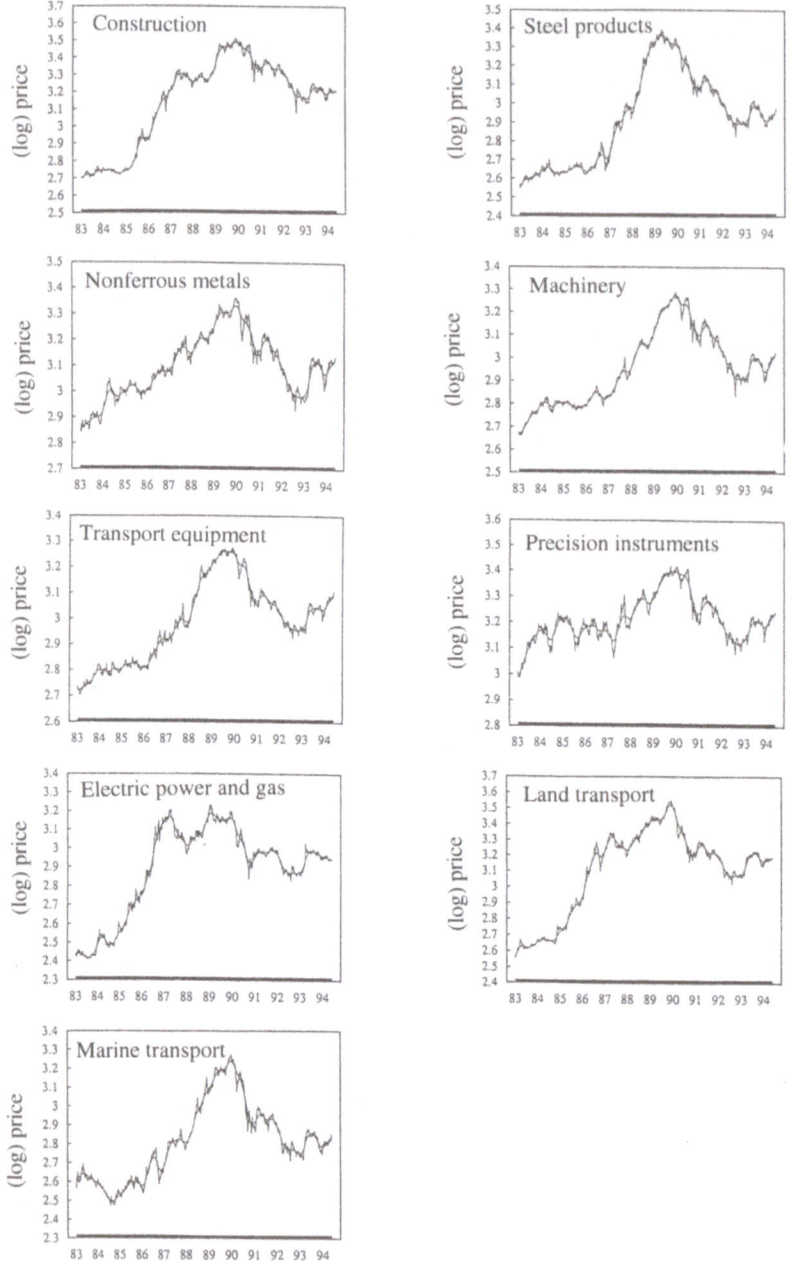

Figure 17.10   Power spectra of cyclical components

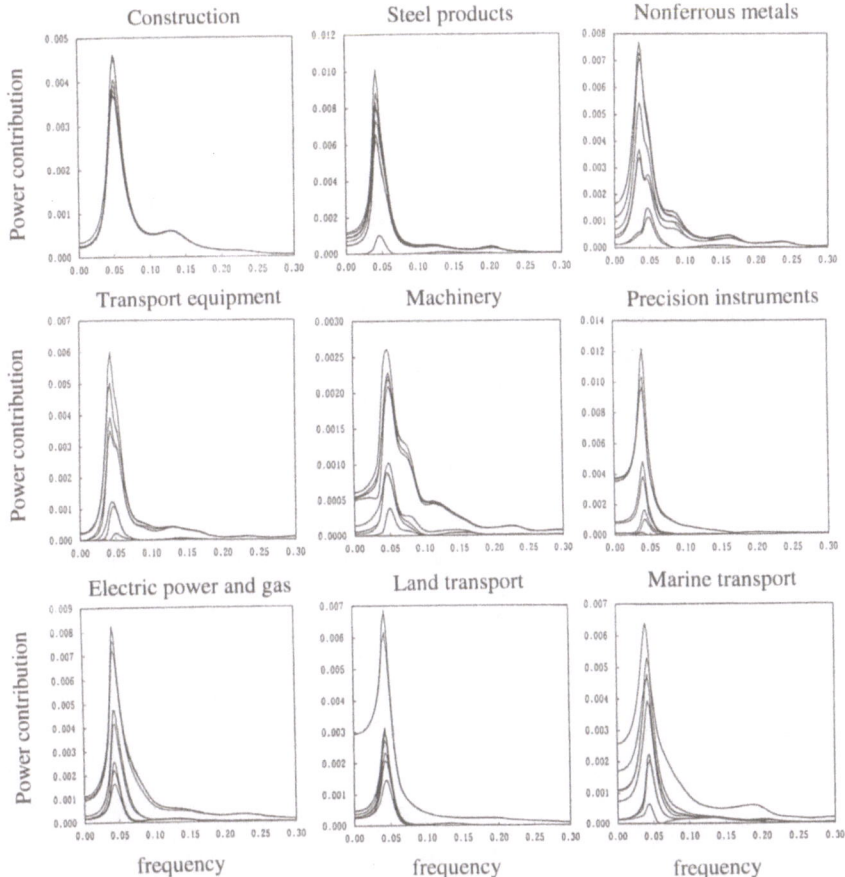

Figure 17.11 Cumulative relative power contribution ratios

Table 17.1 Each industry's relative power contribution ratio at the frequency of maximum power spectrum

	Const.	Steel	Metals	Machin	Trans. equip	Prec. instr.	Elec. gas	Land transp.	Marine transp.
Construction	0.80	0.09	0.07	0.01	0.15	0.01	0.20	0.21	0.10
Steel	0.01	0.53	0.01	0.14	0.17	0.06	0.07	0.09	0.21
Metals	0.00	0.03	0.37	0.05	0.01	0.05	0.04	0.04	0.00
Machinery	0.02	0.07	0.04	0.39	0.07	0.19	0.19	0.06	0.04
Transport	0.00	0.06	0.23	0.07	0.42	0.08	0.07	0.03	0.29
Instruments	0.02	0.03	0.22	0.17	0.04	0.40	0.01	0.02	0.12
Elec. & gas	0.03	0.00	0.03	0.01	0.02	0.00	0.30	0.01	0.02
Land transp.	0.10	0.05	0.05	0.00	0.02	0.05	0.05	0.44	0.08
Marine transp.	0.02	0.12	0.00	0.16	0.11	0.15	0.07	0.10	0.13
Frequency	0.050	0.042	0.035	0.042	0.050	0.040	0.042	0.042	0.042

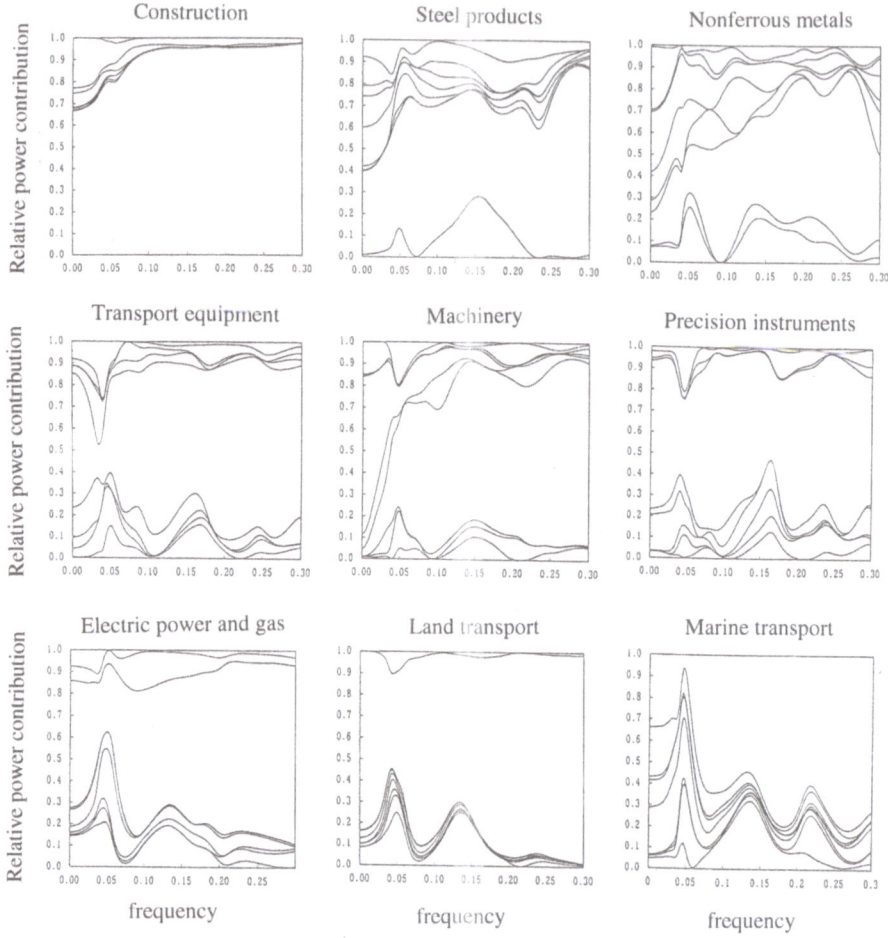

Figure 17.12   Cumulative relative power contribution

weighted market price average of specific stocks, rather than individual stock prices. The reason is that with industry indexes, specific factors affecting individual stocks tend to offset each other, and thus price fluctuations better reflect factors at the industry level. Similar to the analysis of individual stocks in the previous section, we use a nonstationary model to break down fluctuations in the industry indexes into a trend and cyclical component. And then use a VAR model to determine how cyclical fluctuations in each industry index is affected by cyclical fluctuations in other industry indexes.

For each of the nine industry indexes, Figure 17.10 shows actual end-of-week prices from the first week in January 1983 to the fourth weak in May 1994, as well as the respective trend components. Figure 17.11 shows the power spectrum and absolute power contribution ratio of the cyclical component of each industry index. As with the analysis of individual stocks in the previous section, while we again encounter a slight problem with the non-correlation assumption of the white noise component, we

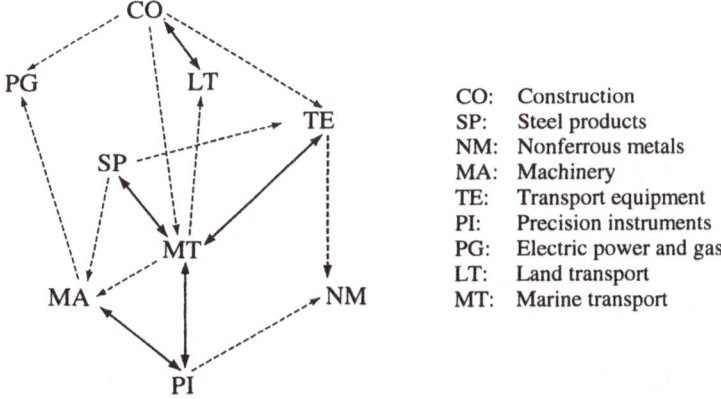

Figure 17.13   Inter-industry influences in fluctuation components

assume that this condition is approximately satisfied and proceed with the analysis. Figure 17.12 shows the cumulative relative power contribution ratios. In each graph, the curves correspond (in order from top to bottom) to marine transport, land transport, electric power and gas, precision instruments, transport equipment, machinery, nonferrous metals, steel products, and construction. Each industry's relative power contribution ratio at the frequency of maximum power spectrum is shown in Table 17.1. Most of the industry indexes attain maximum power spectra in the 0.04–0.05 frequency range. This indicates that the cyclical components of these industry indexes follow a 20–25 week cycle. From the results shown in Figures 17.11 and 17.12 and Table 17.1, we surmise that the industry indexes influence each other in the following ways.

1) Except for marine transport, each industry index's largest fluctuation component is its own particular fluctuation component.

2) The fluctuation components of construction, steel products, machinery, transport equipment, precision instruments and marine transport indexes have a large impact on other industry indexes. On the other hand, the fluctuation components of nonferrous metals and electric power and gas have little impact on other indexes.

3) Based on Table 17.1, in Figure 17.13 we diagram inter-industry influences with relative power contributions of at least 10%. Marine transport is influenced by many of the other indexes, and at the same time influences the other indexes.

4) In addition, transport equipment and precision instruments, which are characterized by high export ratios, both mutually interact with marine transport, and also have a one-way impact on nonferrous metals.

5) There are mutual influences between construction, whose performance is affected by public works spending, and land transport, and between machinery and precision instruments, both of which are part of the processing industry.

The results of the empirical analysis thus show both mutual and one-sided influences among the industry indexes, and in turn among the price fluctuations of

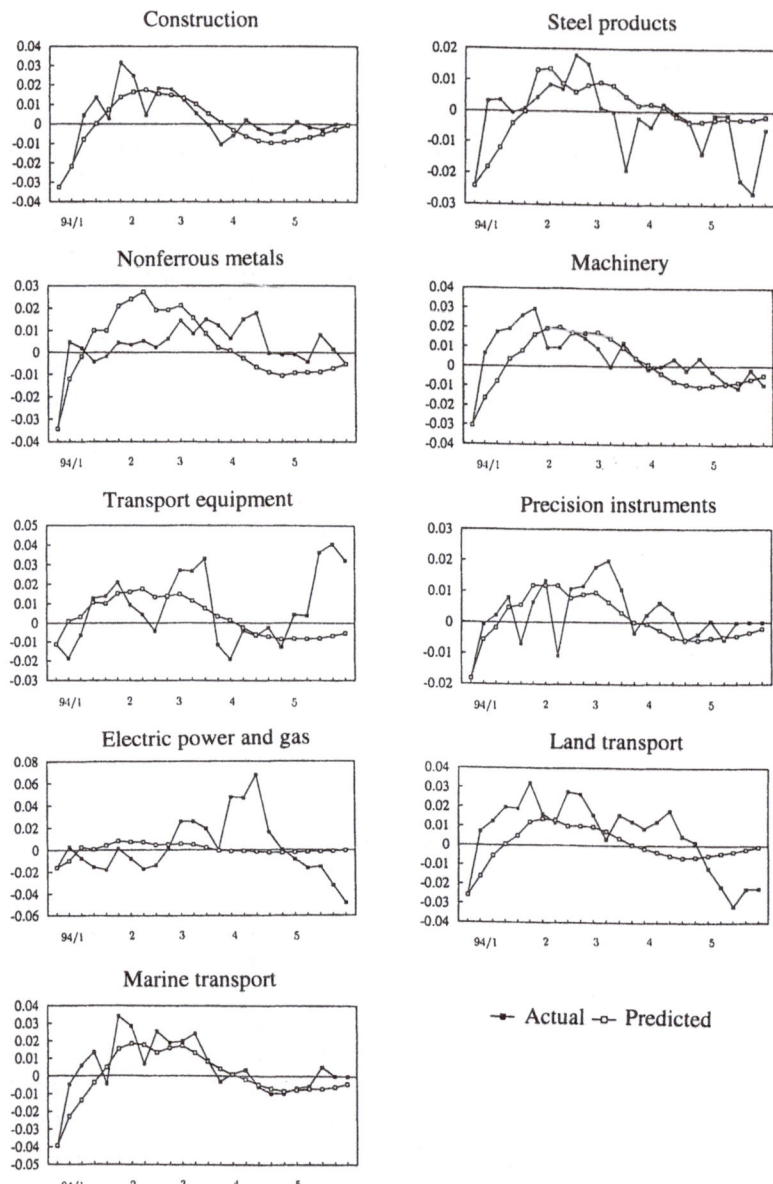

Figure 17.14   The prediction of cycle fluctuations for each industry index

individual stocks comprising the indexes. Confirmation of mutual relationships such as these will contribute to a better understanding of the economics of the stock market.

Finally, Figure 17.14 shows the most interesting practical results of our analysis–the prediction of cyclical fluctuations for each industry index using the VAR model. The simulated prediction begins in the first week of January 1994, and produces good short-term prediction results for industry indexes such as construction, machinery, land transport, and marine transport.

17.4　Conclusion

Until now, time series analyses of financial and securities data have often used autoregressive models that are premised on stationary processes. However, since the prices of most financial assets such as stocks are often nonstationary time series characterized by trends, time series models premised on stationary processes are of limited use. The Kitagawa-Gersch nonstationary model has flexibility for application to a wide variety of models, objectivity with regard to model estimation, and as we have discussed in this paper, compatibility to be combined with VAR models. It is thus an effective approach for explaining the price mechanisms of financial assets.

In the future, the increasing globalization and liberalization of economies and financial markets will pose more diverse and complex risks to asset managers of pensions, life insurance and investment trusts. We expect this to stimulate further fruitful research on time series models of financial asset prices.

References

Akaike, H. (1971), "Autoregressive Model Fitting for Control," *Annals of the Institute of Statistical Mathematics*, Vol.23, 163–180.

Akaike, H. (1974), "A New Look at the Statistical Model Identification," *IEEE, Transactions on Automatic Control*, Vol. AC-19, 716–723.

Akaike, H. (1980), "Likelihood and the Bayes Procedure", *Bayesian Statistics*, J.M. Bernards, M.H. De Griot, D.V. Lindley, and A.F.M. Smith, eds., University Press, Spain, 143–166.

Gersch, W. and Kitagawa, G.: The prediction of time series with trends and seasonalities, *Journal of Business and Economic Statistics*, 1, 253–264, 1983.

Kitagawa, G. and Gersch, W.:A smoothness priors-state space modeling of time series with trend and seasonality, *Journal of the American Statistical Association*, 1984

Tsuda, H. (1992), "The Application of a non-stationary model to the stock movement", Symposium of new time series analysis and application.

Tsuda, H. (1994), "Prediction of stock movement using a non-stationary model", Conference of an applied economic time series.

Tsuda, H. (1994), "Statistics of Equity," Asakura Shoten (in Japanese).

Yamamoto, T. (1988), "The time series analysis of Economics," Sobunsha (in Japanese).

Chapter 18

Dynamic Analysis of Economic Time Series

Sadao Naniwa
Kansai University of International Studies
1-18 Aoyama, Shijimi-cho, Miki-shi, Hyogo
naniwa@kuins.ac.jp

18.1   Introduction

Controversy is liable to be evoked on the matter to what extent economics is of use for enhancement of people's economic welfare. Economics not only should possess the double sides as normative science and positive science but also should make judgment from normative viewpoint on the actual economy, playing a role of providing political suggestions. Meanwhile the empirical evidence of economic activities based on the actual data is laid much stress on as positive science. In consideration of the fact that the economic theory is forced to make development with confrontation to the economic phenomena most characteristic of a period, a high possibility is also pointed out that necessity of the traditional theory being replaced with a new one. That is the reason why the empirical analysis has important role and the economics is called empirical science.

The most of economic data used in the empirical analyses are time series data which are fluctuated irregularly with the lapse of time. Also with the aggregated economic data, the number of the available samples is restricted resulting not a few statistic errors. To use the economic data having such features, statistical methods are employed. A typical example is the econometrics which aims to systematize the method based on the knowledge of the economics, statistics and mathematics. However as described, for example, by Mankiw (1990), a macro-econometric model and traditional macro-economics have given rise to serious breakdown in the double sides of failure in empirical analysis and flaw in theory in recent years, and came into the age of confusion, disruption, and unrest. It is also pointed out that despite the diversification of the economic action or phenomenon which is to be an objective for the analysis and despite the rapid change in the economic theory itself, none of so great a change is seen in the traditional econometric methods. In the policy analyses by large-scale econometric models as well, its credibility has been exposed to criticism.

Under such a situation, the use of time series models based on stochastic process has called attention since the 1970s. A focus of the analysis by the time series model resides in grasping the system's features in a relation of the dynamic feedback and the method shows applicability to the analysis of the economic system. Taking account of the fact that the human activity itself existing behind economic phenomena is greatly influenced by the expectation, desire, etc. at the time in addition to the exogenous shocks, it is natural to intend to grasp the movement of the economic series where economic phenomena are converted into numerical data in a dynamic feedback relation.

To seize the features of the actual economic behaviors, it is necessary to analyze the fluctuation around trend keeping the trends in mind. Conventionally by assuming simply the trend to be linear, analysis of the movement around the straight line has much attempted. However as pointed out by Havenner and Swamy (1978), it also become necessary to understand the longer-term trend itself as a stochastic process. Furthermore in case of the structural changes occurred in the economic system, examining of the period of the changes and study the situation before and after these periods are also an important on the empirical analyses.

In this chapter, the result by applying the time series models recently developed to estimated the trend of the economic time series and the analyses of the dynamical changes around the trend are discussed.

18.2 Trend of the Economic Time Series and the Fluctuation around the Trend

The trend of the economic variables and the fluctuation around the trend reflect the economic activity. Thus, the analyses of the essential features of the major variables are counted for much as the preliminary study to understand the economic condition. In this section, the stochastic trend and the fluctuation around the trend are estimated by the models developed by Kitagawa and Gersch (1984, 1985) considering these features are varying on time. The critical characteristics of the models are the use of prior information in the Bayesian approach (Akaike 1986, Akaike 1980).

18.2.1 Estimation of the Change of the Trend

Consider the observed time series y_n is additively decomposed into the trend factor t_n, the cyclical or shorter-term fluctuation factor x_n, the seasonal factor s_n, and the irregular factor v_n at time n as

$$y_n = t_n + x_n + s_n + v_n. \tag{18.1}$$

The smoothness priors approach original that the trend is assumed to be represented by stochastically perturbed difference equation constraints as

$$\nabla^k t_n = v_{1n}, \quad v_{1n} \sim N(0, \tau_1^2) \quad i.i.d. \tag{18.2}$$

where ∇ is the difference operator $\nabla t_n = t_n - t_{n-1}$, k is the order of the difference, and v_{1n} is assumed to be white noise. The smoothness of the trend is depends on k and τ_1^2. The cyclical factor x_n is assumed to be in the autoregressive model with the order p as

$$\begin{aligned} x_n &= \alpha_1 x_{n-1} + \alpha_2 x_{n-2} + \cdots + \alpha_p x_{n-p} + v_{2n} \\ v_{2n} &\sim N(0, \tau_2^2) \quad i.i.d. \end{aligned} \tag{18.3}$$

The seasonal factor s_n changes gradually and repeatedly every year in almost the same pattern and also assumed to be stochastically perturbed difference equation making use of the smoothness priors that the total sum of the year is close to zero as

$$\begin{aligned} s_n &= -(s_{n-1} + s_{n-2} + \cdots + s_{n-L+1}) + v_{3n} \\ v_{3n} &\sim N(0, \tau_3^2) \quad i.i.d. \end{aligned} \tag{18.4}$$

where L is the seasonal cycle and the change of the seasonality is given by the stochastic term v_{3n}.

These factors are expressed by a state space model as

$$\begin{aligned} z_n &= F z_{n-1} + G v_n \\ y_n &= H z_n + w_n, \end{aligned} \tag{18.5}$$

where z_n is the state vector, F, G and H are coefficient matrices and v_n, w_n are the stochastic terms. These are given as follows;

$$F = \left[\begin{array}{c|c|c} \begin{matrix} C_1 & \cdots & C_{k-1} & C_k \\ 1 & \cdots & 0 & 0 \\ \vdots & \ddots & \vdots & \vdots \\ 0 & \cdots & 1 & 0 \end{matrix} & 0 & 0 \\ \hline 0 & \begin{matrix} \alpha_1 & \cdots & \alpha_{p-1} & \alpha_p \\ 1 & \cdots & 0 & 0 \\ \vdots & \ddots & \vdots & \vdots \\ 0 & \cdots & 1 & 0 \end{matrix} & 0 \\ \hline 0 & 0 & \begin{matrix} -1 & \cdots & -1 & -1 \\ 1 & \cdots & 0 & 0 \\ \vdots & \ddots & \vdots & \vdots \\ 0 & \cdots & 1 & 0 \end{matrix} \end{array} \right],$$

$$G = \left[\begin{matrix} 1 & 0 & 0 \\ 0 & 0 & 0 \\ \vdots & \vdots & \vdots \\ 0 & 0 & 0 \\ 0 & 1 & 0 \\ 0 & 0 & 0 \\ \vdots & \vdots & \vdots \\ 0 & 0 & 0 \\ 0 & 0 & 1 \\ 0 & 0 & 0 \\ \vdots & \vdots & \vdots \\ 0 & 0 & 0 \end{matrix} \right], \quad z_n = \left[\begin{matrix} t_n \\ \vdots \\ t_{n-k+1} \\ x_n \\ \vdots \\ x_{n-p+1} \\ s_n \\ \vdots \\ s_{n-L+2} \end{matrix} \right], \quad v_n = \left[\begin{matrix} v_{1n} \\ v_{2n} \\ v_{3n} \end{matrix} \right], \tag{18.6}$$

$$H = \begin{bmatrix} 1 & 0 & \cdots & 0 & 1 & 0 & \cdots & 0 & 1 & 0 & \cdots & 0 \end{bmatrix},$$

where C_i $(i = 1, \cdots, k)$ reflects trend constraints in (18.2) corresponding to the order k.

18.2.2 Dynamic Expression of the Fluctuation around the Trend

Change of the characteristics of the fluctuation or cyclical movement around the trend is expressed by the time-varying autoregressive model. Suppose that the estimated trend at n from N be $t(n|N)$. Then a time-varying coefficient autoregressive model

$$z_n = y_n - t(n|N) \tag{18.7}$$

is expressed as

$$z_n = \sum_{i=1}^m a_{i,n} z_{n-i} + \varepsilon_n, \quad \varepsilon_n \sim N(0, \sigma^2) \quad i.i.d., \tag{18.8}$$

where $a_{i,n}$ is the time varying autoregressive coefficient at n, m is the order of the model and ε_n is the white noise with mean zero and variance σ^2. The coefficients $a_{i,n}$ are assumed to change gradually with time which is obtained by the perturbed difference equation as

$$\nabla^k a_{i,n} = \delta_{i,n}, \quad \delta_{in} \sim N(0, \tau^2) \quad i.i.d., \tag{18.9}$$

where k is the order of the difference, $\delta_{i,n}$ the stochastic term with mean zero and variance τ^2. The model (18.8) and (18.9) are expressed as the state space model. Suppose the state vector x_n is expressed as ($'$ indicates the transposition)

$$x_n = [a_{1,n}, \ldots, a_{m,n}, \ldots, a_{1,n-k+1}, \ldots, a_{m,n-k+1}]', \tag{18.10}$$

then the state space representation is given as

$$
x_n =
\begin{bmatrix}
a_{1,n} \\
\vdots \\
a_{m,n} \\
a_{1,n-1} \\
\vdots \\
a_{m,n-1} \\
\vdots \\
a_{1,n-k+1} \\
\vdots \\
a_{m,n-k+1}
\end{bmatrix}
=
\begin{bmatrix}
C_1 I_m & \cdots & C_{k-1} I_m & C_k I_m \\
I_m & \cdots & 0 & 0 \\
\vdots & \ddots & \vdots & \vdots \\
0 & \cdots & I_m & 0
\end{bmatrix}
\begin{bmatrix}
a_{1,n-1} \\
\vdots \\
a_{m,n-1} \\
a_{1,n-2} \\
\vdots \\
a_{m,n-2} \\
\vdots \\
a_{1,n-k} \\
\vdots \\
a_{m,n-k}
\end{bmatrix}
+
\begin{bmatrix}
I_m \\
0 \\
\vdots \\
0
\end{bmatrix}
\begin{bmatrix}
\delta_{1,n} \\
\vdots \\
\delta_{m,n}
\end{bmatrix}
$$

$$z_n = [\overbrace{z_{n-1}, \ldots, z_{n-m}}^{m}, \overbrace{0, \ldots, 0}^{(k-1) \times m}] x_n + \varepsilon_n, \tag{18.11}$$

where C_i is the coefficient matrix and I_m is the $m \times m$ unit matrix. In the case of the variance of the system noise $\delta_{i,n}$ is partially increased, the step-like change emerges at that time. Whether the smoothly changing model is adequate or the step-like change model is selected by the information criterion AIC.

Using result of the estimated model (18.8), the changing spectrum on the frequency domain can be intuitionally comprehended. The time-varying spectrum can be shown as

$$p_{f,n} = \frac{\sigma_n^2}{\left|1 - \sum_{j=1}^m a_{j,n} \exp\left(-2\pi i j f\right)\right|^2}, \quad (-1/2 \le f \le 1/2) \tag{18.12}$$

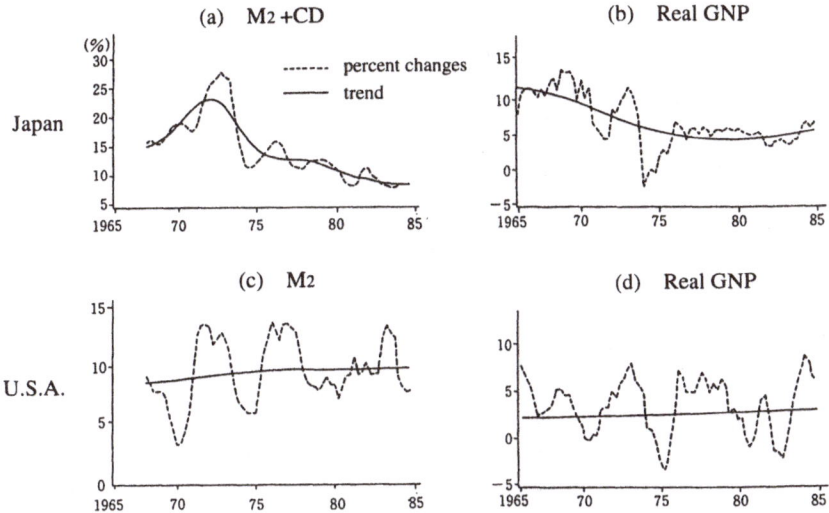

Figure 18.1 Percent changes and trends of the money supply and real GNP in Japan and the US

18.2.3 Examples: Real GNP and Money Supply in Japan and the US

As Friedman (1985) asserted the stability in Japan's economic growth since the 1970s is due to the stable monetary growth, while unstable monetary policy of the US caused instability in the economy. This can be shown as an empirical evidence by examining the characteristics of the fluctuation of the real GNP and the money in both Japan and the US. To analyze, the model (18.1) and (18.2) are applied.

Figure 18.1 depicts the percent changes of the real GNP and money supply and their trends in Japan and the US respectively, from the 1960s to the 1980s. Figure 18.2 shows the time varying spectra. In Figure 18.2, the ordinate indicates the strength of the frequency component and the abscissa denotes the frequency, i.e. a reciprocal of the cycle and the right side one gives shorter cycle. The figure reveals that the greater movement of the cyclical component in the medium or shorter term is noted with Japan's money supply ($M_2 + CD$) to the mid 1970s, but the longer cyclical component is drastically increased in 1973–4 resulting in gradual decrease after that period. The figure also reveals that the real GNP, with which greater cyclical movement until the mid 1970s, is greatly decreased since at the period. The step-like changes in the mid 1970s noticed in Japan's both variables suggest, that structural shift seems to be generated in a direction of the stability of the money supply. On the other hand, no tendency is noted that variation will decrease US money supply (M_2) and a that larger fluctuation can be seen in the real GNP though exhibiting relatively stability in the longer term.

Although these results suggest that structural shift is caused in the real GNP with the stabilization of the money supply since the 1970s, it can not necessarily be said that the stabilization of the money supply expresses causality to stabilize the real GNP. Also, there is a great possibility that there will exist a relationship reverse to the above (Horie-Naniwa 1990). In the meantime, further investigation is required in relation to the fact that the medium or shorter term cycle of Japan's money supply is

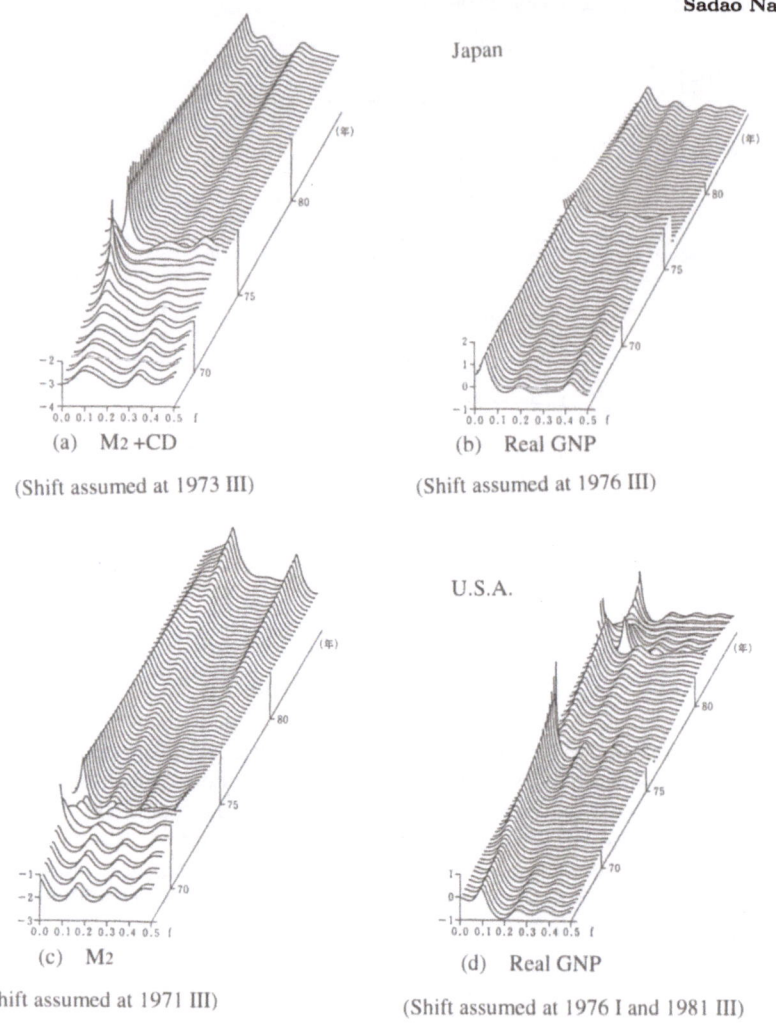

Figure 18.2 Change in the cycle components of the money supply and real GNP in Japan and the US

slightly increased in the 1980s. However in the US, none of the structural shift appears in a direction of the stability of the money supply, and it seems still likely that the real economic activity continues greater variation. However, the results suggest that Friedman's assertion until around the mid 1980s.

18.3 Analysis of Abrupt Change of Trend

18.3.1 Non-Gaussian Model

Kitagawa (1987) introduced a non-Gaussian state space model which is possible to reproduce both abrupt and gradual change of time series. The model is assumed that the stochastic term v_n of the state equation and w_n of the observation equation are independent of each other and are not necessary be Gaussian. On the assumption that the conditional distribution of the state vector when the state x_{n-1} is given at $n-1$ is q and that of the observed value y_n when x_n is given is r, the state space representation is generally expressed as

$$
\begin{aligned}
x_n &\sim q(\cdot|x_{n-1}) \\
y_n &\sim r(\cdot|x_n).
\end{aligned}
\qquad (18.13)
$$

The expression (18.13) can be called a general state space model in which the models described in the previous section are included.

For the non-Gaussian state space model, using the conditional probability, the recursive formulas of the predict or the filter and the smoother are as follows:

One Step ahead Prediction

$$
p(x_n|Y_{n-1}) = \int_{-\infty}^{\infty} q(x_n|x_{n-1})p(x_{n-1}|Y_{n-1})dx_{n-1}
\qquad (18.14)
$$

Filtering

$$
p(x_n|Y_n) = \frac{r(y_n|x_n)p(x_n|Y_{n-1})}{p(y_n|Y_{n-1})}
\qquad (18.15)
$$

where

$$
p(y_n|Y_{n-1}) = \int_{-\infty}^{\infty} r(y_n|x_n)p(x_n|Y_{n-1})dx_n.
\qquad (18.16)
$$

and $Y_m = \{y_1, y_2, \ldots, y_m\}$ is the observed value in consideration of the joint distribution x_n and x_{n+1}

$$
p(x_n, x_{n+1}|Y_N) = \frac{p(x_{n+1}|Y_N)q(x_{n+1}|x_n)p(x_n|Y_n)}{p(x_{n+1}|Y_n)},
\qquad (18.17)
$$

Given the set of the whole observed values Y_N,

Smoothing

$$
p(x_n|Y_N) = p(x_n|Y_n) \int_{-\infty}^{\infty} \frac{p(x_{n+1}|Y_N)q(x_{n+1}|x_n)}{p(x_{n+1}|Y_n)} dx_{n+1}
\qquad (18.18)
$$

is obtained.

If the model is linear and the Gaussian type, then the conditional distribution is normal distribution and Kalman filter can be used for the estimation of the mean and covariance. Kitagawa (1989) proposes the use of the Gaussian-sum filter that is approximated by the mixture of the Gaussian distribution for the high-degree non-Gaussian state space model.

18.3.2   Analysis of the Changes of Stock Price in Japan and the US

In most cases, the changes of the stock prices can be assumed to reflect the efficient market hypothesis where all the obtainable information can be utilized, also uses the information concerning the external financial asset market are assumed. Thus, no stabilization can be seen in the trend of prices. Keeping such a situation in mind, the stock price changes of Japan and the US are analyzed by Kitagawa's non-Gaussian model.

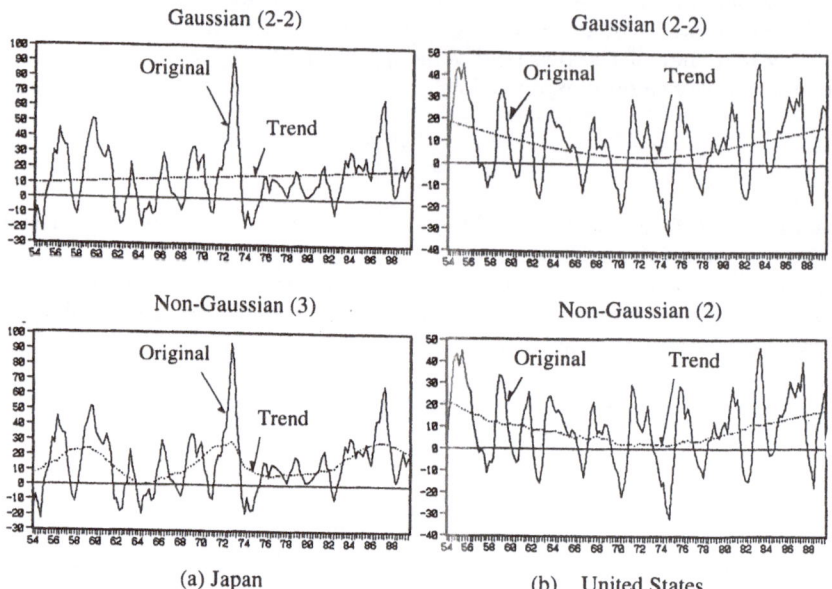

Figure 18.3 Trend by a Gaussian model and non-Gaussian model (with the order of the trend and AR in the parentheses)

Stock prices vary in reflection of the business cycle, but the cyclical movement or amplitude is subject to change depending on time in both Japan and the US. Also in recent years, it is said that concurrence in the variation in Japan and the US has been strengthened. To examine such a pattern, the trend of the stock price variation in Japan and the US was estimated. The results are shown in Figure 18.3.

In the estimation of the non-Gaussian model, the following procedure was taken. (1) The initial values of the parameters such as the order, coefficient, variance, etc. are estimated with a Gaussian model selected by AIC. (2) Since the distribution of the noise of the non-Gaussian model is expressed by the weighted sum of the normal distribution with different variance, estimation is made by changing the combination of the probability distribution having great variance and small variance. (3) To select the non-Gaussian model, AIC is used as the criterion

The results in Figure 18.3 suggest that Japan's stock prices estimated by the Gaussian model show gradual change in the longer term. On the contrary, drastic changes are shown by the non-Gaussian model. This larger changes in the trend suggests that the prices reflect the medium term cycles of fundamental condition of economy. On the other hand, none of great difference is noted in the trend in the non-Gaussian model as well the Gaussian model in the US. The results suggest the US trend is considered to be longer than the one in Japan. The non-Gaussian model can be utilized to estimate the change of the longer term growth power or business condition in the economy, The model gives the possibility to seize the cyclical pattern more explicitly.

18.4   Analysis of the Economic System by a Multivariate Nonstationary Time Series Model

18.4.1   Economic System and Dynamic Characteristics

To analyze the economic system where many of factors are influenced each other, a multivariate time series model should be used. The premise is that the factory composing the economic system are in a relationship of a dynamic feedback where the variables are mutually concerned with themselves, and therefore no assumption should be used in advance such as endogenous variables, internal causes composing the system, or exogenous variables, causes acting on the system from the external side. The exogenous natures of variables are examined by the model analysis.

A practically fitted multivariate model would give an implication which variables are main causes in the system by simulation analyses on cutting off the feedback relationship. Also it can be used to explore policy implication. Oritani (1981), for example, has shown the negative effects of inflation on the economic growth by means of the multivariate autoregressive model based on three Japan's variables of the money supply, prices, and real GNP. The result shows that the higher inflation rate owing to the increase of money supply makes lower the production activity. This evidence is against the discussion that approval of the higher price induces higher the real production activity. Using this model, simulation is also made to examine how the economic activity will be influenced when money supply changes. The results suggest the negative effect to economic growth by the inflation which is caused by the increase of money supply through the dynamic movement of the variables. Horie-Naniwa (1990) applied the multivariate autoregressive model to analyze the relationship between the financial activity and the real economy in the recent economic environment and to the analysis of the stock price variation in the major countries including Japan and the US.

When a multivariate model have higher prediction performance, it can be applied to the control problem in economic systems to find policy effect changing policy instruments. The relationships between the government and the private sector and the relationships among the policies that can also be examined. Examples of control problem are shown in Hiromatsu-Naniwa (1990). An analysis of the structure of the Japanese economy from a point of view of control based on the multivariate autoregressive model is given by Akaike (1989).

18.4.2   Nonstationary Multivariate Model

In the application of the multivariate model introduced above, economic time series are converted into an stationary series by excluding the trend from the original series. This is because there can be hard to find practical non-stationary model at that time. Kato-Naniwa and Ishiguro (1993) try to apply a nonstationary multivariate model as follows.

Yet the k-dimensional observed time series vector be y_n as

$$y_n = (y_{1n}, \ldots, y_{kn})', \quad (n = 1, \ldots, N). \tag{18.19}$$

This vector (18.19) is expanded form of (18.1). Consider here the quarterly series and a 2-variable state vector. Assuming the order of the difference of the trend (d) and the order of the autoregressive factor (m) is 2, respectively and the seasonal cycle L is 4.

Then the state vector is given as

$$z_n = (x_{1n}, x_{2n}, x_{1n-1}, x_{2n-1}, t_{1n}, t_{1n-1}, t_{2n}, t_{2n-1},$$
$$s_{1n}, s_{1n-1}, s_{1n-2}, s_{2n}, s_{2n-1}, s_{2n-2})'. \tag{18.20}$$

The corresponding coefficient matrix of (18.5) is given as:

$$F = \begin{bmatrix}
\begin{matrix} a_{111} & a_{121} & a_{112} & a_{122} \\ a_{211} & a_{221} & a_{212} & a_{222} \\ 1 & 0 & 0 & 0 \\ 0 & 1 & 0 & 0 \end{matrix} & \mathbf{0} & \mathbf{0} \\
\mathbf{0} & \begin{matrix} 2 & -1 \\ 1 & 0 \\ & & 2 & -1 \\ & & 1 & 0 \end{matrix} & \mathbf{0} \\
\mathbf{0} & \mathbf{0} & \begin{matrix} -1 & -1 & -1 \\ 1 & 0 & 0 \\ 0 & 1 & 0 \\ & & & -1 & -1 & -1 \\ & & & 1 & 0 & 0 \\ & & & 0 & 1 & 0 \end{matrix}
\end{bmatrix}$$

$$G' = H = \begin{bmatrix} 1 & 0 & 0 & 0 & 1 & 0 & 0 & 0 & 1 & 0 & 0 & 0 & 0 & 0 \\ 0 & 1 & 0 & 0 & 0 & 0 & 1 & 0 & 0 & 0 & 0 & 1 & 0 & 0 \end{bmatrix}. \tag{18.21}$$

The stochastic noise vector of the state equation is

$$V_n = \begin{pmatrix} v_{x1n} & v_{x2n} & v_{t1n} & v_{t2n} & v_{s1n} & v_{s2n} \end{pmatrix}', \tag{18.22}$$

where a_{ijm} in (18.21) is the i-th variable and the autoregressive coefficients with the order m in variable j, v_{pqn} is the stochastic term of q-th variable on the factor $p(p = x, t, s\!:\!x$ autoregressive cause, t trend factor, s seasonal factor) at time n.

By estimating the nonstationary multivariate model may give new information along the relation ship among the variables comparing with the stationary multivariate model, which is obtained by converting the series into a stationary.

Figure 18.4 illustrates the results of the two variables nonstationary multivariate model of the wholesale price index (WPI) and index of industrial production (IIP).

Table 18.1 AIC of the nonstationary multivariate model of IIP and WPI: Model including the seasonal component in case that the order of the trend and AR are changed

		order of AR		
		1	2	3
	1	-118	-62	-44
trend	2	-136	-173	-163
	3	-91	-141	-153

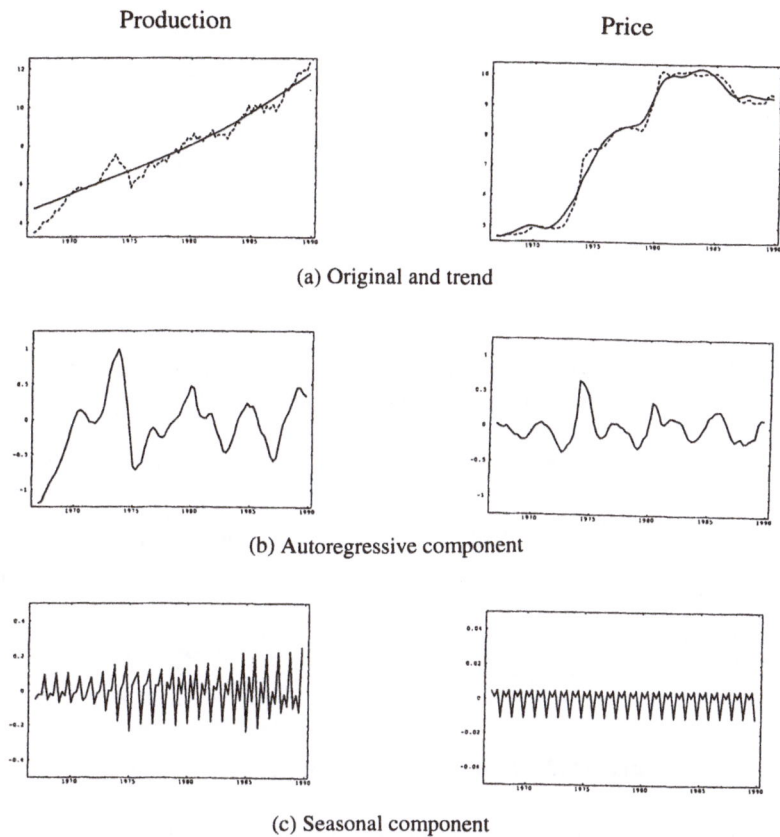

Production   Price

(a) Original and trend

(b) Autoregressive component

(c) Seasonal component

Figure 18.4 Trend, autoregressive component, and seasonal components of the IIP and WPI by a nonstationary multivariate model

Table 18.2 AIC of the nonstationary multivariate model of IIP and WPI: Comparison between the model excluding the seasonal component and the model including the seasonal component

	AIC
Model excluding the seasonal component	−111
Model including the seasonal component	−173

Table 18.1 shows the AIC when the orders of the trend and the autoregressive factor changed. The minimum AIC select when both order is 2. In the original WPI, seasonality is hard to observe, but the model including the seasonal factor is selected as shown in Table 18.2. The AIC suggests the model including the seasonal factor is better than the model excluding the seasonal factor. However as seen in Figure 18.4(c), the seasonality of WPI is, as can be seen in the scales on the ordinate, not

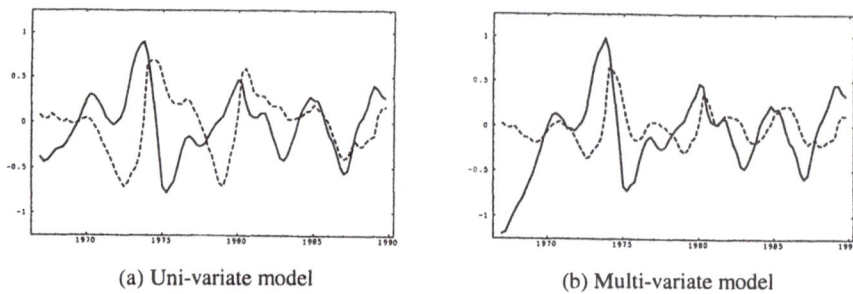

(a) Uni-variate model   (b) Multi-variate model

Figure 18.5 Autoregressive component of IIP (solid line) and WPI (broken line) by the singe-variate model and multivariate model

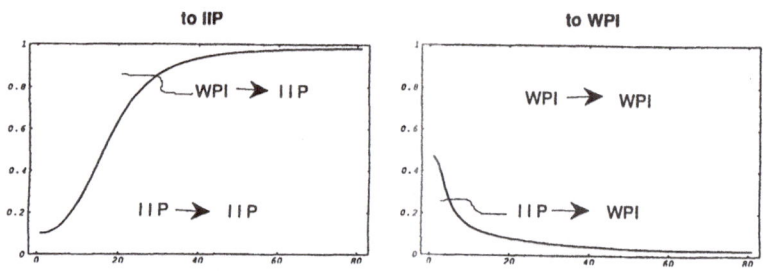

Figure 18.6 Power contribution ratio of IIP and WPI

so remarkable as in the IIP. The autoregressive factors show different movement as in Figure 18.4(b).

The results obtained from a univariate model are also slightly different as seen in Figure 18.5, especially the fluctuation of WPI becomes larger. The model is selected by AIC on the assumption that there exists no relationship among variables in (18.21). In general, the fluctuation of price is relatively smaller than production as seen in original series, the result of nonstationary multivariate model would be adequate which is obtained from estimated series excluding trend.

To examine relationship between IIP and WPI, the power contribution ratio is estimated using the 2-variate autoregressive model as in Figure 18.6. In the lower frequency domain, the influence of the price to the production is stronger. Stabilization of the price is one of the important targets of the monetary policy and the influence of the price to the production activity is larger, as many empirical analyses suggest, the result obtained from this model also reveals explicitly.

The movements of the factors estimated by the nonstationary multivariate model sometimes becomes different by the variables incorporated into the model. This is because the movements are influenced by the mutual relationship of the variables in the system. Thus, the most of care should be made to the use of the estimation results. The stochastic term shown here assumes the normal distribution, which is also to be improved. also with this matter.

18.5 Conclusions

For evaluation of the actual economic policies and comprehension of the policy implication, empirical investigation based on data is of importance to a high degree. The results shown in this chapter reveal that the recent improvement of the time series model may offer effective methods for economic analysis.

It is also important to attempt to make empirical analysis from different viewpoints using a variety of methods. Especially, it is important to pay special attention to the phenomena peculiar to economy, such as peoples' action based on anticipation or expectation and the political or social consciousness of the policy staff which would influence on the economy. Recent development of the stochastic models have made it possible to probe the real activities, and would provide a new prospect for economic analysis.

References

Akaike, H. (1980), "Seasonal adjustment by a Bayesian modeling," *Journal of Time Series Analysis,* Vol. 1, No. 1, 1–13.

Akaike, H. (1980), "Likelihood and Bayes procedure," *Bayesian Statistics,* Bernardo, J. M., De Groot, M. H., Lindley, D. U. and Smith, A. F. M. eds., University Press, Valencia, 143–166.

Akaike, H. (1986), "Information criterion and statistical model," in Japanese, Hoso-Daigaku Kyoiku Shinkokai.

Akaike, H. (1989), "Application of the multivariate autoregressive model," *Advances in Statistical Analysis and Statistical Computing,* Vol. 2, Mariano, R. S. ed., JAI Press Inc., Greenwich, Connecticut, 43–58.

Friedman, M. (1985), "The Feds' Monetarism was never anything but rhetoric," *Wall Street Journal,* December 18.

Havenner, A. and P. Swamy (1978), "A random coefficient approach to seasonal adjustment of economic time series," *Special studies paper,* No. 124, Federal Reserve Board.

Horie, Y. and S. Naniwa (1990), "Japan's financial change and financial policy," in Japanese, Toyo-Keizai Shimposha.

Hiromatsu, T. and S. Naniwa (1990), "Economic time series analysis," in Japanese, Akakura-Shoten.

Hiromatsu, T. and S. Naniwa (1993), "The theory and practice of economic time series analysis," in Japanese, Taga-Shuppan.

Kato, H., S. Naniwa and M. Ishiguro (1993), "A Bayesian method for estimating mutual relationships of multivariate nonstationary time series without individual detrending," *Research memorandom,* No. 471, Institute of Statistical Mathematics.

Kato, H., Ishiguro, M. and Naniwa, S. (1995), "A multivariate stochastic model with nonstationary trend component," Accepted for publication in *Applied Stochastic Models and Data Analysis,* Vol. 10, No. 4.

Kitagawa, G. and W. Gersch (1984), "A smoothness priors-state space modeling of time series with trend and seasonality," *Journal of the American Statistical Association*, Vol. 79, No. 386, 378–389.

Kitagawa, G. and W. Gersch (1985), "A smoothness-priors time varying AR coefficient modeling of nonstationary covariance time series," *IEEE Transactions on Automatic Control*, Vol. AC–30, No. 1, 48–56.

Kitagawa, G. and W. Gersch (1996), "Smoothness priors analysis of time series," Springer.

Kitagawa, G. (1987), "Non-Gaussian state-space modeling of nonstationary time series," *Journal of the American Statistical Association*, Vol. 76, No. 400, 1032–1064.

Kitagawa, G. (1989), "Non-Gaussian seasonal adjustment", *Computers and Mathematics with Applications*, Vol. 18, No. 6/7, 503–514.

Mankiw, N. (1990), "A quick refresher course in macroeconomics," *Journal of Economic Literature*, Vol. , No.

Oritani, Y. (1981), "The negative effects of inflation on the economic growth in Japan," *Discussion paper series*, No. 5, Bank of Japan.

Chapter 19

Processing of Time Series Data Obtained by Satellites

Tomoyuki Higuchi
The Institute of Statistical Mathematics
4-6-7 Minami-Azabu, Minato-ku, Tokyo 106-8569, Japan
higuchi@ism.ac.jp

19.1 Introduction

Mankind, who are full of intelligent curiosity, are always anxious to satisfy such curiosity and are being engaged in development of new observation instruments. The observation utilizing spacecraft such as rockets, balloons, etc. has aimed at exceeding the limitations of the information that can be measured and obtained exclusively on the ground. The human desire to make probe in a direction of height has been expanded and promoted to the outer space, and furthermore to the universe. Thus commencement of such observation methods as to dispatch observation equipment and apparatus has been seen, that is to say, observation by means of artificial satellites, called satellites simply, has been started.

Satellites that have been dispatched into outer space groping for the farther and farther distance for the purpose of strenuously approaching unknown regions bring about a new viewpoint that has never been experienced by the human beings to the mankind enabling them to make observation of the earth from the outer space, once the eyes of the observation instrument are directed to the earth. When things are viewed in perspective, it is pointed out that there exists a side not only allowing the overall variation to be just glanced at but also permitting various types of the information gathered from individual parts behaving as if they are independent of each other to be united to furthermore deepen the comprehension of the objectives. One of the major objectives of the observation using satellites are also found in such an affair. Satellites have become indispensable instruments as an only measure making it possible to observe a wide range of the earth's surface repetitively and continuously for monitoring destruction of the environment and ecological system of the earth that are now attracting global interest of the people. Satellites' mission at present is greatly expanded from qualitative comprehension of the objectives at the primary stage that is recollected by people in a word of "the probe of the unknown regions" to quantitative grasp of the objectives taking up and accumulating a large amount of data with prediction as a target.

In extracting the information obtained from satellite data, three types of problems are found. At the first stage, systematic noise with which a possibility to exercise destructive influence on the inference results is pointed out and none of easy separation can be made is included in the data in addition to simple observation noise. Thus the problems to be solved become ill-posed inverse problems. Secondly, it is very difficult to obtain the data of the fixed area on the earth under the same condition as to a local time and weather condition, etc. Therefore when yearly trend is examined, deliberate analysis is required. Thirdly, many of the sensitivity calibration (adjustment of offset, gain, etc.) of the substances and materials that can be equipped in the satellite have to be procured from the surrounding nature for the objectives for comparison owing to the restriction of the weight that can be equipped on the spacecraft, and, strictly speaking, realization of such calibration is virtually impossible. For comparison of the calibration with the data of the different satellites, care should fully be taken to this matter.

From the above description, it is perceived that procedures of analyzing satellite data require the troublesome and ambiguous matters requiring information processing based on human experiments and instinct. Even with the ability of the computers that have been revealing remarkable progress in recent years, it is very difficult to allow human image and concept to be expressed freely and numerically. However researches of the data analysis methods by means of modeling which is designed to make the most use of the information possessed by the data are in progress recently. By adopting a model rich in flexibility that can freely express features of the data, i.e. a model having so many of parameters and by avoiding falling into strict objectivism aloof from reality, the whole frame of the data analysis is constituted based on the fact that the process itself of the human knowledge acquisition contains a kind of subjective factor. For objective identification of the model chosen among so many of possible models, we use an information criterion into which a frame of the Bayesian approach along a line of AIC is taken. Such a data analysis method provides processing of satellite data with new possibility.

For actual analysis, it is necessary that modeling should be tailored to each type of problems. That is to say, the data analysis method by means of a modern Bayesian model becomes made-to-order. In this chapter, we focus on the problems appearing frequently in the time series analysis of the satellite data, and explain the way of how to solve the problems by means of a modern Bayesian model referred to above.

19.2   Problems to be Dealt With

19.2.1   Flow of the Procedures of Satellite Data Analysis

In processing the satellite data, there exist several steps in the flow reaching an intensive operation of the final high-degree/integrated information. However we give simplified description of such processing hereunder. When some data are transferred from a satellite to the ground using an electromagnetic wave, bit errors are liable to be produced due to some causes. First, calibration of the error should be made in the primary processing for decoding operation of received electromagnetic waves (which is called demodulation). Secondly, data conversion in a generalized sense such as transformation of the coordinate system or calibration of the design of hard system of the observation instrument ahead of launching should be made. Operations to adjust geometric distortion of the image brought about by a variety of causes such as the earth's rotation, curvature, etc., which are called geometric calibration, should also be made (for further details of such operations, refer to Tsuchiya 1990). To the stage of this secondary processing, statistical treatment of the satellite data appears less frequently.

Thirdly, removal of observation noise should be made. "Blot" or "blur" in the images should be removed. The processing that is usually called image processing is the intermediate processing of the above. As observation noise, everything that might possibly prevent the information available at the final stage from being extracted should be included in addition to, needless to say, the noise caused by artificial influence. Vertical, horizontal, or oblique stripes mixed in the images, outliers that were unable to be removed in the primary processing, etc. are exemplified as such a kind of noise. Accordingly the procedure that separates phenomena having evidently nothing to do with the analysis objectives despite the fact of the causes of generation of such phenomena being unknown is also included in this intermediate processing.

The processing at the final stage is high-degree information accumulation corresponding to analysis objects. As processing methods frequently used in terms of the statistical terminology, discriminant analysis, principal component analysis, etc. are designated as representative ones. Detailed interpretation examined from points of view of the individual subjects in the individual scientific fields is applied to the results obtained in the above.

19.2.2   Spin Noise

A regular self-rotational motion (spin) around the spin axis of their body is given to many of satellites for their stabilization of attitude. The spin is likely to bring about cyclical noise to observations and the information is, in some occasions, buried in false information in the worst case. For example when a noise source of which intensity or spatial distribution depends on time is found in a view direction, the observation device measures the noise every time the device's sensor is faced in the direction of the object. As a result, the noise showing an apparent period appears in the obtained data. The noise relating to the spin is just called spin noise for brevity.

Image data from a meteorological satellite to investigate how the cloud is or revealing how natural disaster or human-caused disaster, e.g. forest fires are never free from the influence of this spin noise. Two-dimensional information such as an image is usually composed of combined pieces of one-dimensional information utilizing two-dimensional scans of a normal measurement instrument (Figure 19.1), which is due to the fact that one of scans (in the θ-direction) is usually conducted by the spin motion itself. Even if a scanning has no correlation with the spin motion, the scans themselves induce the noise having periodicity in the data.

None of statistical technique is required with the elimination of the spin noise if the spin noise repeats highly regular cyclic pattern, but it shows a highly temporal/spatial pattern depending on when and where the observation is made. To deal with such a problem, an ad-hoc treatment has been made to allow each data sets to be processed with the instinct based on the application of the existing noise elimination method to so many of the data sets and an abundantly experienced leader's penetrative ability as a judgment criterion. None of sufficient recognition is made at present with not only on-line processing for spin noise elimination but also with a possibility of high-speed large-amount batch processing.

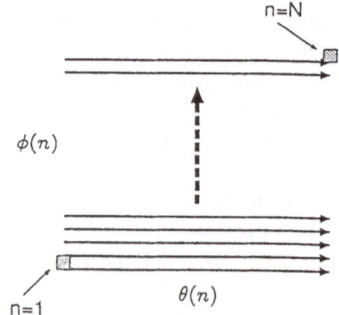

Scanning conducted by the spin motion of the satellite

Figure 19.1 A method of the two-dimensional scan to construct an image

In this chapter, elimination of the spin noise classified in the intermediate process-ing is dealt with as a problem accompanied with the analysis of the time series data obtained by an satellite. Descriptions concerning the above is made as a problem of time series as follows. Unless otherwise specified, let the data y_n $(n = 1, \ldots, N)$ be scalar variable. N is the number of the data. Let the angle around the spindle of the self-rotational motion (called a spin axis) be described with θ_n. The angle θ_n is usually assumed to contain none of errors in the measurement values. Even when the angle is not given as data, we assume that the angle can be defined as a function of n with a few parameters. For example, the angle is given as $\theta_n = \theta_0 + 2\pi n \Delta t/T$ with θ_0 as a parameter representing the initial angle. Here, T is the spin period, and Δt is sampling time.

A problem is summarized such that true signals should be estimated after the spin noise s_n as systematic noise is estimated from the data (y_n, θ_n) $(n = 1, \ldots, N)$ comprised of the scalar-variable observation value y_n and θ_n and is removed from y_n.

19.3   Approach by a Bayesian Model –Simple Model–

Here, we deal with a simple model to eliminate spin noise to give an outline of the satellite data analysis by a Bayesian model.

19.3.1   Composition of an Observation Model

The number of the noise components undesirable in estimating the physical quan-tities that is to be obtained is not always restricted as one. In addition to spin noise, observation noise behaving as the white noise sequences, trend component correspond-ing to the movement of the mean value of the data, etc. are often included in the data. An important matter in eliminating these components are not to proceed the processing in succession in a manner as of elimination of the trend components at first, elimination of the spin noise components at the second stage, and elimination of the white noise at the final stage, but to decompose simultaneously the data into the components that are believed to compose the observation. Only by decompos-ing simultaneously the data, which are complimentarily composed of each other, it becomes possible to do rational elimination of the noise with avoidance of one-sided influence by means of the processing of the other components.

Within a framework of the Bayesian model, we explicitly describe the observations by an observation model. In case that the observation y_n is linearly decomposed into a trend component t_n, spin noise component s_n, and observation noise component w_n, let y_n be formulated by means of the observation model as

$$y_n = t_n + s_n + w_n. \tag{19.1}$$

It is intended that observation y_n is expressed by using the parameters such as t_n, s_n together with the observation noise w_n in an explicit manner.

Describing the observation model in advance using a state vector makes it possible to utilize the algorithm of the Kalman filter for parameter estimation, and convenience is offered thanks to good perspective in the calculation method. The observation model is given as

$$y_n = H_n z_n + w_n, \tag{19.2}$$

using a state vector z_n to express the state of the system at time of n. In case of (19.1), from the system model to be referred to later, the state vector is defined by

$$z_n = [t_n, t_{n-1}, s_n, s_{n-1}]^T. \tag{19.3}$$

The mark T denotes transposition. H_n is $1 \times k$ matrix dependent generally on time n, but a constant matrix $H_n = [1, 0, 1, 0]$ in the example in (19.1). With the linear Gaussian Bayesian model, let the observation noise w_n be one-dimensional normal (Gaussian) white noise with the mean 0 and the variance σ^2.

19.3.2 Composition of the System Model

Suppose the spin noise s_n follows the stochastic difference equation

$$s_n - 2C\, s_{n-1} + s_{n-2} = \xi_n, \tag{19.4}$$

where C is given by $C = \cos(2\pi f_c \Delta t)$ with the spin frequency f_c ($f_c = 1/T$). Furthermore, assume that ξ_n is a one-dimensional normal white noise sequences with the mean 0 and variance τ^2. If $\tau^2 = 0$, then the right-hand side of (19.4) is always 0, bringing about a solution of the equation as a sinusoidal wave, i.e. $s_n = A\sin(2\pi n f_c \Delta t + b)$. Here, A indicates an amplitude, whereas b denotes an initial phase. A behavior of s_n seems far from the sinusoidal wave as τ^2 becomes greater. Equation (19.4) becomes a model of the signal so as to be locally a sinusoidal wave with the frequency f_c. Even if the amplitude is changing gradually as time goes or even if the phase is changed in the course, such a signal (hereafter called a cyclic signal for brevity) can satisfactorily be expressed by (19.4) so long as no change is seen only with the frequency. By setting f_c as the spin frequency this model is one adequate enough to eliminate cyclic signals having the performance of the time-dependent amplitude synchronized to the spin frequency.

In case that $C = 1$, that is to say, $f_c = 0$, the left hand side of (19.4) becomes second-order difference of s_n. This is a signal varying smoothly in a time-dependent manner, that is to say, one of the models to represent the trend component. As for (19.1), the model of the trend component t_n is given by

$$t_n - 2\, t_{n-1} + t_{n-2} = \varepsilon_n, \tag{19.5}$$

where ε_n is a one-dimensional normal white noise sequences with the mean 0 and the variance ν^2. In this occasion, the magnitude of ν^2 determines the smoothness of t_n.

As a model of the trend, the one taking the left-hand side as the first-order difference or higher order difference as well can be considered.

In the framework of the Bayesian model, probability distribution is assumed also for the parameters involved in expressing the observation as shown in (19.4) and (19.5). Equations (19.4) and (19.5) give a system model. The parameter appearing in the system model such as τ^2 or ν^2 is called a hyper-parameter.

The system model is expressed as

$$z_n = F_n z_{n-1} + G_n v_n, \tag{19.6}$$

by using a state vector z_n displaying the state of the system. Together with the above,

$$F_n = \begin{bmatrix} 2 & -1 & 0 & 0 \\ 1 & 0 & 0 & 0 \\ 0 & 0 & 2C & -1 \\ 0 & 0 & 1 & 0 \end{bmatrix}, \quad G_n = \begin{bmatrix} 1 & 0 \\ 0 & 0 \\ 0 & 1 \\ 0 & 0 \end{bmatrix}, \tag{19.7}$$

where z_n is given in (19.3). Furthermore $v_n = [\varepsilon_n, \xi_n]^T$ is a Gaussian distribution with mean vector 0 and the variance covariance matrix

$$\begin{bmatrix} \nu^2 & 0 \\ 0 & \tau^2 \end{bmatrix}. \tag{19.8}$$

19.3.3 Kalman Filter

Combination of (19.2) and (19.6) is called a state space representation of the observations. If state space representation is given, then the estimated value $\hat{z}_n$ of the state vector z_n can be obtained by using Kalman filter and smoother algorithm under the fixed values of hyper-parameters. An optimal value of hyper-parameters can be obtained by AIC minimization. In actual application, the initial value $z_{0|0}$ of the state vector and its variance covariance matrix $V_{0|0}$ should be given. Here, $z_{n|j}$ and $V_{n|j}$ express the mean value of the state vector at the time n and variance covariance matrix under the situation where the data $Y_j = [y_1, \ldots, y_j]^T$ are observed. Usually, it is sufficient to think of a diagonal matrix having 0 vector as $z_{0|0}$, a large value (e.g., 10^6 times the variance of y_n) as $V_{0|0}$ for the sake of convenience.

19.3.4 Application to Rocket Data

We show in Figure 19.2 the results* obtained by applying the procedure based on the aforementioned simple model to the data obtained by a rocket which are subjected to the influence of the spin noise as with the case of an satellite. The data provide the observation $I(z)$ of the intensity of the 5577Å airglow emitted by the excited oxygen atoms at the altitude (height) z. The quasi-cyclic behavior component seen in Panel (a) indicates the rocket's spin noise. The actual observation $I(z)$, is not, the intensity $J(z)$ of the light emitted by the oxygen atoms in the vicinity of the height (altitude) z, but the integrated quantity $I(z) = \int_z^{+\infty} J(z')dz'$ ranging from the infinity to the height (altitude) z. To obtain the estimate of $J(z)$ from $I(z)$, it is necessary, first of all, to eliminate the spin noise together with the observation noise from $I(z)$.

In Figure 19.2(b), the three components t_n, s_n and w_n obtained by decomposing $I(z)$ are shown. The components s_n and w_n are the terms corresponding respectively

---

*The authors are deeply indebted to Professor T. Ogawa of University of Tokyo and Dr. Kazuyuki Kita for the precious information offered by the distinguished scholars as the fruitful outcomes of their joint researches.

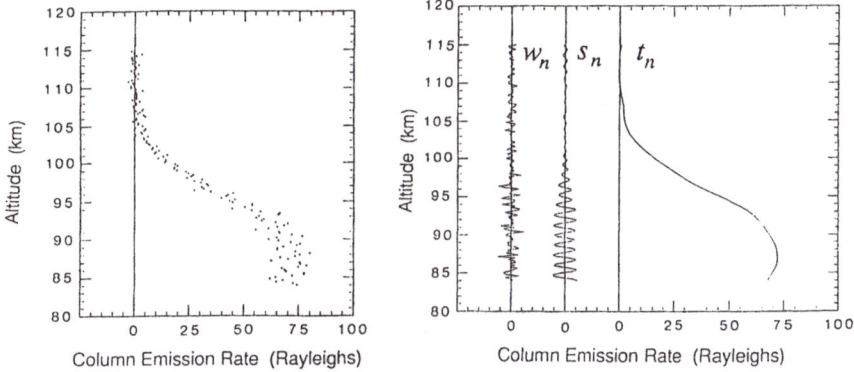

Figure 19.2 (a) Height profile of the intensity of the 5577Å atmospheric light emitted by the excited oxygen atoms. The ordinate is taken as the altitude direction. (b) Profile decomposed into the individual components of the trend t_n, spin noise s_n, and observation noise w_n.

to the spin noise and observation noise, and it is possible to estimate the discretized $J(z)$ by taking the first-order difference of the component t_n obtained as a result of elimination of an effect associated with a spin from the observation $I(z)$. It is clear that even such a simple model makes it possible to realize information processing with extremely high practicality. For details, refer to Higuchi et al. (1988).

19.3.5   Interpretation as a Linear Filter

An estimation of the trend components by using the Bayesian model is regarded as an equivalent to that of low-frequency components. Let the extraction method of the spin noise be compared to the trend estimation method by a classical low-pass filter. When the observation model is given by a simple sum of the several components as seen in (19.1), a simple relationship can be given between the estimates of the obtained components and the data. In case of the above example, it is expressed by

$$T_N = L_t^{\alpha^2} Y_N \qquad (19.9)$$

$$S_N = L_s^{\beta^2} Y_N, \qquad (19.10)$$

where $T_N = [t_1, \ldots, t_N]^T$, $S_N = [s_1, \ldots, s_N]^T$, $Y_N = [y_1, \ldots, y_N]^T$. Furthermore $L_t^{\alpha^2}$ and $L_s^{\beta^2}$ are $N \times N$ matrices corresponding to each t_n and s_n determined by the system model and the value of the hyper-parameter. Here, $\alpha^2 = \nu^2/\sigma^2$ and $\beta^2 = \tau^2/\sigma^2$.

To obtain the expression of (19.9) and (19.10) in a frequency domain, executing discrete Fourier transformation gives

$$\widetilde{T_N} = \widetilde{L_t^{\alpha^2}} \widetilde{Y_N} \qquad (19.11)$$

$$\widetilde{S_N} = \widetilde{L_s^{\beta^2}} \widetilde{Y_N}. \qquad (19.12)$$

The diagonal components of $\widetilde{L_t^{\alpha^2}}$ and $\widetilde{L_s^{\beta^2}}$ determine the frequency response produced when the transformation from the data to each components is viewed as a linear filter.

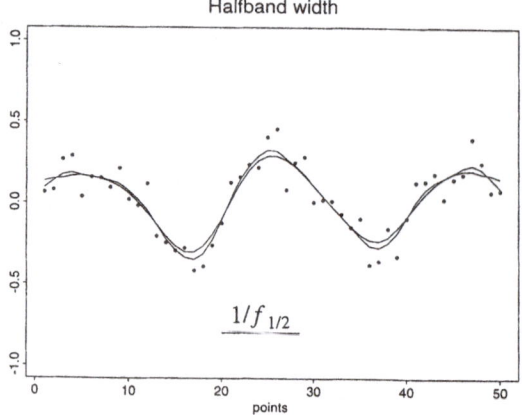

Figure 19.3 Comparison of the smoothing by means of the Bayesian model with the smoothing by means of the moving average. In case of the Bayesian model, the smoothing parameter is automatically determined. $1/f_{1/2}$ is the half-band width obtained from the determined value in accordance with (19.13).

The diagonal component of $\widetilde{L_t^{\alpha^2}}$ describes to what extent of the frequency the component will be allowed to be preferentially passed in the frequency domain, whereas the diagonal component of $\widetilde{L_t^{\beta^2}}$ describes to what extent of the frequency component around f_c is allowed to be selectively passed. The hyper-parameter controls this frequency response.

When the procedure to extract the trend component is regarded as applying a low-pass filter that takes out the low-frequency components, the frequency of which the magnitude of the frequence response attenuates to a half value of that at the frequency 0, i.e. the half-band width $f_{1/2}$ is given approximately as a function of the hyper-parameter α^2 in accordance with the low-pass filter based on the model given by (19.5) (Higuchi 1991)

$$f_{1/2} \simeq \frac{\sqrt{\alpha}}{6} \quad (\text{for } 2^{-12} \le \alpha^2 \le 2^2), \tag{19.13}$$

where the frequency domain is restricted in a range from 0 to $1/2$ on the assumption that the measurement time interval $\Delta t = 1$. Shown in Figure 19.3 are the artificially-generated data y_n, the trend component estimated based on the hyper-parameter automatically determined in accordance with the AIC minimization procedure, and the result of smoothing by means of the simple moving average

$$\hat{t}_n = \frac{1}{2K^* + 1} \sum_{i=-K^*}^{K^*} y_{n+i},$$

having a half-band width which is given by the frequency response from the above expression. K^* is defined by $K^\dagger$ where the frequency response function $H_K(f_{1/2})$ of $(2K + 1)$-point moving average

---

†It is possible to give in a manner as of $K^* = [1/(4f_{1/2}) - 1/2]$, where $[\ \cdot\ ]$ represents the integer part.

$$H_K(f_{1/2}) = \frac{1}{2K+1}\Big[1 + 2\sum_{n=1}^{K}\cos(2\pi n f_{1/2})\Big] \qquad (19.14)$$

takes the nearest value to 0.5. The value of the optimal hyper-parameter obtained for the data shown in Figure 19.3 is $\alpha^2 = 0.25$, and it turns out that $K^* = 2$. From the fact that the trend component almost shows a good agreement with the moving average, $f_{1/2}$ given by (19.13) is appropriate enough.

19.4   Example of a Simple Model

In this paragraph, we show several models having a possibility for application to satellite data processing in a concrete manner to exemplify how to construct models.

19.4.1   Extension to Multi-Components

When a signal is composed of several cyclic components, two models can be considered. One is the model to generalize the observation model as

$$y_n = s_n^1 + s_n^2 + \cdots + s_n^M + w_n, \qquad (19.15)$$

where s_n^m is the m-th quasi-cyclic (wavy) component, and as in the case of (19.4), the system model for each component is given as

$$s_n^m - 2C^m s_{n-1}^m + s_{n-2}^m = \xi_n^m. \qquad (19.16)$$

Furthermore, C^m is a constant given by $C^m = \cos(2\pi f_c^m \Delta t)$ in case that the frequency of s_n^m is f^m. The variance τ_m^2 $(m = 1, \ldots, M)$ of ξ_n^m is also a hyper-parameter for this model.

With another one, the system model is given as

$$\sum_{j=0}^{2M} a_j s_{n-j} = \xi_n, \qquad (19.17)$$

where the coefficient a_j is obtained by comparing the coefficients of s_{n-j} in the equality

$$\sum_{j=0}^{2M} a_j s_{n-j} = \prod_{m=1}^{M}\left(1 - 2C^m B + B^2\right)s_n, \qquad (19.18)$$

where B is the backward operator defined by $s_{n-1} = Bs_n$.

19.4.2   Model of the Decaying/growing Wave

Let it be considered that s_n is a wavy signal locally decaying or growing. When it can be supposed that $s_n = A\exp\{2\pi(g_c + if_c)n\Delta t + ib\}$, the model in (19.4) is generalized as

$$s_n - 2\gamma_c \cos(2\pi f_c)s_{n-1} + \gamma_c^2 s_{n-2} = \xi_n, \qquad (19.19)$$

where $\gamma_c = \exp(2\pi g_c)$, and the positive g_c is called a growth rate of the wave and the negative one is called a damping rate. When $g_c = 0$, (19.19) evidently coincides with (19.4). The model in question expresses the free oscillation in case that friction is found, and is adequate enough to express a wavy behavior globally growing or decaying. For an extension to multiple components, it is recommended that (19.19) be used instead of (19.4) in the discussion in the previous subsection.

19.4.3 Seasonal-Adjustment-Type Model

Long-term economic activity should be investigated in the time series analysis of the economic data, and therefore seasonal adjustment to remove the variation concerned with the season from the data is a very important subject for researches. Because the seasonal adjustment is to remove the variation having one-year cycle from the data, the Bayesian model proposed for the seasonal adjustment can be directly applied to spin noise elimination provided that the cycle of the spin can be given instead of the yearly cycle. When the cycle is r, the simplest model is given as

$$s_n - s_{n-r} = \xi_n .\tag{19.20}$$

In case of monthly data (i.e. the data obtained every month), $r = 12$. This model, which allow s_n to have a non-zero mean component such as $s_n \equiv$ constant, is not an adequate one when the trend component in addition to the spin noise component is intended to be extracted from the data. When (19.20) is re-written as

$$\begin{aligned} s_n - s_{n-r} &= (1 - B^r)s_n \\ &= (1 - B)(1 + B + B^2 + \cdots + B^{r-1})s_n \\ &= \xi_n , \end{aligned}\tag{19.21}$$

it can be seen that the first part of the second line coincides with the first order difference operator used for the model of trend. It is therefore recommended that the condition required by the second pair of the parentheses should be made the model for seasonal adjustment, i.e.

$$\sum_{j=0}^{r-1} s_{n-j} = \xi_n .\tag{19.22}$$

The seasonal adjustment model can be applied to elimination of the spin noise by putting $r = T/\Delta t$, provided that the spin period T is of multiple length of the sampling interval Δt. With this model, it is utterly unnecessary for the spin noise to be of sinusoidal wave, and any kind of signals can be expressed as a fundamental principle provided that the cycle is r. Needless to say, it is possible for the repeating patterns to be varied slowly dependent on time.

19.5 Point Noise Source Model

We hereunder show the approach where the physical structure of the noise source is explicitly taken into the observation model.

19.5.1 Uni-Variate

Suppose the objective generating noise is of the point noise source. At that time, the noise can be written by the product $f(\cdot)I_n$, where I_n is a time-dependent intensity of the point noise source and $f(\cdot)$ can be assumed to be described as a function of the relative angle θ formed between the point noise source and the instrument. Now assume that the relative positioning between the satellite and the point noise source remains unchanged during the period of the observation, then we can get $f(\cdot) = f(\theta)$.

Under the assumption of such a noise source, let a problem to estimate the physical quantity $x = x(t, \theta)$ depending on the time t and the space (which is equivalent to the angle θ in this occasion) be considered. Now suppose that the time dependency

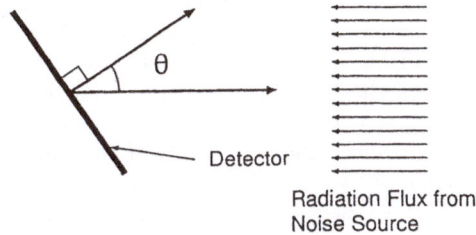

Detector

Radiation Flux from
Noise Source

Figure 19.4 Exemplified schematic relation between the observation and the instrument. The observation is received by the sensor of parallelly-placed plates. Around the satellite, the observation is in proportion to the flux from light sources almost parallel with each other in the distance.

of x is almost negligible within Δt ($|\Delta t \cdot \partial x/\partial t| \simeq 0$). Also suppose that the x within this interval is a physical quantity determined exclusively by the angle. That is to say, suppose that the component x_n obtained by removing an effect of the point noise source and observation noise from the data y_n at time $t = n$ is given by $x_n \simeq x(\theta_n)$ using the angle $\theta = \theta_n$. Furthermore, by using a small angle $\Delta\theta$, let $x(\theta + \Delta\theta)$ be approximated as

$$x(\theta + \Delta\theta) \simeq x(\theta) + x(\theta) - x(\theta - \Delta\theta). \tag{19.23}$$

At that time, assuming that the θ_n is so designed as to allow $\Delta\theta = \theta_n - \theta_{n-1}$, we have

$$x(\theta_{n+1}) \simeq x(\theta_n) + x(\theta_n) - x(\theta_{n-1})$$
$$x_{n+1} \simeq 2x_n - x_{n-1}. \tag{19.24}$$

In short, a trend model is adopted for x_n assuming that the second-order difference shows small variation.

The intensity I_n of the point noise varies slowly dependently on time, and we adopt a trend model assuming that the first-order difference shows small variation. At that time, $z_n = [I_n, x_n, x_{n-1}]^T$ with respect to the state vector, whereas H_n of the observation model is given as $H_n = [f(\theta_n), 1, 0]$. As $f(\theta)$, let a function having small parameters be considered corresponding to the purpose of the application. For example, when the obtained data are so designed as to be in proportion to the flux coming out from the noise source, we obtain (Figure 19.4)

$$f(\theta) = \begin{cases} \cos\theta & |\theta| < 90° \\ 0 & 90° \le |\theta| \le 180°. \end{cases} \tag{19.25}$$

In addition to the above, the factors such as $\cos^2\theta_n$, $\exp(-\sin^2\theta_n)$, quadratic polynomial, etc. are considered as candidates. $f(\theta)$ should be designed so as to express the characteristics of the instruments in dealing with practical problems.

19.5.2 Expansion to Multi-Variate

To examine space-distribution structure of the object, several instruments steered into different directions are occasionally mounted on a plane onto which a spin axis is included. At that time, the observation becomes the multi-variate $y_n = [y_n^1, y_n^2, \ldots, y_n^M]^T$.

Furthermore, y_n^m is the observation at time n obtained with the m-th instrument (channel-m). Here, M is called the channel number. When the angle of the channel m from the spin axis in the plane including the spin axis is described as ϕ^m, the relationship $f(\cdot)$ between the channel m and the point noise source is written by θ_n and ϕ^m. That is to say, $f(\cdot) = f(\theta_n, \phi^m)$. At that time, not by considering a model with respect to every channels, that is, for y_n^m of the individuals but by thinking of an extensive model for multi-variate time series y_n, an estimation of a function form of $f(\theta_n, \phi^m)$ can be made based on the more rational assumption. Furthermore, it becomes possible to estimate features such as the difference among the equipment characteristics among channels from the data.

Suppose the component x_n^m which is obtained by removing the influence of the point noise source at time n and the observation noise from the observation of the channel m along line of the solving method in case of uni-variate, is in accordance with the second-order difference model. On the other hand, suppose that I_n is expressed by the first-order difference model as in the case of uni-variate. From this it is understood that the state vector becomes the $(1+2M) \times 1$ vector $z_n = [I_n, x_n^1, x_{n-1}^1, \ldots, x_n^M, x_{n-1}^M]^T$. On the other hand, H_n of the observation model becomes the $M \times (1+2M)$ matrix

$$
\begin{bmatrix}
f(\theta_n, \phi^1) & 1 & 0 & & & & \\
f(\theta_n, \phi^2) & & & 1 & 0 & & \mathbf{0} \\
\vdots & & & & & \ddots & \\
f(\theta_n, \phi^M) & \mathbf{0} & & & & 1 & 0
\end{bmatrix}.
\tag{19.26}
$$

The structure of the variance/covariance matrix of the vector of the observation noise $w_n = [w_n^1, w_n^2, \ldots, w_n^M]^T$, should be determined by taking into account the factors including the characteristics among the channels.

19.5.3   Expansion of the Point Noise Source Model

When the influence from the noise source cannot be expressed by the point noise source referred to above, it is necessary to express $I_n \cdot f(\theta_n)$ in a form with higher flexibility. Let it be considered that the θ dependency of the spin noise should be expressed by a first-order spline function $R_n(\theta)$ at each time n. First of all, let node points θ_j $(j = 1, \ldots, J)$ be determined by equally dividing $[0, 360°]$ by d_θ. Here, $J = 360/d_\theta$. Secondly, let values of $R_n(\theta)$ at each node points be denoted as r_n^j. $R_n(\theta)$ in a range of each regions $[(j-1)d_\theta, \ jd_\theta]$ is given by

$$
R_n(\theta) = (1-a)r_n^j + ar_n^{j+1},
\tag{19.27}
$$

where $a = \theta/d_\theta - [\theta/d_\theta]$ with $[\theta/d_\theta]$ indicating the integer part of θ/d_θ. Also on the assumption that $R_n(\theta)$ is a periodic function, $r_n^{J+1} = r_n^1$.

By assuming that $R_n(\theta)$ varies slowly dependently on time, let it be assumed that r_n^j is in accordance with the first-order difference model. It is possible for the variance of the first-order difference to be in common with j or to be optimized so that the likelihood will individually be maximum. In general, some kinds of prior information are available, and modeling so designed as to effectively utilize such information will be made. As in the case of uni-variate point noise source, assume that x_n to be measured can be expressed by the second-order difference model. Accordingly the state vector becomes $z_n = [r_n^1, r_n^2, \ldots, r_n^J, x_n, x_{n-1}]^T$. H_n of the observation model

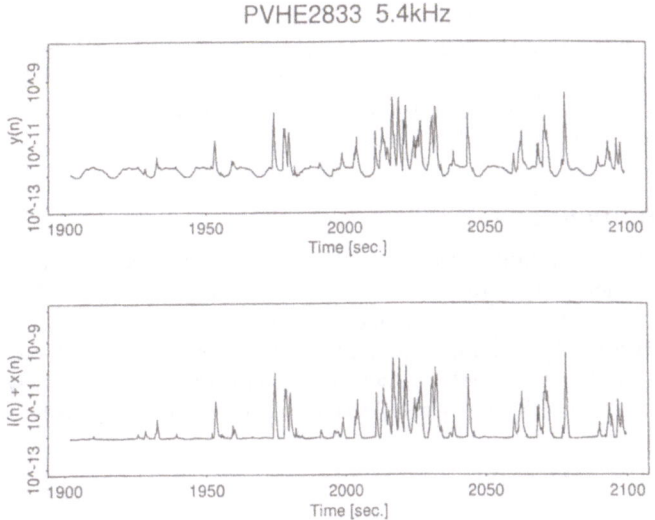

Figure 19.5 (a) Intensity of the electric field observed by Pioneer Venus Orbiter. (b) Data after systematic noise caused by spin noise is eliminated.

becomes $1 \times (J+2)$ vector defined by

$$H_n = [\overbrace{0,\ldots,0}^{j_n \text{ times}}, 1-a, a, \overbrace{0,\ldots,0}^{J-j_n-2 \text{ times}}, 1, 0], \tag{19.28}$$

where j_n is an integer defined by $[\theta_n/d_\theta]$.

When there locally exist the generation sources of the signal x to be measured in space, no approximation of (19.23) can be established. In this occasion, we decompose the observation y_n as

$$y_n = w_n I_n f(\theta_n) + x_n, \tag{19.29}$$

in accordance with selection (by combining the judgment of $x_n = 0$ or no, with estimation of I_n and $f(\theta)$ (Higuchi 1993)).

Assume I_n varies slowly, and is expressed by the first-order difference model. At the same time, assume $f(\theta)$ shows a spatially smooth behavior. These assumptions can be accomplished by applying the trend model taking the second-order difference of f_j $(j = 1, \ldots, 360/d_\theta)$, as an infinitesimal amount, where f_j is defined by discretizing $f(\theta)$ by d_θ. The model in question, which is a model of the observation depending on both time and space, is, in short, one of the simplest statistical model called generally space-time model. Let a search of the optimal x_n, I_n, and f_j be made by the procedure to do independently the judgment as to whether $x_n = 0$ or not, smoothing the time domain to obtain I_n, and smoothing the space domain to obtain f_j.

The results[*] obtained by applying this method to real satellite data are shown in Figure 19.5. Shown in (a) is just part (792 points) of the 20394 data y_n observed.

---

[*]The authors are deeply indebted to Professor C. T. Russell, Dr. R. J. Strangeway, and graduate student Mr. G. K. Crawford of Los Angeles School, University of California for the precious information offered by the distinguished researchers as the fruitful outcomes of their joint researches.

The cyclic variation is equivalent to the spin noise. When the signal corresponding to x_n is observed, y_n takes extremely large values. In (b), the data $I_n + x_n$ obtained by removing and calibrating the influence of $f(\theta_n)$ from the observed data y_n are shown. It is found that the systematic noise evidently concerned with the spin noise seen in (a) is removed in (b). I_n in the interval shown in this figure is almost constant, but varies when viewed through all of the data.

19.6   Conclusions

In the above, we explain several models to remove spin noise in the satellite time series data. Although the frame of the Bayesian model is extremely of the unified style, it is necessary to think of a model specialized into the individual problems in case of application. To develop a new model for each actual problem takes a time for the users, but the new analysis method based on the new model will enable the analyzers' ideas or knowledge to be effectively utilized to the utmost.

References

Higuchi, T., K. Kita, and T. Ogawa (1988), "Bayesian statistical inference to remove periodic noises in the optical observation aboard a spacecraft," *Applied Optics*, Vol. 27, No. 21, 4514–4519.

Higuchi, T. (1991), "Frequency domain characteristics of linear operator to decompose a time series into the multi–components," *Annals of the Institute of Statistical Mathematics*, Vol. 43, No. 3, 469–492.

Higuchi, T. (1993), "A method to separate the spin synchronized signals using a Bayesian approach," *Proceedings of the Institute of Statistical Mathematics*, Vol. 41, No. 2, 115–130.

Higuchi, T. (1994), "Separation of spin synchronized signals using a Bayesian approach," *Proceeding of The Frontiers of Statistical Modeling: An informational approach*, (eds. H. Bozdogan), Kluwer Academic Publishers, 193–215.

Tsuchiya, K. (1990), *An Introduction to Remote Sensing*, Asakura Publishing Co., Tokyo (in Japanese).

Chapter 20

Analysis of Earth Tides Data

Yoshiaki Tamura
National Astronomical Observatory,
Division of Earth Rotation
Mizusawa Astrogeodynamics Observatory
Mizusawa, Iwate-ken 023-0861, Japan
tamura@miz.nao.ac.jp

20.1 What are Earth Tides?

20.1.1 Phenomenon of Earth Tides

The sun and the moon existing near the earth generate tidal forces on the earth. The force is produced because of the fact that the attraction from the heavenly bodies are slightly different from each other on the center and on the surface of the earth. Also the force is in proportion to the gradient of the attraction, and becomes larger in proportion to the size of the body to which the attraction is affected. The magnitude of the attraction is in proportion to the mass of the heavenly body and in inverse proportion to the square of the distance according to the law of gravitation. The tide-generating force is obtained by differentiating the gravitational attraction in space. Consequently, the force is in proportion to the mass of the heavenly body and in inverse proportion to the cube of the distance between the heavenly body and the earth. From this reason, the tide-generating force by the moon is more than twice larger than that by the sun, while the gravitation of the sun is by far larger than that of the moon.

The ocean tide is widely known as a tidal phenomenon. The tidal force affects influences not only on the ocean as a fluid on the earth surface but also on the solid earth itself. The solid earth is not a perfect rigid body, but an elastic one. Therefore the solid earth slightly deforms when an external force is applied to. The tidal force on the solid earth and the earth's deformation caused by tides is called earth tides, and its scale is 0.2–0.3m in case of the vertical displacement, approximately 10^{-6}ms^{-2} in the change of gravity acceleration, and approximately 5×10^{-8} in the strain change (see the schematic diagram in the upper side on the left in Figure 9.2). In addition, the change of the tilts or plumb line and periodical change of the earth's rotational rate are observed as earth tides. With the frequency of the earth tides, the diurnal and semidiurnal components exceed as well as the ocean tides. However, terdiurnal-to-quatriurnal short period components and fortnightly-to-semiannual long period components also exist.

Several type of spring gravimeters have widely been used for gravity tide measurements. In recent years, superconducting gravimeters are in use. The latter gravimeter measures fine positional change of a niobium sphere caused by the gravitational acceleration change, which sphere is levitated by a superconducting magnetic field at the temperature of liquid Helium 4K. In the observation of the strain change, an invar bar or a quartz tube with the length of 10–20m is used as a reference of the length. An end of such bar or tube is fixed onto the ground, and relative deviation of the position from the ground is measured on the opposite side of the bar. Such a strainmeter is settled in a pit to avoid temperature change. The vertical and horizontal tidal displacements are directly measured by using the so-called space geodetic techniques such as VLBI (very long baseline interferometry) or SLR (satellite laser ranging).

With any of the phenomena of the earth tides, the magnitude is just of the 10^{-7} order compared with the gravity acceleration of the earth or the earth's radius. Even for detecting the earth tides, resolution of 10^{-7} order is required in the observation. Furthermore to obtain significant information from tidal phenomena, resolution to detect the relative change of 10^{-9}–10^{-10} is required in the observation of earth tides. In the case of ocean tides, their amplitude differs considerably depending on regions. However since a phenomenon of the amplitude of 50cm–1m is observed at the resolution of 1cm, thus the signal to noise ratio of the observation is relatively good. Also since it is quite seldom that a reference bench mark on the land vertically varies a great deal compared with the amplitude of the tides, large drift is hardly included in the record. However in the case of the earth tides, it is not usual that large instrumental drift or irregular drift caused by natural phenomena is contaminated in the record. Moreover, variations of the measurement environment such as atmospheric pressure or temperature change perturb tidal signals.

For an analysis method of the earth tides, it is requested that the tidal constants (amplitude ratio and phase lag to a theoretical value), which express response characteristics of the earth against external force, should be precisely determined excluding such drifts and perturbations from observation data. Furthermore, occurrence of missing data and data offsets are often seen in actual observations, and flexible countermeasures to the above are also required.

20.1.2 Purpose of the Analysis of the Earth Tides

The purpose of the analysis of the earth tides phenomena is to examine the response of the earth to the tidal force caused by the moon and the sun. From the observation of the earth tides, analysis of the fluid core resonance existing in the inside of the earth and improvement of an earth model in the tidal frequency band are carried out as global problems for examples. Also as a regional/local problem, locality of the tidal factors reflecting the difference of the elastic constants depending on the difference of the ground structure is investigated. Furthermore in places such as volcanic districts, estimation of the time variation of the internal physical properties is under way from the time change of the tidal factors. With the analysis of the oceanic tides, not only the explanation of an oceanic physical phenomenon itself but also fulfilling the daily requirement such as sea level forecast at harbors or tidal current forecast in gulfs and channels is one of the major purposes. With the data analysis of the strains or tilt changes with the purpose of the earthquake prediction or crustal movement detection, interpretation of the drift components obtained by removing the tidal component from the observation data is sometimes more important than the precise determination of the tidal factors themselves.

In examining the response of the solid earth or oceans to the tidal force, it is necessary to know well the tidal force as a source of external force in advance. As described at first, the tide-generating force is derived from the motion of the moon and the sun. Since the motion of the moon and the sun is known accurately, the calculation of the tidal force can be made with the accuracy satisfactory in practice. Two types of calculation methods are available. One method is to obtain tidal force directly from the position of the heavenly bodies, and other method uses harmonic expansion tables expanded on a frequency domain (a method to obtain the force as the sum of sinusoidal waves). In latter case, the individual components expanded on the frequency domain is called tidal components or tidal waves, and their amplitude and phase on the rigid earth are precisely obtained. Approximately 1200 terms of tidal components including very tiny ones are known (Tamura 1987). Principal tidal waves and the ones with closer angular velocity are dealt with as a wave group. The 1200 components can be separated into a few tens wave groups. Observation days required for separating the wave groups are determined from the difference of the angular velocity between principal waves. To separate principal tidal waves, one month of the observation period is required at least. Principal tidal waves representing the wave group are listed in Table 9.1. The argument in the table is the coefficient relating with the motion of the moon and the sun, and determines the frequency and phase of the individual components. The amplitude in the table is the one normalized in some method. For an actual calculation method of the tidal force, refer to Tamura (1987) and Harrison (1985).

In the tidal analysis, obtaining amplitude ratio and phase difference between observations and the harmonically expanded theoretical values becomes a controversial matter. Great difference is noted at this point with the power spectrum analysis method which obtains the distribution of the power density in the frequency domain ignoring phase information.

20.2   Analysis Model

20.2.1   Sampling Interval

About the sampling interval of the tidal data, one hour interval is an adequate one because the period of target tidal phenomena is 1/3 days to one day. When dense sampling data with a short time interval is available, re-sampled time series data must be prepared excluding the short period fluctuation by employing an adequate digital low-pass filter.

It is not advisable to apply a model to the dense sampling data without any pre-processing from a view point of calculation time. When dense observation data is at hand, the procedure shown below is applied in general. First of all, do a preliminary analysis and make predicted values using the result. Secondly, examine the residuals from the original observation. From the comparison with the predicted values, remove the outlying observations by using a suitable threshold level. From the corrected data by removing outliers, make the "normal point" data with adequate time intervals (adequate space intervals depending on problems) to be used as the input data for the analysis.

Table 20.1 Harmonic table of principal tidal waves.

Component	Degree n	m_1	m_2	m_3	m_4	m_6	At J2000 Angular velocity($°$/h)	Amplitude
Long Period Tide								
Sa	2	0	0	1	0	1	0.04106668	0.011549
Ssa	2	0	0	2	0	0	0.08213728	0.072732
Mm	2	0	1	0	−1	0	0.54437471	0.082569
Mf	2	0	2	0	0	0	1.09803304	0.156303
Diurnal Tide								
Q_1	2	1	−2	0	1	0	13.39866089	0.072136
O_1	2	1	−1	0	0	0	13.94303560	0.376763
M_1	2	1	0	0	1	0	14.49669393	−0.029631
P_1	2	1	1	−2	0	0	14.95893136	0.175307
S_1	2	1	1	−1	0	1	15.00000196	−0.004145
K_1	2	1	1	0	0	0	15.04106864	−0.529876
J_1	2	1	2	0	−1	0	15.58544335	−0.029630
OO_1	2	1	3	0	0	0	16.13910168	−0.016212
Semidiurnal Tide								
$2N_2$	2	2	−2	0	2	0	27.89535483	0.023009
μ_2	2	2	−2	2	0	0	27.96820848	0.027768
N_2	2	2	−1	0	1	0	28.43972953	0.173881
ν_2	2	2	−1	2	−1	0	28.51258319	0.033027
M_2	2	2	0	0	0	0	28.98410424	0.908184
L_2	2	2	1	0	−1	0	29.52847895	−0.025670
S_2	2	2	2	−2	0	0	30.00000000	0.422535
K_2	2	2	2	0	0	0	30.08213728	0.114860
Terdiurnal Tide								
M_3	3	3	0	0	0	0	43.47615636	−0.011881

In the description made hereafter, the sampling interval of one hour is supposed. Also with the observation period, it is assumed that an observation period of more than one month, which is necessary to resolve principal tidal components, is available.

20.2.2   Modeling of the Drift

In constructing an analysis model, one of the most difficult problems is how to deal with the drift contained in the observation values. In case of the analysis of the crustal movement data, proper estimation of the drift itself is in some occasions a very important purpose. It is a very rare case that the drift can be expressed as a polynomial of the time t or a constant for a certain period. Although such a modeling is possible when the secular sea level change is considerably small as seen in the oceanic tides, it is very hard to express the drift with a polynomial of the time t or with a sum of the sinusoidal waves in case of the ordinary earth tides data.

In order to obtain the drift, we may use a digital filter or a method to obtain continuous drift by dividing the analysis interval every several days and by applying polynominals or a spline functions to the individual intervals. Actually, the tidal analysis has been made conventionally in such a manner mentioned above. However in applying such a filter, it is difficult to separate perfectly the required information from the drift components. Similarly to this, there is a restriction in the estimation of the drift by dividing the analysis period into sections. In consideration of such matters, a new method to estimate simultaneously the tidal factors and an arbitrary form of the drift was proposed (Ishiguro *et al.* 1984; Tamura *et al.* 1991).

In the proposed method, it is assumed that the drift only changes smoothly in time without assuming a specific function on the form of the drift. That is to say, the tidal parameters are estimated after adding the binding condition that the squared sum of the second difference of the drift (or the third difference at an option) is adequately smaller.

The value of the drift at every observation, d_n, is estimated by assuming the condition that

$$d_n - 2d_{n-1} + d_{n-2} \approx 0 . \tag{20.1}$$

In case of the tidal parameter estimation, it is an important problem to what extent the binding condition defined referred to above should be strengthened. The estimation problem of this drift d_n is equivalent to the smoothing problem of the one-dimensional data. By introducing a suitable type of prior distribution under the expectation that $d_n - 2d_{n-1} + d_{n-2}$ distributes around 0, the problem can be solved by applying the Bayes method proposed by Akaike (1980).

20.2.3   Modeling of Perturbations

In order to remove the influence of the atmospheric pressure change or temperature change affected on the observation of the tides, the response method can be applied in the model. Denoting the observed data such as atmospheric pressure, temperature, etc. by x_n, it is assumed that the perturbation R_n is expressed as

$$R_n = \sum_{k=0}^{K} b_k x_{n-k} , \tag{20.2}$$

by using the response coefficient (response weights) b_k, where K indicates the maximum number of the lags. When $K = 0$, it follows that a simple proportional relation is assumed. When there are several associated observation data sets , it is recommended to consider the sum of corresponding response coefficients.

With the influence of the atmospheric pressure change, a physical model can be constructed as the change of the mass distribution possessed by the atmosphere. Actually, the influence of the atmospheric pressure change is, in the observation of the gravity tides, expressed as a sum of the change of the attractive force possessed by the atmosphere and the change of the load onto the crust. When the perturbations are removed by a stochastic model such as response method, the response coefficients to be obtained from the analysis comprises an amount of the influence reflecting such physical processes and a value including the instrumental response characteristics. The response estimated from the actual data analysis is observed as decrease of $3 \times 10^{-9}\text{ms}^{-2}$ of the gravity per the increase of the pressure 1hPa, and is well consistent with the value by the physical model.

When seveeral associated observation data sets are available, we should pay attention to the interpretation of the obtained response coefficients if a strong correlation exists among the associated observation data. In the observation, for example, of the strain change in a pit, adiabatic expansion and compression are caused by the change of the atmospheric pressure, and then a considerable strong correlation is often seen in the changes of the atmospheric pressure and the temperature in the pit in the short period variation of less than several hours. If a pair of the associated observation data sets having very strong correlation are used for the analysis, then the individual response coefficients are different a great deal from the ones in case of using only one associated observation data set. For the purpose of removal of the influence of the disturbance noise, only the total amount R_n obtained by summing up both the cases has a meaning.

20.2.4   Observation Model

The observed time series y_n can be expressed by

$$y_n = \sum_{m=1}^{M}(\alpha_m C_{mn} + \beta_m S_{mn}) + \sum_{k=0}^{K} b_k x_{n-k} + d_n + \varepsilon_n \,, \tag{20.3}$$

in consideration of the response and drift of the associated observation data mentioned in previous subsection, where m represents the number of the tidal component group (M is the total number of the wave group), α_m, β_m are unknown tidal constants, C_{mn}, S_{mn} are the theoretical values of the tidal component number m. They express the theoretical values of in-phase and 90° out-phase components respectively. Furthermore ε_n is the irregular observation error component. The parameters of the tidal constants α_m, β_m and the response coefficients b_k are estimated by minimizing following equation $J(d)$ applying least squares method.

$$J(d) = \sum_{n=1}^{N}\left\{ y_n - \sum_{m=1}^{M}(\alpha_m C_{mn} + \beta_m S_{mn}) - \sum_{k=0}^{K} b_k x_{n-k} - d_n \right\}^2$$

$$+ v^2 \sum_{n=1}^{N} \{d_n - 2d_{n-1} + d_{n-2}\}^2, \tag{20.4}$$

where v^2 is a coefficient to control the smoothness of the drift. It must be noted that all d_n of every observation epoch are treated as unknowns in the model. This makes it possible to deal with the data accompanying complicated drift which cannot be expressed with a polynomial or sinusoidal waves.

Even when data missing is occasionally occurred in the observations, no problem is caused in the calculation of the above equation $J(d)$. Moreover, when the offset of the zero-level is produced in the observations, the estimation procedure of its offset amount is converted into a problem to estimate the response coefficient to a step function if its position is specified. Thus there is no difficulty in modeling of observational data.

If the (9.4) is schematically expressed, it is given as

$$J(d) = \sum_{n=1}^{N}\{\text{irregular component(residual)}\}^2 + v^2 \sum_{n=1}^{N}\{\text{second difference of drift}\}^2.$$

$$\tag{20.5}$$

The first half of the equation is the same as the general least squares method allowing the square sum of the errors to be minimized. The second half expresses the condition imposed on the drift. With this constraint, the solution never becomes undetermined even if the total number of the unknowns is larger than the number of the observations by putting the drift values at every epoch as unknowns.

To determine the tidal constants α_m, β_m and response coefficients b_k by minimizing $J(d)$, the value of v^2 should be given. If the value of v^2 is taken as exceedingly large, then the freedom of the drift form is made smaller resulting in the assumption of the drift whose form is closer to a straight line. If the value of v^2 is taken as small contrarily to the above, the freedom of the drift form becomes larger. In an extreme case of $v^2 = 0$, the form of the drift becomes entirely free and the residuals become 0. However this result will be just non-sense solutions. The parameter v^2 referred to above is called a hyperparameter, which is the parameter to determine the prior distribution of the parameters (the value d_n at each epoch).

20.2.5 Bayes Model and ABIC

To determine the form of the drift and to determine the tidal parameters, etc., the choice of suitable values of the hyperparameter v^2 is an essential problem. The value of v^2 can be chosen by introducing an adequate Bayes model with the use of the Bayesian information criterion proposed by Akaike (1980). Its is briefly summarized as follows.

Based on the assumption that the distribution of the irregular component ε_n is of the normal distribution with a zero mean and variance σ^2, the distribution of the observations, when the parameters including d_n are known, are given by the density function

$$L = \left(\frac{1}{2\pi\sigma^2}\right)^{N/2} \exp\left\{-\frac{1}{2\sigma^2}\sum_{n=1}^{N}(\text{irregular component})^2\right\}. \tag{20.6}$$

With the drift d_n, the prior distribution having a density function given by

$$P = \left(\frac{v^2}{2\pi\sigma^2}\right)^{N/2} \exp\left\{-\frac{v^2}{2\sigma^2}\sum_{n=1}^{N}(\text{second difference of drift})^2\right\} \tag{20.7}$$

is assumed. Let the hyperparameter v_2 be chosen so that the integral concerning the parameter d of the product of the two density functions referred to above will be maximized. Otherwise by taking natural logarithm of the integral and defining ABIC as

$$\text{ABIC} = -2\log\left\{\int LP \, dd\right\}. \tag{20.8}$$

Let v_2 be chosen so that this ABIC will be minimized. Since both the observation distribution L and the prior distribution P are of the normal distributions, ABIC can be obtained analytically. To obtain the v giving the minimum ABIC, we use a discrete search by $\sqrt{2}$-time step starting from an initial value of an adequate v. Also corresponding with AIC, ABIC* is defined by correcting with the number of parameters adjusted to minimize ABIC. Finally, ABIC* is calculated by

$$\text{ABIC}^* = N\log 2\pi + N\log\hat{\sigma}^2 + N + \log\det(I + v^2 D^t D)$$

$$- N\log v^2 + 2(\text{number of parameters}), \tag{20.9}$$

where $\sigma^2 = J(\hat{d})/n$, $\hat{d}$ expresses the estimated parameter, and I is a unit matrix and D is the $n \times n$ matrix defined as

$$D = \begin{bmatrix} 1 & & & & 0 \\ -2 & 1 & & & \\ 1 & -2 & 1 & & \\ & \ddots & \ddots & \ddots & \\ 0 & & 1 & -2 & 1 \end{bmatrix}. \tag{20.10}$$

In the calculation of ABIC*, the initial values d_0, d_{-1} of the drift are necessary. They can be estimated by considering as unknowns. Or they can be taken from the drift of the pre-analysis period, in case that the analysis period is shifted successively.

In supplemental notes, some kinds of outliers are often included in the real observation data, and the irregular component ε_n in the observations is in some occasions not in accordance with the normal distribution which is a basic assumption in the analysis model. When the minimum ABIC is searched for such data, abnormally small v_2 is sometimes obtained or abnormally large v_2 is obtained contrarily. Therefore let the adequate lower and upper limit values of v be settled emperically in the search of minimum ABIC. In such an occasion, re-analysis should be made after removing the outliers.

20.3   Tidal Analysis Program BAYTAP-G

20.3.1   Functions

A tidal analysis program BAYTAP-G (Bayesian Tidal Analysis Program-Grouping Model) has been developed as a joint research of the Institute of Statistical Mathematics and International Latitude Observatory of Mizusawa (National Astronomical Observatory, Mizusawa at present) (Ishiguro et al. 1984; Tamura et al. 1991), in which program the analysis model mentioned in previous section is adopted. The program is widely used not only for the analysis of the earth tides data but also for the analysis of the crustal movement observation data at present.

BAYTAP-G has the following functions.

1) Determination of the tidal constants.
2) Determination of the drift and the calculation of its power spectrum.
3) Response calculation of the associated observation data such as atmospheric pressure data.
4) Interpolation of the missing values, estimation of an amount of the step.
5) Detection of the outliers.
6) Calculation of ABIC* to check whether the adopted model is an adequate one or not.

For classification of the tidal component groups, 12–31 groups can be selected considering both of the observation period and the observation accuracy. Also in case that the analysis period is as short as one week with the purpose of the outliers detection, using only 3–5 tidal components is possible.

It is a considerably difficult problem to detect strictly the outliers. BAYTAP-G adopted quite a simple method to detect outliers. If the estimated irregular component (residual) becomes more than 4 times of a mean value, and if an absolute value of the second difference of the drift likewise becomes more than 4 times of a mean value, they are listed up as the candidates of the outliers. Program users are advised to make the re-analysis after removing the outliers referring to this list. The newest version of BAYTAP-G removes outlying data from input data set automatically and re-analyzes the data iteratively. The judgment of outlier is made by using two threshold levels given a priori by the program user.

20.3.2   Example of Analysis

Here we show an example of tidal analysis. The observation data used here is the north-south components of the crustal strain observed with a quartz-tube extensiometers installed at the Esashi Earth Tides Station, National Astronomical Observatory (39.1°N, 141.3°E). Figure 9.1 shows the Fourier spectrum of the strain data covering

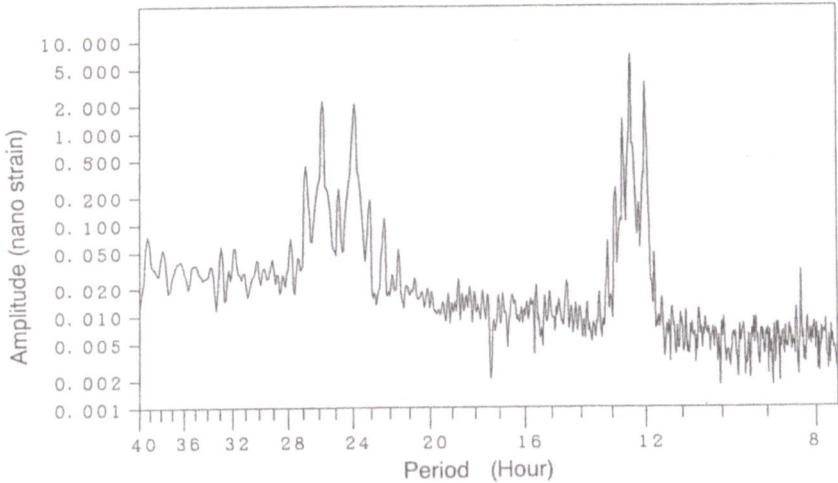

Figure 20.1 Fourier spectrum of the crustal strain NS component observed at the Esashi Earth Tides Station.

one year in 1993. The unit of the amplitude is 10^{-9} strain. From the diurnal band to the terdiurnal band, peaks of the spectrum corresponding to the principal waves listed in Table 9.1 appear.

Figure 9.2 gives an example of the analysis with BAYTAP-G for one-month period data in April 1993. In the upper part of the left side of the figure, raw observation data (unit: 10^{-9} strain; expansion of the ground is taken to be positive) are shown and missing observations are included in several places. In the left side of the middle row, the tidal components synthesized from the estimated tidal constants are shown. In the bottom of the left side, atmospheric pressure response parts are shown. In the upper part of the right side of the figure, the drift components are shown. Even in the periods of missing observations, the drift is continuously obtained. Since the tidal components of the diurnal and the shorter bands are removed, and the response components to the atmospheric pressure are removed, it becomes clear how the drift is in detail to the extent of the order of 10^{-9} strain. In the right central part, irregular parts (residuals) are shown. To show hereby as an example, the parts having slightly large residuals are not removed. In case of the analysis aiming precise determination of tidal constants, re-analysis should be carried out after removing such outliers, if any. In the right bottom, the drift components obtained when no response components are considered are shown for the sake of comparison. It is obvious that almost all of the irregular drift fluctuation comes from the variation of the atmospheric pressure. It is exceedingly difficult to express such drift form seen in right bottom of Figure 9.2 in a polynomial of the time t or sinusoidal waves. In the study that precursory phenomena of an earthquake are to be found by watching the detailed change of the crustal strain, it might be understood that it is effective enough to estimate the drift at every observation epoch by removing tidal components and disturbances such as atmospheric pressure change.

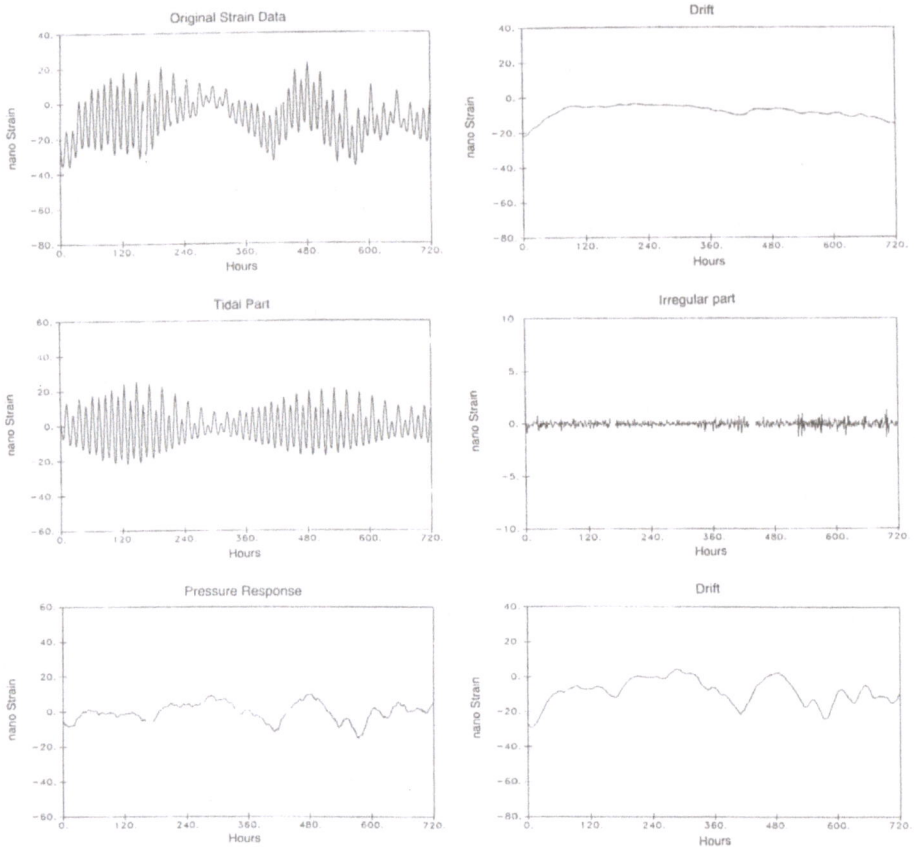

Figure 20.2 Example of the analysis of the crustal strain data using BAYTAP-G. Left side in the upper part: Observation values including missing data, Left side in the center: Tidal components, Left side in the bottom: Response components with atmospheric pressure, Right side in the upper part: Drift components, Right side in the center: irregular parts (residuals), Right side in the bottom: Drift components not in consideration of the response of the atmospheric pressure.

20.3.3 Special Use

BAYTAP-G can be used, as one of the special uses, for smoothing the observation data employing a Bayes model without estimating the tidal constants. Moreover, a use for detecting any kinds of outliers, not for mainly aiming at smoothing the data, is also possible.

Also since response calculation of the associated observation data such as atmospheric pressure can be made, BAYTAP-G can be used as a general response analysis program utterly independent of the tidal analysis problems.

20.4   Focal Points in the Analysis

With supplementary explanation concerning BAYTAP-G, several notices are described in this section.

The first thing to be taken up is that BAYTAP-G is not an "almighty tool" to determine the tidal constants by any means. Provided that no tidal components are included in the data to be analyzed, determination of the tidal constants is utterly meaningless. We strongly recommend to apply any kind of spectrum analysis to the data before carrying out tidal analysis in order to confirm presence or absence of tidal components, and to examine the observational noise level.

For that purpose, the spectrum analysis by simple FFT shown in Figure 9.1 might be satisfactory. As the analysis program is becoming a gigantic one and its contents are growing as a black-box-like tool, users are required to investigate the quality of the input data on their own responsibility.

In the general least squares method, a model giving the minimum residual squared sum is adopted as a good model. Despite this fact, BAYTAP-G selects a model allowing the form of the drift to be as smooth as possible in parallel with adequate minimization of the residual sum of squares. For the judgment of the goodness of the model, ABIC is used instead of the residual sum of squares. When comparison is made between the case that the associated observation data are not used for the analysis and the case that the associated data are used, it occasionally appears at a glance that the residual squared sum is not improved so well in latter case. However as shown in the example in Figure 9.2, the drift closer by far to a straight line is obtained when the associated observation data are used in the analysis, i.e. the value of the hyperparameter v^2 becomes larger. Thus smaller values of ABIC are usually obtained. Also in this case, it is usual that the estimation errors of the tidal constants become smaller even if the residual sum of squares is not reduced so much.

The use of ABIC must be paid attention to. ABIC is a criterion to check the goodness of the different models for the same data set. The use of ABIC is utterly not adequate for the judgment as to which quality of the observation data of January or February is better. For such comparison, how large the hyperparameter v^2 to determine the straightness of the drift can be used as an index if no outliers are included.

In BAYTAP-G, assumption that the second difference of drift becomes close to 0 is adopted as a fundamental option of the restricting condition of the drift form. For the sake of effective constraints that the drift is linear (straight-line-like), adequate sampling intervals are required for the phenomenon to be analyzed. For example when response of the rather slowly-changing atmospheric pressure variation with several-day cycle is to be obtained, it is occasionally advisable for the sampling interval to be taken as 2–3 hours instead of one hour.

The Bayes model may be introduced to another part in addition to the determination of the drift in the tidal analysis. It is considered that the relations

$$\alpha_1 \approx \alpha_2 \approx \alpha_3 \approx \cdots \approx \alpha_M, \qquad (20.11)$$

$$\beta_1 \approx \beta_2 \approx \beta_3 \approx \cdots \approx \beta_M \qquad (20.12)$$

are allowed to be satisfied in order to express smooth frequency response characteristics by introducing the prior distribution to the tidal constants α_m, β_m. It is recommended that such processing be applied to the case that actual observation made just half a year, notwithstanding that separation of the tidal constants of S_1 and K_1 waves requires a 1-year long observation period.

20.5   Concluding Remarks

The tidal analysis program BAYTAP-G is written with the standard FORTRAN77 language. We considered that the program would be used on a main frame computer when it was developed. It is now usable on various types of workstations or personal computers because machine dependence is avoided as far as possible. Approximately 2MB of memory size is required for a practical use. This size of the program is compact enough to be executed in almost all computers including personal ones.

The program is distributed either independently or as a part of the time series analysis program package TIMSAC-84 from the Institute of Statistical Mathematics (Ishiguro and Tamura 1985). At present, maintenance and management of the program version are left in the hand of National Astronomical Observatory, Mizusawa, and can be distributed with suitable media. Both institutions are the inter-university research institutes, and researchers are allowed to visit the institutes as inter-university-basis use carrying their own data and can be provided with required programs after conducting several tentative analysis. Also the program and sample data sets can be obtained via Internet at present.

Using the tidal constants determined by BAYTAP-G, a program to predict the tides have been developed. The program, which is relatively a compact one, is available with satisfaction on such as personal computers simple enough to be handled by anyone. By applying the program on the personal computers used for data-acquisition systems, it might be possible to eliminate the tidal components from continuous observation data of the crustal movement at semi-real time.

A variety of tidal analysis methods have been developed till now, but most importance was placed on a matter how the amount of the calculation should be minimized in the age of hand-operated calculation. Even from the time when use of computers was started, it has been still important problem to cut down the volume of the data to be handled and to reduce the amount of the calculation. Today, such restriction is seldom seen in most of computers, and the physical phenomena appear in the observations have been modeled as faithfully as possible. It becomes possible to solve simultaneously many of unknowns from huge number of observation data. It has been quite unnecessary to display craftsman-like techniques, and heightening analysis accuracy has been possible by conducting straightforward modeling the observing phenomena.

References

Akaike, H. (1980), "Likelihood and the Bayes procedure," *Bayesian Statistics*, eds. J. M. Bernardo, M. M. DeGroot, D. V. Lindley and A. F. M. Smith, University Press, Valencia, Spain, 143–166.

Ishiguro, M., T. Sato, Y. Tamura and M. Ooe (1984), "Tidal data analysis –an introduction to BAYTAP–," *Proceedings of the Institute of Statistical Mathematics*, Vol. 32, 71–85 (in Japanese with English abstract).

Ishiguro, M. and Y. Tamura (1985), "BAYTAP–G" in TIMSAC–84, *Computer Science Monographs*, Vol. 22, 56–117.

Harrison, J. C. (1985), *Earth Tides*, Van Nostrand Reinhold Company, New York, 98–106.

Tamura, Y. (1987), "A harmonic development of the tide-generating potential," *Marées Terrestres Bulletin d'Informations*, Vol. 99, 6813–6855.

Tamura, Y., T. Sato, M. Ooe and M. Ishiguro (1991), "A procedure for tidal analysis with a Bayesian information criterion," *Geophysical Journal International*, Vol. 104, 507–516.

Chapter 21

Detection of Groundwater Level Changes Related to Earthquakes

Norio Matsumoto
Geological Survey of Japan
1-1-3 Higashi, Tsukuba, Ibaraki 305-8567, Japan
norio@gsj.go.jp

21.1   Introduction

An earthquake is a waveform representing catastrophic release of strain energy that was gradually accumulated in the crust. Crustal strain is considered to be changed by some causes immediately before the huge destruction, and leveling, tilt, volumetric strain, groundwater level and radon concentration in groundwater are continuously observed with an aim of detecting precursory changes mainly in the South Kanto and the Tokai districts. The most famous example of crustal deformation preceding a large earthquake is the change of level in Kakegawa, Shizuoka Prefecture before the 1944 To-Nankai Earthquake (M7.9; Mogi, 1985).

The following three reasons can be considered for changes in the groundwater level related to earthquakes:

1) Deformation of aquifers,
2) New and/or changing path of the groundwater flow to or from aquifers caused by fault movement and/or strong ground motion,
3) Surface waves from distant earthquakes.

It is considered that the changes in groundwater level caused by 1) and by 2) might occur before earthquakes and the changes caused by 1) are related to changes in crustal strain. To assume that the aquifer (layer saturated with groundwater) is a kind of large volumetric strain meter and groundwater level in the aquifer is responsive to crustal strain, the aquifer is required to be confined (aquifer whose well water level is shallower than the depth of the aquifer). Barometric pressure and earth tide greatly influence the groundwater level in such an aquifers (Roeloffs, 1988). Also, when the aquifer is relatively shallow, the influence of rainfall cannot be neglected.

From the reasons described above, elimination of effects of barometric pressure and earth tide from groundwater level is necessary in isolating information concerning crustal strain from the water level of the aquifer which is sensitive to the crustal strain. Furthermore, the effect of rainfall should also be adjusted. For that purpose, it is necessary to establish an objective method to adjust for the effects of barometric pressure, earth tide, and rainfall in evaluating qualitatively the water level change in earthquakes. However, the barometric and the tidal effects has conventionally be estimated in accordance with simple linear model using one coefficient determined by the least squares method (Roeloffs, 1988). Rainfall data themselves have been shown exclusively as reference data without the rainfall effect being objectively quantified.

In this chapter, a method will be described based on a more realistic state space model, which makes it possible to extract the water level change related to earthquakes by separating the effect of barometric pressure, earth tide and rainfall, and noise components from the observed groundwater level time series. Moreover, I would like to describe the remarkable changes detected immediately after thirteen earthquakes in the water level at Haibara Observation Well, Shizuoka Prefecture during a period from April 1981 to December, 1989.

21.2   Observation Data

Groundwater level, discharge rate and radon concentration in groundwater are observed by the Geological Survey of Japan in 15 wells at 7 observation sites in the South Kanto and the Tokai Districts as of March 1994. Groundwater level is observed in 10 wells at 5 sites among them. In this chapter, groundwater level data obtained at Haibara Observation Well is analyzed for the period from April 1981 to December 1989.

Haibara Observation Well is 170m deep, and groundwater level, barometric pressure, and rainfall are observed. The data are sent to the Geological Survey of Japan, Tsukuba, Ibaraki Prefecture by wire in real time. The observation is continued from February 1981 with accuracy of 1mm in groundwater level, 1hPa in barometric pressure and 0.5mm in rainfall (Takahashi, 1993). The observed water level, barometric pressure, and rainfall together with theoretical earth tide are shown in Fig. 10.1. The observed water level is related to the barometric pressure, rainfall and earth tide, and the response to barometric pressure is especially large.

Several percent of the observations are missing from February 1981 to March 1983 owing to trouble of telemetry system. Furthermore, there are some outliers in the water level data because of incorrect synchronization between the data sampling system and the water level sensor. The state space model analysis was applied after interpolation of missing data and elimination of the outliers, which is described in chapter 22 in detail.

21.3   Data Analysis Method

21.3.1   Conventional Data Analysis Method

In this section, the conventional analysis method is described for adjustment of effects of barometric pressure and earth tide in groundwater level data.

Suppose groundwater level observed at time n is y_n, barometric pressure is p_n, and theoretical earth tide (changes of crustal strain calculated from the gravitational attraction of the sun and the moon) is e_n. Let the observed water level be expressed

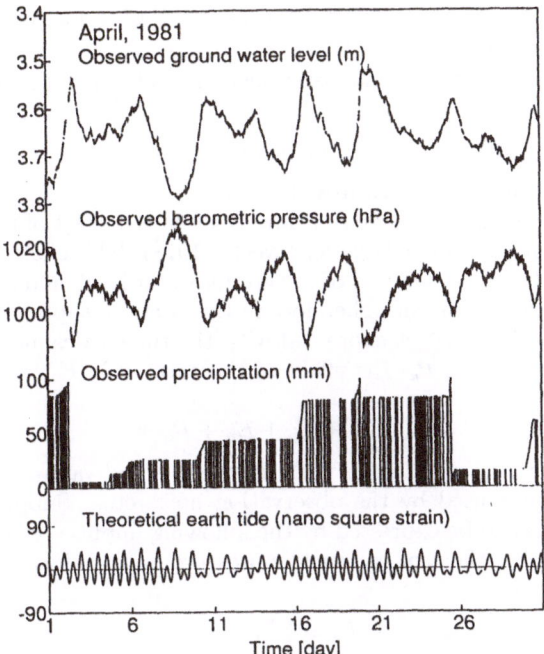

Figure 21.1 The observed groundwater level, barometric pressure, rainfall data and theoretical earth tide at Haibara Observation Well in April 1981. The rainfall data is denoted in accumulated values. Breaks-off occasionally seen in the graphs indicate occurrences of missing observations.

with the following multi-regression model using the barometric pressure and the earth tide during a period when the rainfall is scanty:

$$y_n = ap_n + be_n + \varepsilon_n , \qquad (21.1)$$

where ε_n is a residual. Coefficients a and b can be estimated by conventional least squares methods, such as Householder method (Sakamoto *et al.*, 1986).

This method cannot take into account influence of time delay of barometric pressure, and furthermore cannot estimate effect of rainfall. Thus the analysis method shown in the next section is developed.

21.3.2 Analysis by a State Space Model

In this section a method is described which estimates barometric, tidal and rainfall effects, and noise components using a state space model, and which estimates water level changes related to earthquakes by eliminating these effects and the noise component from the observed water level.

The outline of this method is as follows:

1) Construction of a groundwater level model expressing adequately the barometric, the tidal and the rainfall effects included in the observed water level.

2) Expression of time series model by a state space model and calculation of the likelihood by the Kalman filter using the time series model.

3) Searching for the maximum likelihood estimate of model parameters using the nonlinear optimization method.

4) Choice of the optimal groundwater level model using the AIC minimization method.

Below, each step of the method is explained in detail.

Time Series Model of Groundwater Level   Owing to influence of unsaturated layer above the aquifer, there is a time delay in the effect of barometric pressure on the water level of the confined aquifer (Weeks, 1978). This effect has been neglected in the conventional analysis method of the groundwater level, and furthermore rainfall effects are not considered at all. Therefore in this section, suppose that the observed water level y_n $(n = 1, \ldots, N)$ is expressed with the time series model using the effect of the barometric pressure P_n, the effect of the earth tide E_n, and the effect of the rainfall R_n.

$$y_n = t_n + P_n + E_n + R_n + \varepsilon_n \tag{21.2}$$

In (21.2), ε_n is Gaussian white noise with the mean 0 and variance σ^2, which denotes accidental variation caused by the observation noise etc.. Suppose the barometric and the tidal effects can be expressed by the following impulse response models which include the time delay effect of barometric pressure p_n and the earth tide e_n:

$$P_n = \sum_{i=0}^{\ell} a_i p_{n-i},$$
$$E_n = \sum_{i=0}^{m} b_i e_{n-i}. \tag{21.3}$$

Furthermore, suppose that the rainfall effect can be expressed by a linear autoregressive model, taking the rainfall r_n as input, as follows:

$$R_n = \sum_{i=1}^{k} c_i R_{n-i} + \sum_{i=1}^{k} d_i r_{n-i}. \tag{21.4}$$

Equation (21.4) can express the rainfall effect during a long period with a small number of coefficients. However, it is unusual for the response of groundwater level to rainfall to be oscillatory, so adequate restrictions must be placed on the autoregressive coefficients. Suppose the trend t_n $(n = 1, \ldots, N)$ changes gradually according to the following random walk model;

$$t_n = t_{n-1} + v_n, \qquad v_n \sim N(0, \tau^2) \tag{21.5}$$

This trend, which is the residual variation after eliminating the barometric, the tidal and the rainfall effects, and noise component from the observed water level, shall be hereafter called the corrected water level.

Estimation of Corrected Water Level by State Space Representaion and Kalman Filter   When the observed water level, barometric pressure, theoretical earth tide and rainfall $(y_n, p_n, r_n, e_n, \ n = 1, \ldots, N)$ are given, let us consider how to obtain the corrected water level t_n based on the models (21.2)–(21.5). For this purpose, the coefficients of the barometric and the tidal effects $a_0, \ldots, a_\ell, b_0, \ldots, b_m$, the coefficients of the rainfall effect $c_1, \ldots, c_k, d_1, \ldots, d_k$, and the rainfall effect itself

$R_1, \ldots, R_N$, and the variances τ^2, σ^2, that is, $2 \times (N + k) + \ell + m + 4$ unknown parameters should be estimated in addition to N corrected water level values. Thus let the time series models in (21.2)–(21.5) be expressed by the following state space model:

$$\begin{aligned} x_n &= Fx_{n-1} + Mr_n + Gv_n \\ y_n &= H_n x_n + w_n \end{aligned} \qquad (21.6)$$

where the state vector x_n is defined as follows:

$$x_n = (t_n, a_0, \ldots, a_\ell, b_0, \ldots, b_m, R_n, \ldots, R_{n-k+1})^t \qquad (21.7)$$

Furthermore, $v_n \sim N(0, \tau^2)$, $w_n \sim N(0, \sigma^2)$

$$F = \begin{bmatrix} 1 & & & & & \\ & I_{\ell+1} & & & & \\ & & I_{m+1} & & & \\ & & & c_1 & 1 & \\ & & & \vdots & & \ddots \\ & & & c_{k-1} & & 1 \\ & & & c_k & & \end{bmatrix}, \quad G = \begin{bmatrix} 1 \\ 0 \\ 0 \\ 0 \\ \vdots \\ 0 \\ 0 \end{bmatrix}, \quad M = \begin{bmatrix} 0 \\ 0 \\ 0 \\ d_1 \\ \vdots \\ d_{k-1} \\ d_k \end{bmatrix}, \quad (21.8)$$

$$H_n = (1, p_n, \ldots, p_{n-\ell}, e_n, \ldots, e_{n-m}, 1, 0, \ldots, 0) \qquad (21.9)$$

Here, the state vector x_n can be effectively estimated using the following Kalman filter and smoothing algorithm when parameters σ^2, τ^2, $c_1, \ldots, c_k$, $d_1, \ldots, d_k$ are given.

[One-step-ahead prediction]

$$\begin{aligned} x_{n|n-1} &= Fx_{n-1|n-1} + Mr_n \\ V_{n|n-1} &= FV_{n-1|n-1}F^t + \tau^2 GG^t \end{aligned} \qquad (21.10)$$

[Filter]

$$\begin{aligned} K_n &= V_{n|n-1}H_n^t (H_n V_{n|n-1} H_n^t + \sigma^2)^{-1} \\ x_{n|n} &= x_{n|n-1} + K_n(y_n - H_n x_{n|n-1}) \\ V_{n|n} &= (I - K_n H_n) V_{n|n-1} \end{aligned} \qquad (21.11)$$

where V_n is the covariance matrix of x_n, and $x_{n|n-1}$ is the estimate of x_n based on the data obtained by the time $n - 1$.

Furthermore, the following smoothing algorithm can allow us to estimate the n-th state vector $x_{n|N}$ and the covariance matrix $V_{n|N}$ based on N observations.

[Fixed-interval smoother algorithm]

$$\begin{aligned} A_n &= V_{n|n} F_{n+1}^t V_{n+1|n}^{-1} \\ x_{n|N} &= x_{n|n} + A_n(x_{n+1|N} - x_{n+1|n}) \\ V_{n|N} &= V_{n|n} + A_n(V_{n+1|N} - V_{n+1|n}) A_n^t \end{aligned} \qquad (21.12)$$

Because the state vector $x_{n|N}$ includes the corrected water level t_n, the rainfall effect R_n, barometric coefficients a_i, and tidal coefficients b_i, the estimates of those values can be obtained after processing the Fixed-interval smoother algorithm.

If $x_{0|0}$, $V_{0|0}$, τ^2, σ^2, $c_1, \ldots, c_k$, $d_1, \ldots, d_k$ are given in advance as initial values, then the log likelihood, the barometric, the tidal and the rainfall effects, and the corrected water level $t_1, \ldots, t_N$ can be calculated by means of the algorithms described above.

Estimation of Parameters and Choice of Models   The best estimates of τ^2, σ^2, $c_1, \ldots, c_k$, $d_1, \ldots, d_k$ are obtained by the maximum likelihood method. The log likelihood $\ell(\theta)$ of the groundwater level model is given by the following equation:

$$\ell(\theta) = -\frac{1}{2}\left\{N\log 2\pi + \sum_{n=1}^{N}\log D_{n|n-1} + \sum_{n=1}^{N}\frac{(y_n - y_{n|n-1})^2}{D_{n|n-1}}\right\}, \tag{21.13}$$

where

$$y_{n|n-1} = H_n x_{n|n-1}, \quad D_{n|n-1} = H_n V_{n|n-1} H_n^t + \sigma^2, \tag{21.14}$$

and these can be calculated using the results of the Kalman filter procedure.

According to Kitagawa and Gersch(1996), if the ratio τ^2/σ^2 of the variances τ^2 and σ^2 in the state space model is given, σ^2 is automatically determined. Hence the number of hyperparameters (the parameters of the prior distribution) to be estimated by optimization is decreased by 1 and becomes $2 \times k + 1$.

When the orders of the barometric, the tidal and the rainfall effects (ℓ, m and k, respectively) are fixed, $2 \times k + 1$ of the hyperparameters can be estimated. The likelihood function obtained by the Kalman filter is nonlinear with respect to the hyperparameter, and therefore nonlinear optimization must be used to find the maximum likelihood estimate $\hat{\theta}$. In this chapter, the simplex method (for example, Kowalik and Osborne, 1968) was used.

Choice of Orders   The optimal orders ℓ, m, and k of the barometric, the tidal and rainfall effects are determined using AIC minimization method. In case of the groundwater level model, AIC is given as follows:

$$\text{AIC} = -2\ell(\hat{\theta}) + 2(\ell + m + 2k + 4) \tag{21.15}$$

We choose the orders of the model that minimize the AIC.

21.4   Analysis of Actual Data

Approximately 70,000 data during a period of 8 years and 8 months analyzed. For this amount of data, the computation time to evaluate all combinations of orders in the effects of barometric pressure, earth tide and rainfall (for example, $25 \times 4 \times 6 = 600$ trials) is prohibitive. Therefore we actually perform the analysis according to the following procedure.

21.4.1   Determination of Orders of Barometric and Tidal Effect

First of all, let us consider the estimation of the orders ℓ, m and the coefficients $a_0, \ldots, a_\ell$, $b_0, \ldots, b_m$ of the barometric and the tidal effects exclusively. A state space expression of this model is given as follows, eliminating the rainfall effect from equation (21.8).

$$y_n = t_n + \sum_{i=0}^{\ell} a_i p_{n-i} + \sum_{i=0}^{m} b_i e_{n-i} + \epsilon_n \tag{21.16}$$

$$F = \left[\begin{array}{c|c|c} 1 & & \\ \hline & I_{\ell+1} & \\ \hline & & I_{m+1} \end{array}\right], \quad G = \left[\begin{array}{c} 1 \\ 0 \\ 0 \end{array}\right] \tag{21.17}$$

Table 21.1 The order of the effect of barometric pressure ℓ, the order of the effect of earth tide m and AIC by the analysis in section 21.4.1. $\ell = 22$ and $m = 2$ minimizes AIC.

$\ell \setminus m$	0	1	2	3
18	−48439.9	−49229.7	−51405.2	−51089.4
19	−48601.5	−49201.9	−51453.3	−51266.8
20	−48706.7	−49013.5	−51474.0	−51357.7
21	−48753.1	−50455.2	−51477.9	−51384.1
22	−48738.2	−50448.2	−51485.8	−51372.3
23	−48654.4	−50407.9	−51404.6	−51265.8
24	−48521.1	−50305.4	−51256.9	−51171.0
25	−48492.5	−50226.2	−51202.5	−51198.0

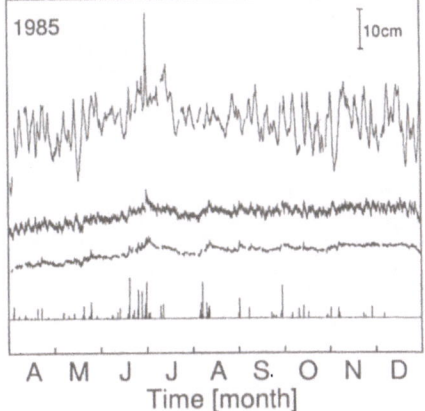

Figure 21.2 Observed water level, the residual analyzed by the conventional method in section 21.3.1, the trend of the analysis in section 21.4.1 and rainfall per hour during the period from April to December, 1985.

$$H_n = (1, p_n, \ldots, p_{n-\ell}, e_n, \ldots, e_{n-m}), \tag{21.18}$$

where initial values are taken as $x_{0|0} = (t_0, 0, \ldots, 0)^t$, $V_{0|0} = I$. Meanwhile, t_0, the initial value of the trend, is taken as an unknown. That is, t_0 and the hyperparameter τ^2/σ^2 are the parameters we estimate using nonlinear optimization in this section.

In order to estimate the optimal model, we calculate AIC for many trial of combination of orders of the barometric and tidal effects ℓ, m using the 6,600 data samples obtained between April and December, 1985 when rainfall was relatively small and no great earthquakes occurred.

As a result, the AIC is minimized for $\ell = 22$ and $m = 2$ as listed in Table 21.1. In Figure 21.2, the trend t_n when $\ell = 22$, $m = 2$ is shown together with the residual analyzed using the conventional analysis method shown in section 21.3.1.

The trend obtained by the analysis in this section is smoother than the residual in the conventional method, and the influence of the rainfall is clearer. This is because the noise component is removed and the time delay effect of barometric pressure is

Table 21.2   Orders of the rainfall effect k and AIC by the analysis mentioned in section 21.4.2, where $\ell = 22$ and $m = 2$.

k	AIC
1	−58371.7
2	−58374.3
3	−58538.2
4	−58534.7
5	−58530.4
6	−58536.0

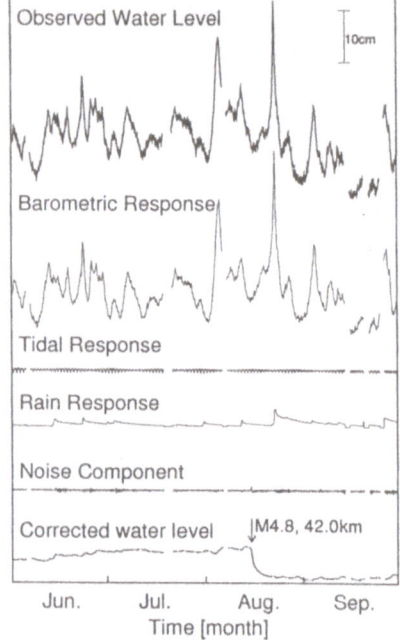

Figure 21.3   Observed water level, the barometric, the tidal and the rainfall effects, noise component, and corrected water level during the period from April to September, 1981 by the analysis mentioned in section 21.4.2.

sufficiently estimated in the state space analysis.

21.4.2   Estimation of Rainfall Effect

Next, the order k and the coefficients c_i, d_i $(i = 1, \ldots, k)$ of the rainfall effect in equation (21.4) are estimated by 6,600 data observed from April to December 1981, fixing the orders ℓ and m of the barometric and the tidal effects at $\ell = 22$ and $m = 2$ as obtained in the previous section 21.4.1. We reestimate a_i and b_i simultaneously with estimating k, c_i and d_i.

If k is large value, it is difficult to choose the initial values of the $2k + 1$ hyper-parameters that will lead to the globally optimum solution. Therefore it becomes

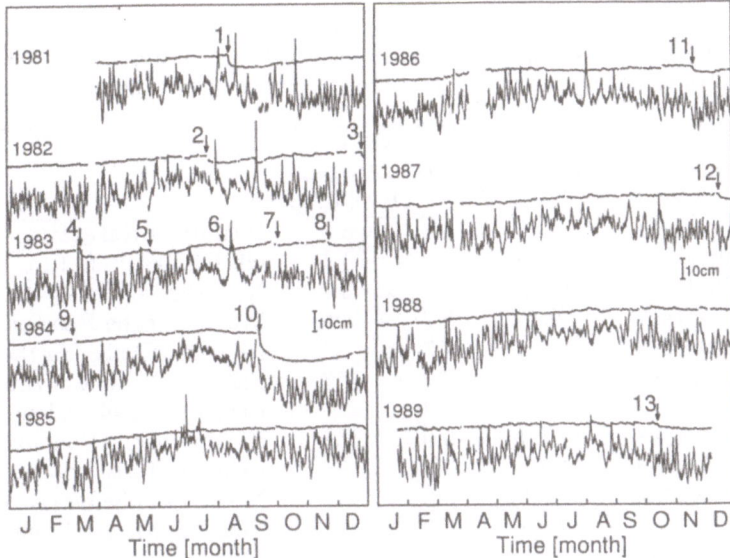

Figure 21.4   The heavy line indicates the corrected water level after eliminating the barometric, the tidal and the rainfall effects, and the noise component by the analysis in section 21.4.3. The thin line indicates the observed water level. Both are during the period from February, 1981 to December, 1989.

problems how to take initial values at a nonlinear optimization. In this section, we take the initial values of the coefficients calculated by the method of Matsumoto (1992) which allows the coefficients of the rainfall effect to be estimated relatively easily, and in which the effect of rainfall is separately estimated from the barometric and the tidal effects. The other initial values were chosen as in the previous section 21.4.1. Here, the optimization and the calculation of AIC were made for the order k ($k = 1, \ldots, 6$) of the rainfall effect. The results are listed in Table 21.2.

When $k = 3$, AIC is minimized. The trend t_n estimated using this model is shown in Figure 21.3. It is revealed from the figure that the barometric, the tidal and the rainfall effect can be adequately estimated. Furthermore, we can now see that the corrected water level changes after the earthquake of M4.8 at August 15th, although we cannot distinguish whether the raw water level changes or not in that time.

21.4.3   Estimation of Corrected Water Level during the Entire Period

Finally, we estimate barometric, tidal and rainfall effects, noise component, and corrected water level in the entire period from April 1981 to December 1989, substituting the optimal orders $\ell = 22$, $m = 2$, and $k = 3$ obtained in the previous section into the model of equation (21.8) In this calculation, the number of the hyperparameters to be estimated is also $2 \times k + 1$ $(c_1, \ldots, c_k, d_1, \ldots, d_k, \tau^2/\sigma^2)$.

The data sets, excluding missing observations includes 68,237 observations. The corrected water level analyzed in this section is shown in Figure 21.4. This analysis reveals, in the corrected water level, 13 instances of remarkable changes following earthquakes. The magnitudes of the earthquakes, hypocentral distances from the

Table 21.3 The list of the earthquakes with which the corrected water level drops in Figure 21.4 . In the amounts of the water level drops, "–" indicates the amount of the drop cannot be determined because of missing observations.

No.	Date	Hypocentral distance	M	Amount of water level drop	Epicenter
1	8/15/1981	42.0	4.8	6.3	Vicinity of Kakegawa
2	7/23/1982	374.9	7.0	4.4	Off Ibaraki Prefecture
3	12/28/1982	155.8	6.4	3.4	Near the Miyake Island
4	3/16/1983	66.0	5.7	4.6	Vicinity of the Hamanako Lake
5	5/26/1983	621.9	7.7	1.5	Nihonkai-Chubu-Earthquake
6	8/8/1983	113.1	6.0	2.5	Eastern Yamanashi Prefecture
7	10/3/1983	150.4	6.2	–	Near the Miyake Island
8	11/24/1983	57.1	5.0	2.1	Vicinity of the Hamanako Lake
9	3/6/1984	741.5	7.9	–	Near the Torishima Island
10	9/14/1984	128.0	6.8	14.1	Nagano-ken-Seibu Earthquake
11	11/22/1986	126.1	6.0	3.0	Near the Niijima Island
12	12/17/1987	226.6	6.7	2.7	Chiba-ken-touhou-oki Earthquake
13	10/15/1989	122.0	5.7	1.7	Near the Izu-Oshima Island

Haibara Observation Well, amount of water level change etc. are listed in Table 21.3.

A relationship between hypocentral distance 'Dis' from Haibara Observation Well and the magnitude 'M' of the earthquakes caused remarkable changes in the corrected water level is shown in Figure 21.5. For the Haibara Observation Well, the line $M = 2.45 \log Dis + 0.69$ appears to be boundary in that if the magnitude M of the earthquake is large enough or if the hypocentral distance Dis is short enough that the earthquake plots left of this line, then the corrected water level will change immediately after the earthquake.

21.5 Conclusions

Many reports have been released since earlier times purporting that groundwater level changes greatly before and after earthquakes (Rikitake, 1976). However, groundwater level changes owing to many causes, and therefore it has been difficult to evaluate quantitatively the effect of earthquakes on water level. In this chapter, it is shown that eliminating the barometric, the tidal and the rainfall effects together with the noise component from the observed groundwater level makes it possible to detect distinct groundwater level changes related to earthquakes. The corrected water level might include possible precursors of earthquakes such as, for example, gradual decrease before earthquake No.10 in Figure 21.4. Further analysis and evaluation are in progress for this matter at present.

In prosecuting earthquake prediction research, it might be necessary to use this method to analyze not only groundwater level, but also strain, tilt and gravity, and it might be indispensable to make those observations and evaluate them with close relationships.

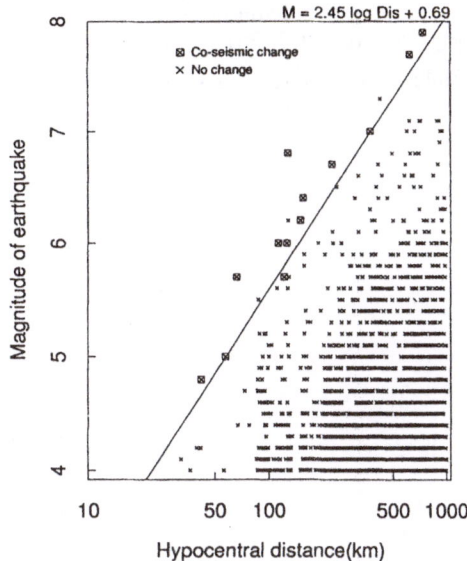

Figure 21.5 Relationship between the magnitude of the earthquakes occurring between April 1981 and December 1989 and the hypocentral distance from Haibara Observation Well

References

Kitagawa G. and Gersch W. (1996), *Smoothness priors analysis of time series*, Springer-Verlag, New York.

Kowalik, J. and Osborne, M. R. (1968), *Methods for unconstrained optimization problems*, Elsevier Publishing Company, Inc.

Matsumoto, N. (1992), "Regression analysis for anomalous changes of ground water level due to earthquakes," *Geophysical Research Letters*, Vol. 19, 1193–1196.

Mogi. K. (1985), *Earthquake Prediction*, Academic Press, Tokyo.

Rikitake, T. (1976), *Earthquake prediction*, Elsevier Publishing Company, Inc., Dordrecht.

Roeloffs, E. A. (1988), "Hydrologic precursors to earthquakes: a review," *Pure and Applied Geophysics*, Vol. 126, 177–206.

Sakamoto, Y., Ishiguro, M. and Kitagawa, G. (1986), *Akaike Information Criterion Statistics*, D.Reidel Publishing Company, Inc.

The author wishes to express his appreciation Prof. G. Kitagawa for his many suggestions and comments on this work started from 1988. The author is grateful to Dr. E. Roeloffs of U. S. Geological Survey for her careful reading of English draft version. This research carried out under the ISM Cooporative Research Program from 1988 to 1997 under Grants 97-ISM-CRPA-57.

Takahashi, M.(1993), "Groundwater telemetering system for earthquake prediction",
    Journal of Geography, Vol. 102, 241–251 (in Japanese).

Weeks, E. P. (1978), "Field determination of vertical permeability to air in the un-
    saturated zone," *U. S. Geological Survey Professional Paper*, No. 1051, 1–41.

Chapter 22

Processing of Missing Observations and Outliers in Time Series

Genshiro Kitagawa
The Institute of Statistical Mathematics
4-6-7 Minami-Azabu, Minato-ku, Tokyo 106-8569, Japan
kitagawa@ism.ac.jp

22.1 Missing Observations and Outliers

Thanks to rapid and remarkable development of electronic devices such as computers and measurement instruments, a variety of time series has been available automatically and continuously, resulting in minute-by-minute accumulation of a large amount of data. However when time series is observed for a long period, part of the time series is sometimes unable to be observed owing to some accidental causes such as malfunction of instrument or some physical restrictions of the observation system. In such a case, unobserved data is called missing observations. Even with only several percent of the missing observations, if they are studded with the data, the length of actually available data may become very short if only a part observed continuously is taken out.

Figure 22.1 shows the underground water level data observed in a well of the depth 170m, Haibara, Shizuoka by Geological Survey of Japan. On this observation point, the underground water level, atmospheric pressure, precipitation, etc. have been observed at an interval of 2 minutes since 1979 (Kitagawa and Matsumoto 1996). The resolution and accuracy of the water level gauge have been 1mm and ±2mm, respectively. In the analysis described hereunder, the data obtained by sampling the observed values since April 1981 at a 10-minute interval are used.

The initial problem in the analysis of this underground water level data is that so many missing observations are included. In Figure 22.1, the missing observations are shown by giving the minimum value of the ordinate. Accordingly the vertical lines indicate the portions having missing observations, and it is understood that such portions are studded with the entire observation interval. The time series analysis was conventionally applied to the portions where continuous data were able to be obtained. Even after sampling the data with one hour interval, the maximum length

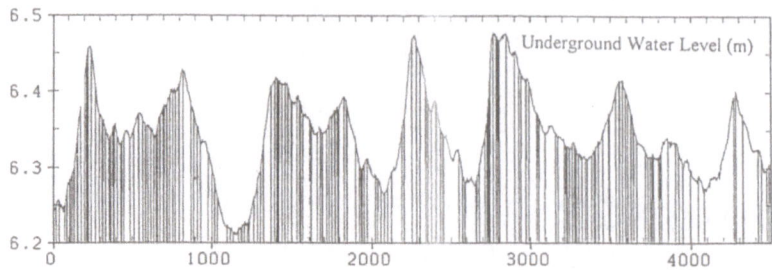

Figure 22.1　Underground water level in meter. Sampling interval: 10min

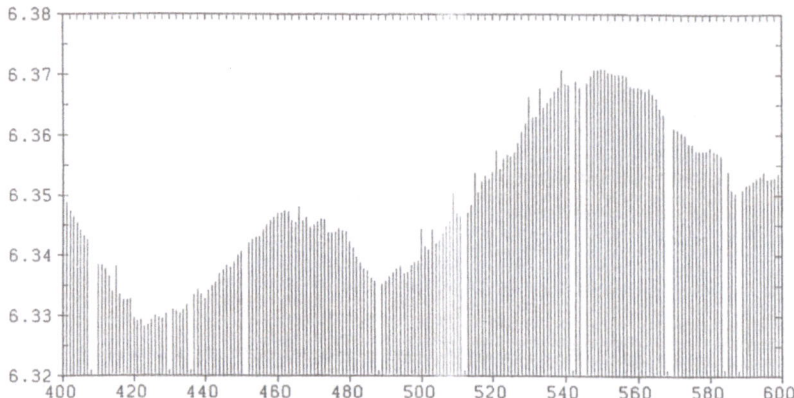

Figure 22.2　Part of the underground water level data in meter. Sampling interval: 10min.

of the continuously observed data was only 200 hours by March 1983 and 600 hours since that time on.

In such a case, we sometimes analyze the data by regarding all the observations are obtained after replacing the missing observations with a mean value of the data or by estimating them with linear interpolation. Such processing is equivalent to assuming an arbitrary model for a time series, and there is a possibility that great bias is produced in subsequent analysis. Therefore in order to continuously monitor the abnormality of ground water level by effectively using the observed data, it is necessary to establish a method to handle a model or to estimate the missing observations by automatically processing the data with many missing observations.

Furthermore, Figure 22.2 which shows a part of the ground water level data in Figure 22.1, reveals that leaps in an upward direction exist. The leap is produced due to non-synchronism of the sensor with the telemeter at the time of the measurement of the water level, and continues to be seen until 1990 when a new measurement method is introduced.

The observed values remarkably deviated from the ordinary variation of the data are called outliers. In case of Figure 22.2, the cause of the abnormality is known and most of them are easily detectable for human observers. However to detect or to treat reasonably the outliers that are contained as much as several percent in the data of approximately 520,000 observations and to correct them, development of an automatic processing method is required.

In this chapter, we first show a method to estimate a parameter of a time series model from the data containing missing observations by using a state space model and Kalman filter together with a method to estimate missing observations by using an estimated model. Secondly, we introduce a method of treating the data with both outliers and missing observations by using non-Gaussian distribution with heavier tail than the normal distribution for the observation noise. These methods allow the modeling by eliminating adequately the influence of the outliers. However since the Kalman filter cannot take into account the non-Gaussianity, it is necessary to use a non-Gaussian filter.

22.2   Processing of Missing Observations

22.2.1   State Space Model of Time Series

Suppose y_n is an ℓ variate time series. A state space model of the time series is given by

$$
\begin{aligned}
x_n &= F_n x_{n-1} + G_n v_n & \text{(system model)} & \quad (22.1) \\
y_n &= H_n x_n + w_n, & \text{(observation model)} & \quad (22.2)
\end{aligned}
$$

where x_n is a k-dimensional vector that cannot be directly observed, and is called a state. v_n is called system noise, which is an m-dimensional Gaussian white noise with the mean 0 and the variance covariance matrix Q_n. On the other hand, w_n is called observation noise, which is an ℓ-dimensional Gaussian white noise with the variance covariance matrix R_n. F_n, G_n, and H_n are $k \times k$, $k \times m$, and $\ell \times k$ matrixes, respectively. Most of the linear models used in the time series analysis can be dealt with in a unified manner by expressing them in this state space model form.

For example, when a time series y_n is expressed by the AR model

$$
y_n = \sum_{i=1}^{k} a_i y_{n-i} + v_n, \qquad (22.3)
$$

a state space representation of the model can be obtained by defining the state vector as $x_n = (y_n, y_{n-1}, \ldots, y_{n-k+1})^T$ and by putting F, G, and H to

$$
F = \begin{bmatrix} a_1 & a_2 & \cdots & a_k \\ 1 & \cdots & 0 & 0 \\ \vdots & \ddots & \vdots & \vdots \\ 0 & \cdots & 1 & 0 \end{bmatrix}, \qquad G = \begin{bmatrix} 1 \\ 0 \\ \vdots \\ 0 \end{bmatrix}, \qquad (22.4)
$$

$$
H = [\, 1 \ 0 \cdots 0 \,],
$$

with $Q_n = \sigma^2$ and $R_n = 0$. On the other hand, putting $R_n = \sigma^2$ gives the expression of the time series where observation noise with the variance σ^2 is added to AR process.

22.2.2   Estimation of the State by Kalman Filter

By using a state space model of a time series, the exact likelihood can be calculated even for the data with missing observations which enables the maximum likelihood estimation of the parameters. Let I_n be a set of the time points until the time n with which the time series was actually observed. In case that there are no missing observations, it is given by $I_n = \{1, \ldots, n\}$. We now consider a problem of estimating the state x_n based on the observed value $Y_m \equiv \{y_i | i \in I_m\}$. Especially in case that $m < n$, the problem is called prediction, whereas it is called a filter if $m = n$, and a smoothing if $m > n$.

Let the conditional mean of x_n given the observations Y_m be expressed as $x_{n|m}$. Also let the variance covariance matrix of x_n be expressed as $V_{n|m}$. In case that there are no missing observations, both the one-step-ahead prediction $x_{n|n-1}$, $V_{n|n-1}$ and the filter $x_{n|n}$, $V_{n|n}$ can be calculated recursively by the following Kalman filter:

One-step-ahead prediction

$$
\begin{aligned}
x_{n|n-1} &= F_n x_{n-1|n-1} \\
V_{n|n-1} &= F_n V_{n-1|n-1} F_n^T + G_n Q_n G_n^T.
\end{aligned}
\tag{22.5}
$$

Filter

$$
\begin{aligned}
K_n &= V_{n|n-1} H_n^T (H_n V_{n|n-1} H_n^T + R_n)^{-1} \\
x_{n|n} &= x_{n|n-1} + K_n(y_n - H_n x_{n|n-1}) \\
V_{n|n} &= (I - K_n H_n) V_{n|n-1}.
\end{aligned}
\tag{22.6}
$$

In case that y_n is missing, it can be seen that $I_n = I_{n-1}$ and $Y_n = Y_{n-1}$ holds. Therefore if y_n is missing, we have $x_{n|n} = x_{n|n-1}$ and $V_{n|n} = V_{n|n-1}$. From this it is suggested that the filter step be neglected. Formally, it is equivalent to apply the filter step with the Kalman gain $K_n = 0$ by putting $R_n = \infty$. When $x_{n|n-1}$ and $V_{n|n-1}$ are obtained, from equation (22.2), the predicted value of y_n and its variance covariance matrix can be easily obtained by

$$
\begin{aligned}
y_{n|n-1} &= H_n x_{n|n-1} \\
d_{n|n-1} &= H_n V_{n|n-1} H_n^T + R_n.
\end{aligned}
\tag{22.7}
$$

22.2.3   Likelihood Calculation and Estimation of the Parameter of Time Series Model

Suppose a time series model with a parameter θ is available and its state space model is given. Also suppose, when a time series $Y_n = \{y_j | j \in I_n\}$ is given, the joint density function of Y_n specified by the time series model is expressed as $f_N(Y_N|\theta)$. Here, since $Y_n = \{Y_{n-1}, y_n\}$ holds when y_n is observed, $f_n(y_n|\theta)$ can be decomposed as

$$
f_n(Y_n|\theta) = f_n(Y_{n-1}, y_n|\theta) = f_{n-1}(Y_{n-1}|\theta) g_n(y_n|Y_{n-1}, \theta).
\tag{22.8}
$$

On the other hand, if y_n is not observed, from $Y_n = Y_{n-1}$, we have

$$
f_n(Y_n|\theta) = f_{n-1}(Y_{n-1}|\theta).
\tag{22.9}
$$

The likelihood of the model is defined by

$$
L(\theta) = f_N(Y_N|\theta).
\tag{22.10}
$$

Therefore by applying

$$
f_n(Y_n|\theta) = \begin{cases} f_{n-1}(Y_{n-1}|\theta) g_n(y_n|Y_{n-1}, \theta) & \text{if } y_n \text{ is observed} \\ f_{n-1}(Y_{n-1}|\theta) & \text{if } y_n \text{ is missing} \end{cases}
$$

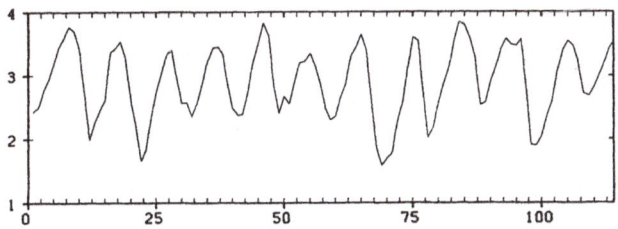

Figure 22.3   Canadian Lynx Data

repeatedly for $n = N, N - 1, \ldots, 2$, the likelihood of the time series model can be expressed as

$$L(\theta) \;=\; \prod_{n \in I_N} g_n(y_n|Y_{n-1}, \theta), \tag{22.11}$$

by the product of the conditional density functions satisfying $n \in I_N$. Here for the sake of simplicity, let $Y_0 = \phi$ (empty set) and $f_1(y_1|\theta) \equiv g_1(y_1|Y_0, \theta)$. The log-likelihood is then given by

$$\ell(\theta) \;=\; \log L(\theta) \;=\; \sum_{n \in I_N} \log g_n(y_n|Y_{n-1}, \theta). \tag{22.12}$$

Meanwhile, $g_n(y_n|Y_{n-1}, \theta)$ is the predictive distribution of y_n given the observation Y_{n-1}. Since the distribution becomes ℓ-dimensional normal distribution with the mean $y_{n|n-1}$ and the variance covariance matrix $d_{n|n-1}$, it can be expressed as

$$g_n(y_n|Y_{n-1}, \theta) \;=\; \left(\frac{1}{\sqrt{2\pi}}\right)^{\ell} |d_{n|n-1}|^{-\frac{1}{2}} \exp\left\{-\frac{1}{2}\varepsilon_n^T d_{n|n-1}^{-1}\varepsilon_n\right\}, \tag{22.13}$$

where the vector of prediction error is defined by $\varepsilon_n = y_n - y_{n|n-1}$. Therefore by substituting this into equation (22.11), the log-likelihood of the time series model can be obtained by

$$\ell(\theta) \;=\; -\frac{1}{2}\left\{\ell N \log 2\pi + \sum_{n \in I_N} \log |d_{n|n-1}| + \sum_{n \in I_N} \varepsilon_n^T d_{n|n-1}^{-1}\varepsilon_n\right\}. \tag{22.14}$$

Many of the stationary time series models such as an AR model, ARMA model, etc. and non-stationary models such as a trend model, seasonal adjustment model, etc. can be expressed in a linear-Gaussian state space model. Accordingly an unified algorithm to calculate the log-likelihood has successfully been obtained with the use of Kalman filter even when the missing observations are contained in the data. The maximum likelihood estimates of the parameters of the time series model can be obtained by maximizing the log-likelihood function by using the numerical optimization procedure (Dennis and Schnabel 1983).

To examine the influence of the missing observations in model estimation, an experiment as shown below was conducted. Illustrated in Figure 22.3 are the Canadian Lynx Data widely known in the time series analysis, revealing catch of lynx of every year in Mckengie river area, Canada. The data length, $N = 114$. With this data, AR models with various orders were estimated by the maximum likelihood method and the goodness of the models were compared by AIC (Sakamoto et al. 1986). By this criterion, the order 11 was selected as the best order for the complete data. For the sake of brevity, the order is hereunder fixed to this 11, and consider the case when $\alpha = 10\%$, 20%, 30%, 40%, and 50% of the observations are missing. Three cases as shown below were considered.

1) Shortening: Let a terminal portion of the data be the missing observations. In this case, it is equivalent that the data number is made fewer.

2) Random: Let the location of the missing observations be chosen at random.

3) Continuity: Let the two intervals with length $\alpha/2\%$ each from $n = 21$ and $n = 71$ are assumed to be the missing observations.

The goodness of the estimated model was evaluated by the Kullback-Leibler information quantity (Sakamoto *et al.* 1986).

$$I(g; f) = \int_{-\infty}^{\infty} \log \left\{ \frac{g(x)}{f(x)} \right\} g(x) dx. \tag{22.15}$$

Here g and f are the distributions specified by the AR models estimated from $N = 114$ complete data and the data with missing observations. Assuming that the parameters of the models are respectively given by $(\sigma^2, a_1, \ldots, a_{11})$ and $(\hat{\sigma}^2, \hat{a}_1, \ldots, \hat{a}_{11})$, $I(g; f)$ can be calculated by

$$I(g; f) = -\frac{1}{2} \left\{ 1 + \log \left(\frac{\sigma^2}{\hat{\sigma}^2} \right) + \sigma^2 \left(C_0 - 2 \sum_{m=1}^{11} \hat{a}_m C_m + \sum_{m=1}^{11} \sum_{l=1}^{11} \hat{a}_m \hat{a}_\ell C_{m-\ell} \right) \right\}, \tag{22.16}$$

where C_k is an autocovariance function determined by the model $(\sigma^2, a_1, \ldots, a_{11})$ and can be obtained by solving the Yule-Walker equation,

$$C_0 = \sum_{i=1}^{11} a_i C_i + \sigma^2$$

$$C_k = \sum_{i=1}^{11} a_i C_{k-i}, \qquad k = 1, \ldots, 11. \tag{22.17}$$

Table 22.1 shows the change of the Kullback Leibler information quantity $I(g; f)$, when the value of α was altered. Even in the case that the missing observations are scattered at random, deterioration is of the same order as the case when the data number is $\alpha\%$ decreased. In case of the same data number, contrarily to the

Table 22.1 Influence of the missing observations to the goodness of the estimated model evaluated by $I(g; f)$

α	shortened	random	continuous
10%	0.00262	0.00476	0.00701
20%	0.00504	0.00843	0.01998
30%	0.00731	0.01192	0.17355
40%	0.07128	0.04692	0.08226
50%	0.09432	0.10036	0.18144

Table 22.2 Change of $I(g; f)$ by the AR order

order	$I(g; f)$	order	$I(g; f)$	order	$I(g; f)$
1	0.5100	7	0.0982	13	0.0103
2	0.1598	8	0.0899	14	0.0104
3	0.1544	9	0.0857	15	0.0107
4	0.1342	10	0.0608	16	0.0181
5	0.1254	11	0.0000	17	0.0182
6	0.1230	12	0.0093	18	0.0249

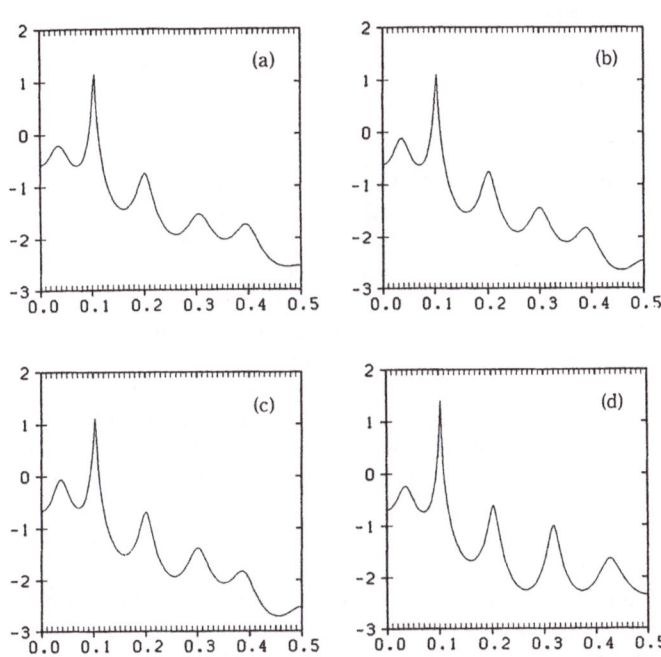

Figure 22.4 Change of the spectra by the missing observations; (a) Original data, (b) 10% missing, (c) 30% missing, (d) 50% missing

expectation, random missing case may yield better models rather than the shortened case.

Table 22.2 shows the information quantity $I(g; f)$ with respect to the difference between the model with various orders of the AR model and the model with order 11. From the table, it can be seen that the models estimated from the data with up to 40% random missing observations are better than the model with the order of less than 11 estimated from the complete data.

Figure 22.4 shows how the spectrum obtained from the estimated AR model changed when missing observations appear at random. The abscissa indicates the frequency, whereas the ordinate denotes the logarithmic values of the power spectrum. (a) shows estimated values based on the original data free from missing observations. On the other hand, (b), (c), and (d) show the case when 10%, 30%, and 50% of the observations are missing at random. It can be seen that even with many missing observations, very good estimates of the spectrum are obtained.

22.2.4   Interpolation of Missing Observations by Smoothing

The problem of smoothing is to estimate the state $x_n, 1 \leq n \leq N$ when a time series $Y_N = \{y_n | n \in I_N\}$ is given. With respect also to this smoothing, recursive algorithm called fixed-interval smoothing can be used as with the case of Kalman filter. While the filter estimates of x_n is obtained by using the observations up to the time n, the algorithm of smoothing uses all the observations. Therefore by the smoothing, we can obtain estimates better in accuracy than the filter.

Fixed-Interval Smoothing

$$
\begin{aligned}
A_n &= V_{n|n} F_n^T V_{n+1|n}^{-1} \\
x_{n|N} &= x_{n|n} + A_n(x_{n+1|N} - x_{n+1|n}) \\
V_{n|N} &= V_{n|n} + A_n(V_{n+1|N} - V_{n+1|n})A_n^T.
\end{aligned}
\tag{22.18}
$$

In the algorithm of smoothing, the results of Kalman filter, i.e. $x_{n|n-1}$, $x_{n|n}$, $V_{n|n-1}$, and $V_{n|n}$ are used for calculation. Hence to do smoothing, $\{x_{n|n-1}, x_{n|n}, V_{n|n-1}, V_{n|n}\}$, $n = 1, \ldots, N$ should be obtained by the Kalman filter at first. After that, the smoothed values can be obtained by the backward recursion, starting from $x_{N-1|N}, V_{N-1|N}$ to $x_{1|N}, V_{1|N}$. In this connection, the initial values $x_{N|N}, V_{N|N}$ are required in order to start the backward algorithm of smoothing, but it should be noted that these can be obtained by the Kalman filter.

The interpolation of the missing observations is easily realized from the smoothed values of the state x_n. That is to say, since a relation between the state x_n and the time series y_n is given by the observation model (22.2), the interpolation of the missing observation y_n and its variance covariance matrix can be obtained by

$$
y_{n|N} = H_n x_{n|N} \tag{22.19}
$$

$$
d_{n|N} = H_n V_{n|N} H_n^T + R_n. \tag{22.20}
$$

Shown in Figure 22.5 is the result of the estimation of the missing observations, when missing observations of 40% are assumed. (a) indicates the case when the missing observations appeared at random, whereas (b) shows the case when they appear continuously. The solid line shows the estimated observations and their standard error bound. On the other hand, o shows the assumed missing observations which are not used for calculation. Since observations are usually available both in front and in the rear in case that (a) is at random, the estimation accuracy of the missing observations is considerably good. On the other hand, in the continuously missed case (b), the estimated error bound is large on the central portion of the missing interval. Even in that case, however, periodical variation of the time series is well reproduced.

22.3   Processing of Outliers

22.3.1   Non-Gaussian State Space Model

For processing of the outliers included in the data and the missing observations as shown in Figures 22.1 and 22.2, consider a simple non-Gaussian state space model

$$
\begin{aligned}
x_n &= x_{n-1} + v_n \\
y_n &= x_n + w_n,
\end{aligned}
\tag{22.21}
$$

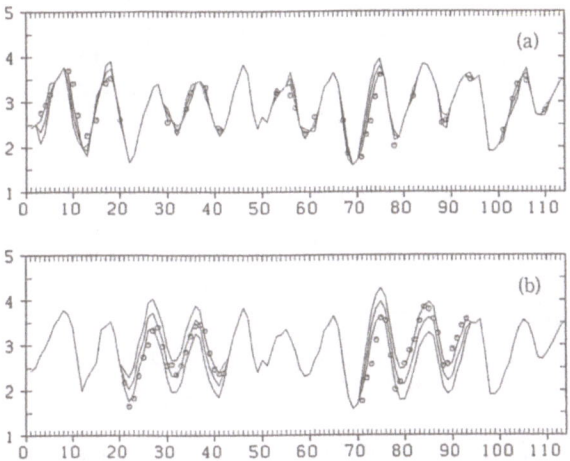

Figure 22.5 Interpolated data when 40% missing observations are assumed, (a) In case of randomly missing observations, (b) In case of continuously missing observations

where v_n is distributed as the normal distribution with mean 0 and the variance τ^2, $q(v|\tau^2)$, but w_n is not necessarily distributed as normal distribution and its density function is given by $r(w|\theta)$. In general, in the state space model of (22.2), the Gaussian observation noise w_n can be replaced by a non-Gaussian noise.

For processing of the data with outliers as shown in Figure 22.2, we use a heavy-tailed distribution than the normal distribution for observation noise. As an example of such a distribution, Cauchy distribution

$$r_c(w|\sigma^2) = \frac{\sigma}{\pi(w^2 + \sigma^2)} \tag{22.22}$$

is frequently used. The density function, $r_c(w|\sigma^2)$, of the Cauchy distribution takes a large value in the vicinity of 0, not so much rapid attenuation as in the normal distribution is seen when the absolute value of w becomes large. Therefore if the Cauchy distribution is used, then it takes a value close to 0 at a high probability. However it is possible to express the existence of a possibility that exceedingly large variation will appear at a low probability, and appearance of outliers can very easily be incorporated into a model.

Despite the above, the abnormality in case of the underground water level data is due to a problem of the measurement method, and always appears in the positive side. Thus in this case, we may also consider a Gaussian-mixture distribution

$$r_m(w|\alpha, \sigma^2, \mu, r^2) \sim (1-\alpha)N(0, \sigma^2) + \alpha N(\mu, r^2), \qquad \mu > 0 \tag{22.23}$$

where $N(0, \sigma^2)$ exhibits the normal distribution with the mean 0 and the variance σ^2, and corresponds to the observation noise during the normal hours. On the other hand, $N(\mu, r^2)$ expresses the distribution during the abnormal hours.

22.3.2 Non-Gaussian Filter and Smoothing

By using a distribution given in (22.20) or (22.21) for the observation noise of the trend model in (22.19), the smoothing of the data with outliers as shown in Figure

22.2 can be made.

However, since Kalman filter cannot yield good estimates for state space model with non-Gaussian noise, it is necessary to use the following algorithms of the non-Gaussian filter and smoothing which directly calculate the conditional density function $p(t_n|Y_m)$ of the state t_n.

Non-Gaussian Filter and Smoothing

$$p(t_n|Y_{n-1}) = \int_{-\infty}^{\infty} q(t_n - t_{n-1}|\tau^2)p(t_{n-1}|Y_{n-1})dt_{n-1}$$

$$p(t_n|Y_n) = \frac{r(y_n - t_n|\theta)p(t_n|Y_{n-1})}{p(y_n|Y_{n-1})} \tag{22.24}$$

$$p(t_n|Y_N) = p(t_n|Y_n)\int_{-\infty}^{\infty} \frac{q(t_{n+1} - t_n|\tau^2)p(t_{n+1}|Y_N)}{p(t_{n+1}|Y_n)}dt_{n+1},$$

where $p(y_n|Y_{n-1}) = \int r(y_n - t_n|\theta)p(t_n|Y_{n-1})dt_n$. Since p and r generally become non-Gaussian distribution, they cannot be determined just by obtaining a mean vector and variance covariance matrix as did in Kalman filter. In order therefore to realize the algorithms of non-Gaussian filter and smoothing, a method to approximate a general density function is required. As one of such methods, we approximate it by a step function by dividing the range of the density into a certain number of intervals, for example 400 intervals. Then, the right hand sides of equations in (22.3) can be calculated numerically by using the product or sum of the values in the intervals. For further details, refer to Kitagawa (1987) and Kitagawa and Gersch (1996).

22.3.3   Processing of Outliers

Table 22.3 shows AIC of the fitted models and the maximum likelihood estimates of the parameters. Obviously, it is comprehended that the model with Gaussian-mixture distribution is the best one. It is noted that AIC of the Cauchy distribution model is worse than the normal distribution model. This is because that Cauchy distribution has heavy tails on both the sides, notwithstanding that the outliers in Figure 22.2 appeared exclusively in the positive side.

In Figure 22.6, just a slight portion of the trend estimated by the non-Gaussian smoothing is shown. The thin line indicates the estimated values by the normal distribution model, whereas the bold line denotes the estimated values by the Gaussian-mixture distribution model. Because the observation noise is exceedingly small at the normal time, the maximum likelihood estimate of the dispersion of the normal distribution becomes small.

The estimates by the Gaussian model are strongly affected by the outliers and the great leap of the trend in an upward direction is noticed in the part of the outliers.

Table 22.3 Values of 3 models and AIC supposed on the observation noise together with the maximum likelihood estimates of the parameters

Model	AIC	$\hat{\tau}^2$	$\hat{\sigma}^2$	$\hat{r}^2$	$\hat{\alpha}$	$\hat{\mu}$
Normal	−8741.1	0.28×10^{-5}	0.18×10^{-6}			
Cauchy	−8654.9	0.19×10^{-5}	0.18×10^{-6}			
Gaussian-mixture	−8936.1	0.16×10^{-5}	0.13×10^{-6}	0.10×10^{-5}	0.05	0.004

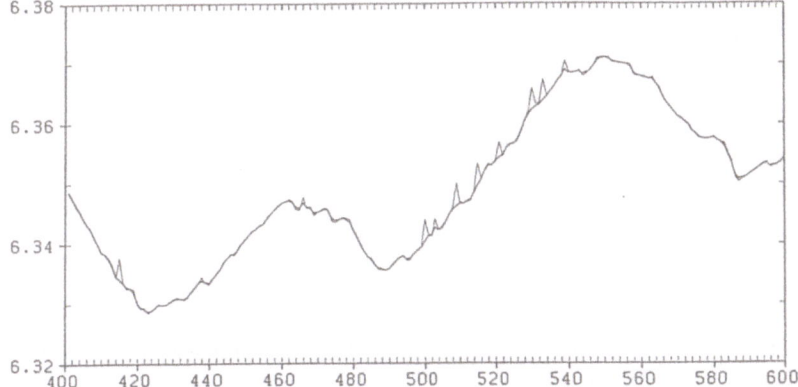

Figure 22.6 Trend obtained by the non-Gaussian smoothing; thin line: normal distribution model, bold line: Gaussian-mixture distribution

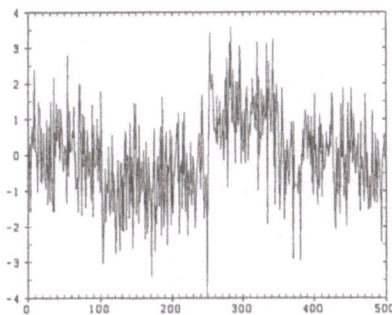

Figure 22.7 Artificially generated data with suddenly shifted mean value

On the other hand, the estimates by the Gaussian-mixture model are very smooth and adequately ignored the outliers.

22.3.4 Detection of Structural Change

By using a non-Gaussian distribution for the system noise, it is possible to treat an occasion when the state of the system suddenly changes. Figure 22.7 shows an artificially generated data, in which the mean value changes midway on the 3 points drastically. Complying with such data, the trend is estimated by assuming that both the system noise and the observation noise are normally distributed. The estimated trend is undulated as shown in Figure 22.8(a), and furthermore it cannot follow the drastic changes of the mean value and the delay of response is seen.

For this case, the Pearson family of distributions was applied to the system noise. The density function of the Pearson family of distributions is given by

$$q(v|b, \tau^2) = \frac{c}{(v^2 + \tau^2)^b}, \tag{22.25}$$

where $0.5 < b < \infty$, c is the normalizing constant to allow the integral in the whole space of $q(v)$ to be 1, resulting in $c = \tau^{2b-1}\Gamma(b)/\Gamma(\frac{1}{2})\Gamma(b - \frac{1}{2})$. Meanwhile, b is a

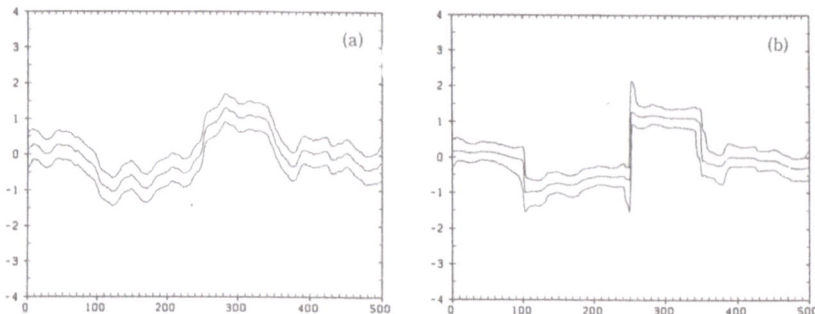

Figure 22.8 Median and 90% error bound of the estimated trend (a) Gaussian model, (b) Non-Gaussian model

Table 22.4 Change of AIC for various value of the shape parameter of the system noise density

b	$\hat{\tau}^2$	$\hat{\sigma}^2$	log-likelihood	AIC
0.6	1.30×10^{-8}	1.024	-741.99	1487.98
0.75	2.20×10^{-7}	1.022	-741.94	1487.89
1.0	3.48×10^{-5}	1.022	-742.25	1488.50
2.0	1.05×10^{-2}	1.018	-745.25	1495.50
∞	1.22×10^{-2}	1.043	-748.52	1501.03

parameter expressing the shape of the distribution. The distribution becomes the Cauchy distribution if $b = 1$, the t-distribution for $b = (k+1)/2$, $k = 1, 2, \cdots$, and the normal distribution if $b = \infty$. In addition, τ^2 is a parameter concerning the dispersion of the distribution. Table 22.4 shows change of the value of AIC as the value of b is changed and the maximum likelihood estimates of the parameters. From the values of AIC, it can be seen that the model is the best when $b = 0.75$ and is the worst when $b = \infty$.

Figure 22.8(b) shows the trend estimated by non-Gaussian smoothing with the best shape parameter, $b = 0.75$. In comparison with the case by the Gaussian distribution, the estimated values smooth enough as a whole is obtained, and the jumps on the 3 places are well detected.

22.4   Conclusions

With the use of the state space model and the Kalman filter, parameters can be estimated even when missing observations are included in a time series. On the other hand, the missing observations can be estimated by using the estimated model. Furthermore by employing an adequate non-Gaussian distribution for the observation noise, processing of the data with outliers can be made.

In subsection 22.3, a method to interpolate the missing observations and outliers is shown. In case that the data is not so long and the order of the model is low, the models such as the AR model, ARIMA model, etc. can directly be fitted to the data with missing observations or outliers, or can be used for smoothing.

References

Dennis, J. E. Jr. and Schnabel, R. B. (1983), *Numerical Methods for Unconstrained Optimization and Nonlinear Equations*, Prentice Hall, New Jersey.

Sakamoto, Y., Ishiguro, M. and Kitagawa, G. (1986), *Akaike Information Criterion Statistics*, D. Reidel Publishing Co, Dordrecht.

Kitagawa, G. (1987), "Non-Gaussian state space modeling of nonstationary time series", *Journal of the American Statistical Association*, Vol. 82, 1032–1063 (with discussion).

Kitagawa, G. and Gersch, W. (1996), *Smoothness Priors Analysis of Time Series*, Lecture Notes in Statistics, No. 116, Springer-Verlag, New York.

Kitagawa, G. and Matsumoto, N. (1996), "Detection of coseismic changes of underground water level", *Journal of the American Statistical Association*, Vol. 91, No. 434, 521–528.

Chapter 23

Mental Preparation for Time Series Analysis

Hirotugu Akaike
1-7-14-204 Toride, Toride, 302 Japan

23.1 Introduction

The philosophical or psychological frame of the researcher is reflected on the way of the development of a scientific research. Since the statistical method is adopted in case where the structure of the objective is inherently undetermined, particular adjustment of the mental framework is required in its application. In ordinary textbooks of statistics, this aspect is not usually discussed. The examples of real time series analysis included in this book provide precious clues to find out how fruitful outcomes have been obtained by approaching concrete problems with particular frames of mind.

The present author has participated in several projects for the handing of practical problems and experienced some success by developing necessary new methods. Based on these experiences, several matters which are supposedly concerned with the mental preparation for the time series analysis will be discussed in this chapter, where the handling of statistical models is treated as the main subject.

23.2 Time Series Analysis and Statistical Science

Since the word statistics is historically adopted to denote the source material for the explanation of the state or condition of a country, it is generally accepted to mean the mass of data or its characteristic values, and the statistics as a science started as the science of the method of handling a large amount of data. With the refinement of the theory, the discussion of the handling of the data of small size has followed, but there yet remains the tendency to accept statistics as the science for the discussion of the procedure of the handling of data.

Such understanding simply represents the conceptualization of statistics corresponding to a particular phase of its development. The essence of the statistical method lies in the creation of the required information by using the data. The material that is required for the action under the uncertainty is the information, and this is given in the form of the resources for the construction of a prediction or more generally of an inference. The statistical science, which is adopted as the title of this series of the books, is defined as the science of intelligent activities of human beings concerning such processes of creation of information. For further discussion of the concept of statistical science, refer to Akaike (1994). It is obvious that for the people who live through the continuous process of making selection or decision under uncertainty statistical-scientific study is indispensable.

The analysis of the data of the births and deaths in Breslau, now a Polish city, made by an astronomer Halley is widely known as the first example of the mathematical analysis of social statistical data. This was for the purpose of assessing the sum of rational installments for pensions and it was aimed at grasping the dynamics for the prediction of the movement of the social masses. Historically viewed, it is plainly observable that the problem of prediction was playing fundamental role at the starting of a statistical-scientific study. Time series analysis intends to grasp the characteristics of the temporal movement or the dynamics of an object and its final purpose is the realization of an adequate prediction. Thus, it can be said that time series analysis offers a typical field of study that explicitly shows the essence of the statistical-scientific study. Time-dependent consideration lies at the origin of statistical thought.

23.3   Prediction and Expectation

If the task of obtaining a value concerning the future with a particular procedure at present is called foreknowing, prediction is required when the process of foreknowing contains undetermined or uncertain elements.

The environment which contains uncertain elements places the people under various types of mental strain. This is a hard environment, but it is also a reality of human being to find some pleasure in there. A great feeling of satisfaction is obtained when various expectations entertained for the future have been fulfilled as the result of actions following the expectations. This is the reason why the plays or games with random elements have old histories and that they are continuously newly devised and enjoyed.

From the above observation, it can be seen that a systematic approach to the uncertainty can be realized only by expressing the structure of our expectation properly. Historically, mathematical discussion of probability started with the problem of evaluation of the gains and losses of bets. The concept of probability is variously explained, but its essence is the formal expression of the structure of our expectation. The problem of prediction can also be effectively handled by adopting a stochastic structure that adequately expresses the expectation.

To cope with the uncertainty for the future, it is necessary to think deeply about what to do by applying all the available knowledge or experience. Effective use of stochastic structures is generally impossible without such efforts.

23.4   Ultimate Truth and Models

Discussion of conventional statistical method treated mainly the inference where the object to be considered has a particular structure with some unknown constants. This is the formulation of ordinary statistical estimation. The theory of statistical test handles the problem in the form of the discussion of whether a hypothesis is true or not. At the bottom of these kinds of discussion lies the idea of the so-called "true structure" or "truth".

If we follow the observation in the previous section, the statistical method to cope with the uncertainty is related to the expression of our expectation. Here, except for the special case where the object is a stochastic mechanism whose structure is already confirmed, the way of expressing expectation depends greatly on the use of our knowledge or experience. Accordingly there is nothing like an absolutely true structure. Furthermore, since an expectation is composed to lead the activity to yield effective results in an objectively defined environment, the expression of the expectation is effectively realized by using a model which represents the characteristics of the environment or object. Accordingly we approach the ultimate truth or true structure through the process of the search for better models.

This corresponds well to the view that, taking into account the fact that even an exceedingly fundamental scientific knowledge has deepened its content through the accumulation of human experiences, the truth we seek is just a matter expressed by a model giving an approximation to the object and is relative to the present knowledge, except for the case of the discussion of the truth or false of a simple situation such as the content is confirmed by opening a box.

The model which is used in the application of statistical method, including time series analysis, is a statistical model. This model uses the stochastic structure which is a mathematical expression of the expectation and represents the frame of the data utilization in a form useful for the action for the future. Naturally, there is nothing like an ultimate model, and advances are continuously made for better models. A model is the expression of an hypothesis, and the proposal of an hypothesis is the fundamental activity of intelligence. The meaning of a set of data is generated and the information is created only by the proposal of an hypothesis. Without the proposal of an hypothesis, based on the wide knowledge and deep experience of the analyst, there is no hope of realization of a good model and its proper use. More importantly, the basic problem of the choice of the data to be used for the analysis cannot be settled without the proposal of an hypothesis.

23.5   Evaluation of a Model and Information Criterion

To propose a single hypothesis and discuss its truth has been a major problem of the conventional theory of statistical test. Judging from the discussions in the preceding section, this is a curious way of handling hypotheses. Attention should have been paid to the process of proposing various hypotheses and searching for better hypotheses or models through the comparisons based on the data.

The basic problem here is the rule or the criterion for the comparison. When this is arbitrary, effective accumulation of experiences by a person or society cannot be realized. In the case of a statistical model, the model is given in the form of a stochastic structure determining the probability or probability density $f(X)$ of yielding the data X. The likelihood of the model with respect to a specific data x observed is defined by $f(x)$. From the point of view of the information criterion, it is observed that the log likelihood, $\log f(x)$, obtained by taking the logarithm of the likelihood provides an adequate basis for the comparative evaluation of the model.

The quantity called the amount of information is defined by

$$I(g : f) = E_g \log g(X) - E_g \log f(X). \tag{23.1}$$

This provides a measure of how far the distribution $g(\cdot)$ is separated from the distribution $f(\cdot)$. E_g denotes the expected value, or mean value, under the assumption that X follows the distribution $g(\cdot)$. This quantity is nonnegative and the smaller $I(g : f)$ or the greater $E_g \log f(X)$ represents the better approximation of $f(\cdot)$ to $g(\cdot)$. By taking the log likelihood $\log f(x)$ as an unbiased estimate of $E_g \log f(X)$, the point of view of the information criterion adopts the log likelihood as the criterion for the evaluation of the goodness of the model given by $f(\cdot)$.

The advantage of this point of view is that it clarifies the fact that the log likelihood has substance that fits perfectly to the discussion developed in the preceding section. The distribution $g(\cdot)$ corresponds to the true distribution in the conventional theory and the above explanation shows that the fact that the distribution is unknown causes no obstacle for practical use of the log likelihood. Especially it clearly demonstrates the intersubjectivity of the log likelihood. It assures the fact that the meaning of the log likelihood as the relative evaluation of the model remains unchanged, even if $g(\cdot)$ represents the ideal distribution that is subjectively entertained as ultimate, but may be conceived differently, by each individual.

In the case the model includes an unknown parameter A such as $f(\cdot|A)$, when the data x is given, the model given by the value of A that maximizes $\log f(x|A)$, the maximum likelihood estimate, is usually regarded as a good one.

However, for the comparison of models, bias is generated with the log likelihood of the model by the adjustment of the parameter with the data. Rectifying the bias and expressed in a form corresponding to the ordinary concept of error, the information criterion AIC is defined by

$$\text{AIC} = (-2) \, (\text{maximum log likelihood}) + 2(\text{number of parameters}), \tag{23.2}$$

where the logarithm denotes natural logarithm. This is originated in the consideration of the problem of determination of the order of the autoregressive model of time series and its application is extended to the general statistical models. This shows that time series analysis contains something that touches the fundamental aspect of statistical-scientific thinking. A detailed discussion of the relation between the information criterion and prediction, and the study of entropy by Boltzmann, is given in Akaike (1985).

23.6   Confirmation of Validity

Even if it is considered that use of the information criterion has lead to a good model, it does not assure the reliability of the model for practical use . Confirmation of the validity in the situation of practical application is indispensable. In the case of the time series analysis, confirmation of the validity of the method or model is possible by the implementation of prediction or control. Generally, in the case of the analysis by a statistical model, it can be checked whether the performance of the method is an expected one or not by making simulations following the stochastic structure conceived by the model. This is one of the advantages of the method of analysis by statistical model.

The point of view of prediction easily provides ideas for the improvement of the model. When the present author discussed the procedure for the systematic development of practical application of the Bayes model, the problem of seasonal adjustment was taken up as an example and a model which assumed the second order difference of the trend to be white noise was adopted for the representation of the trend. This model fits to the idea of the conventional seasonal adjustment procedure that tries the confirmation of the state of the trend in the past and is easy to handle. However from the point of view of the prediction of a fluctuating trend, it must be considered that this model which essentially assumes the movement of the trend to be linear is basically inadequate. It is obvious that adopting an autoregressive model of low order, such as the first or second order, would be much more practical.

Thus the discussion of practical applicability clearly shows that, without being satisfied by a simple application of existing methods, it is necessary to continue the effort to achieve the practical applicability through the improvement of the use of a model or of the model itself, until a positive result is obtained that is quite acceptable for the researchers engaged in the handling of the practical problem. This shows that so long as the statistical science is concerned with the clarification of the structure of our expectation it can never be separated into the two types of activities of the development of mathematical theory and its application.

23.7 Conclusions

Even if the convenience of handling data is increased by the development of computers and other information-related equipments, it does not follow from this that it will automatically produce useful outcomes. When the development of a new statistical method or its application is contemplated, the success or failure depends on how deeply we possess a concrete feeling of the present problem. This is not restricted to the case of the time series analysis. There is no guarantee of success in mechanically imposing a man-made frame of outlook, like a statistical model, to an objective object. Only the continual process of the improvement of the model and its interpretation, realized by the process of thinking the objectives deeply, making observations from a variety of angles, and applying concrete or imaginative actions to the object to check the validity of the hypothesis, can lead to a good result.

References

Akaike, H. (1985), "Prediction and entropy," in *A Celebration of Statistics*, A. C. Atkinson and S. E. Fienberg, eds., Springer Verlag, New York, 1–24.

Akaike, H. (1994), "What is statistical science," *Proceedings of the Institute of Statistical Mathematics*, Vol. 42, No. 1, iii–ix. (in Japanese)

Appendix

Genshiro Kitagawa

ABIC   An information criterion developed for the evaluation of the goodness of the hyper-parameters of Bayesian models. When the data distribution is $p(y|\theta)$ and the prior distribution of the parameter θ is given by $\pi(\theta|\lambda)$ with a hyper-parameter λ, it is defined by

$$\text{ABIC} = -2\log \int p(y|\theta)\pi(\theta|\lambda)d\theta + 2k , \tag{A.1}$$

where k is the dimension of the hyper-parameter (Akaike 1980; Sakamoto 1991).

AIC   Akaike Information Criterion. A criterion introduced by Akaike (1973) for the evaluation of the statistical models and is defined by

$$\text{AIC} = -2(\text{the maximum log-likellihood}) + (\text{number of parameters}). \tag{A.2}$$

The model with smaller value of AIC is considered as better model. In the case of univariate autoregressive model with order m, AIC is given by

$$\text{AIC} = N(\log 2\pi\hat{\sigma}^2 + 1) + 2(m+1), \tag{A.3}$$

where N is the data length, $\hat{\sigma}^2$ is the maximum likelihood estimate of the innovation variance and log denotes the natural logarithm. For the k-variate autoregressive model with order m, it is given by

$$\text{AIC} = kN(\log 2\pi + 1) + N\log|\hat{\Sigma}| + 2mk^2 + k(k+1), \tag{A.4}$$

where $|\hat{\Sigma}|$ is the determinant of the maximum likelihood estimate of the variance covariance matrix of the innovation process (Akaike 1973; Sakamoto *et al.* 1983; Kitagawa 1993).

AR model   A time series model representing the current value y_n using the weighted average of the past values as

$$y_n = \sum_{j=1}^{m} a_j y_{n-j} + v_n , \tag{A.5}$$

is called an AR (autoregressive) model. Here v_n is a Gaussian white noise sequence with mean 0 and variance σ^2 which is independent on the past values y_{n-j}. m, a_j and σ^2 are called the AR order, AR coefficient and the innovation variance, respectively.

Autocorrelation Function   The function of k defined as the correlation between the stationary time series y_n and y_{n-k} is called the autocorrelation function and is denoted by R_k. R_k is easily obtained from the autocovariance function by $R_n = C_n/C_0$.

Autocovariance Function   The function of k defined as the covariance between the time series y_n and the k-lag series y_{n-k},

$$C_k = \frac{1}{N} \sum_{n=k+1}^{N} (y_n - \mu)(y_{n-k} - \mu), \qquad (A.6)$$

where μ is the mean, is called the (sample) autocovariance function. k is called the lag. By replacing the N in the denominator of (A.6) by $N - k$, we can obtain an unbiased estimator of the true autocovariance function. However to assure the positive-definiteness of the variance covariance matrix of (A.27), N rather than $N - k$ is usually used in the definition (A.6).

Coherency   Assume that the power spectra and the cross spectrum of the time series x_n and y_n are given by $p_x(f)$, $p_y(f)$ and $p_{yx}(f)$, respectively. Then the coherency $\gamma^2(f)$ is defined by

$$\gamma^2(f) = \frac{|p_{yx}(f)|^2}{p_y(f)p_x(f)}. \qquad (A.7)$$

The coherency can be interpreted as the square of the correlation between two time series y and x, at frequency f.

Cross-correlation Function   Correlation of two stationary time series y_n and x_{n-k}, considered as a function of the lag k, is called the cross-correlation function, and is denoted by R_k^{yx}. R_k^{yx} can be obtained from the cross-covariance of y_n and x_n, C_k^{yx} and the variances of y_n and x_n, C_0^{yy} and C_0^{xx}, by $R_k^{yx} = C_k^{yx}/\sqrt{C_0^{yy}C_0^{xx}}$.

Cross-covariance Function   Covariance of two stationary time series, y_n and x_{n-k}, considered as a function of the lag k,

$$C_k^{yx} = \frac{1}{N} \sum_{n=k+1}^{N} (y_n - \mu_y)(x_{n-k} - \mu_x), \qquad (A.8)$$

where μ_y and μ_x are means of y_n and x_n, respectively, is called the cross covariance function.

Cross-spectrum   Fourier transform of the cross covariance function C_k^{yx}, $k = 0, \pm 1, \ldots$ of a stationary time series y_n,

$$p_{yx}(f) = \sum_{k=-\infty}^{\infty} C_k^{yx} \exp(-2\pi i k f), \quad -1/2 \le f \le 1/2 \qquad (A.9)$$

is called the cross spectrum. Here i denotes the imaginary number defined by $i^2 = -1$.

FPE   Final Prediction Error. A criterion for the selection of the order of AR model proposed by Akaike (1969). For the univariate AR model with order m, it is defined by

$$\text{FPE} = \frac{N+m+1}{N-m-1}\hat{\sigma}^2 , \qquad (A.10)$$

where N is the data length and $\hat{\sigma}^2$ is an estimate of the innovation variance. The model with smaller value of FPE is considered better. There is a relation between FPE and AIC that $N \log \text{FPE} \approx \text{AIC}$. MFPE and FPEC are the extensions of the FPE criterion to the multi-variate (vector) AR model and vector AR model with control inputs, respectively.

Frequency Response Function   Fourier transform of the impulse response function $\{h_k\}$,

$$H(f) = \sum_{k=-\infty}^{\infty} h_k \exp\{-2\pi i k f\}, \qquad (A.11)$$

is called the frequency response function. It is a complex valued function and is expressed by

$$H(f) = \alpha(f) \exp\{i\Phi(f)\}. \qquad (A.12)$$

Here $\alpha(f)$ and $\Phi(f)$ are called the amplitude and the phase, respectively. There is a relation between the frequency response function and the spectra such as

$$p_{yy}(f) = H(f)p_{yx}(f) = |H(f)|^2 p_{xx}(f). \qquad (A.13)$$

Householder Method   An algorithm to obtain the least squares estimate precisely based on the triangularization of the design matrix without solving the normal equation. Since this Householder method facilitate a efficient computation for the inclusion or removal of a set of variables and the computation of the likelihood of the model, it is sometimes used for fitting autoregressive model or regression models (Sakamoto *et.al.* 1986; Kitagawa 1993).

Impulse Response Function   Suppose that there is a relation between two time series y_n and x_n such as

$$y_n = \sum_{k=0}^{\infty} h_k x_{n-k} + v_n , \qquad (A.14)$$

then $\{h_k\}$ is called the impulse response function of y_n to x_n. The impulse response h_k expresses the response of y_{n+k} with time-lag k when impulse with magnitude 1 is given to x_n.

Kalman Filter   Kalman filter is the algorithm for the computationally efficient recursive estimation of the unknown state x_n of the state space model. The estimators of the state x_n given the observations up to time $n-1$ and n are respectively called the (one-step-ahead) predictor and the filter.

Least Squares Method A method to estimate the parameters of the model by minimizing the sum of squares of the differences between the actual observations and the predicted values obtained from the assumed model.

Likelihood Criterion which is frequently used for the estimation of the parameters of a statistical model. Given the model $p(y|\theta)$ and N independent observations $y_1, \ldots, y_N$, it is defined by

$$L(\theta) = \prod_{n=1}^{N} f(y_n|\theta). \tag{A.15}$$

The logarithm of the likelihood $\ell(\theta) = \log L(\theta)$ is called the log likelihood.

Maximum Likelihood Method A method of obtaining the parameters of the model by maximizing the likelihood or the log likelihood of the model. The parameters estimated by the maximum likelihood method is called the maximum likelihood estimates. For an AR model, approximations to the maximum likelihood estimates can be obtained by the Yuke-Walker method or the least squares method. For an ARMA model or more general state space models, the likelihood function becomes nonlinear and a numerical optimization is used to obtain the maximum likelihood estimators.

Mean The (sample) mean of the stationary time series $y_1, \ldots, y_N$ is defined by $\mu = \frac{1}{N} \sum_{n=1}^{N} y_n$.

Multivariate AR model A model for multivariate time series which expresses the current value as the weighted average of the past ones and the innovation,

$$y_n = \sum_{j=1}^{m} A_j y_{n-j} + v_n \tag{A.16}$$

is called the Multivariate (or Vector) Autoregressive (MAR or VAR) model. v_n is a multivariate white noise sequence with mean 0 and the variance covariance matrix Σ and is independent of the past value y_{n-j}. The parameters of the model, m, A_j and Σ are called the order, AR coefficient matrix and the innovation variance covariance matrix, respectively. TIMSAC-78 contains a program that can determine different order for each (i, j) component. In this case we have to determine k^2 orders.

Optimal Control Assume that the control system with control inputs u_n and controlled outputs y_n is expressed by a state space model

$$\begin{aligned} x_n &= F x_{n-1} + G u_n + v_n \\ y_n &= H x_n. \end{aligned} \tag{A.17}$$

Then the optimal control input which minimizes the quadratic objective function

$$I = E[x_n S x_n^t + u_n R u_n^t], \tag{A.18}$$

where E denotes expectation, is given by

$$u_n^* = K x_n. \tag{A.19}$$

Here K is called the optimal control gain and can be computed by using the dynamic programming (Akaike and Nakagawa 1988).

Power Contribution   If the power spectrum of the i-th variate is expressed as the sum of the effect of independent noise inputs as

$$p_i(f) = \sum_{j=1}^{k} q_{ij}(f),$$

then the quantity $r_{ij}(f)$ defined by

$$r_{ij}(f) = \frac{q_{ij}(f)}{p_i(f)} \tag{A.20}$$

is called the relative power contribution (or relative noise contribution) of the variable j to the variable i. If the k-variate time series is expressed by an AR models and if its innovation variance and covariance matrix is of diagonal form $\Sigma = \text{diag} \{\sigma_1^2, \ldots, \sigma_k^2\}$, then it is obtained by

$$q_{ij}(f) = |b_{ij}(f)|^2 \sigma_j^2, \tag{A.21}$$

where $b_{ij}(f)$ is (i, j) element of the inverse matrix of the frequency response function of the AR operator defined by

$$A(f) = I - \sum_{k=1}^{m} A_k \exp\{-2\pi i k f\}. \tag{A.22}$$

In the use of the relative power contribution, we should carefully check the independence of the noise process, namely the diagonality of the innovation variance covariance matrix (Akaike and Nakagawa 1988).

Smoothing   In general, smoothing is an operation to fit a smooth curve to a data set which contains observation errors. In the state estimation for a state space model, it means to estimate a state x_n using the past, present and future observations.

Spectrum   Fourier transform of the autocovariance function C_k, $k = 0, \pm 1, \ldots$ of the stationary time series y_n,

$$p(f) = \sum_{k=-\infty}^{\infty} C_k \exp\{-2\pi i k f\}, \tag{A.23}$$

is called the spectrum (power spectral density function). If the time series is expressed by an AR model

$$y_n = \sum_{j=1}^{m} a_j y_{n-j} + v_n, \quad v_n \sim N(0, \sigma^2), \tag{A.24}$$

its power spectrum is given by

$$p(f) = \frac{\sigma^2}{|1 - \sum_{j=1}^m a_j \exp\{-2\pi i j f\}|^2} \cdot \qquad \text{(A.25)}$$

State Space Model   State space model represents the behavior of the time series y_n by using a state vector x_n as

$$\begin{aligned} x_n &= F_n x_{n-1} + G_n v_n \\ y_n &= H_n x_n + w_n \,, \end{aligned} \qquad \text{(A.26)}$$

where v_n and w_n are white noise sequences and are called the system noise and the observation noise, respectively. The first model in (A.26) is called the system model and the second one the observation model. AR and ARMA models and various nonstationary time series models can be treated samely by using the state space representation. The models that do not assume the normality of the noise v_n or w_n are called the non-Gaussian state space model.

Stationary Time Series   Time series whose stochastic characteristics such as the mean, the variance and the autocovariance function do not change with time is called stationary. Otherwise it is called nonstationary.

The Institute of Statistical Mathematics   Inter-university research institute in Japan for the research on the statistical science. Established in 1947 as a research organization of the Ministry of Education, Science, Sports and Culture. (Home page address, *http://www.ism.as.jp/*)

Time Series   The observations of a randomly fluctuating phenomenon are called time series. It is usually observed at a equally spaced time interval and is denoted by $\{y_1, \ldots, y_N\}$. When more than one related phenomena are observed simultaneously, e.g., temperature and air pressure, it is called multivariate (or vector) time series.

TIMSAC   <u>Ti</u>me <u>S</u>eries <u>A</u>nalysis and <u>C</u>ontrol program package. Program package for the analysis, prediction and control of time series developed by Prof. H. Akaike and other members of the Institute of Statistical Mathematics, Tokyo. There are four mutually inclusive versions, TIMSAC, TIMSAC-74, TIMSAC-78 and TIMSAC-84. Academic users can obtain the software freely from the Center for the Development of Statistical Computing, the Institute of Statistical Mathematics.

White Noise   The time series y_n which is independent on the past values and is characterized by $C_0 = \sigma^2$ and $C_k = 0$ for $k \neq 0$ is called the white noise with the variance σ^2. If y_n is distributed as Gaussian distribution, then it is called a Gaussian white noise. The power spectrum of the white noise with variance σ^2 is a constant over frequency f and is given by $p(f) = \sigma^2$.

Yule-Walker Method   The popular method of estimating the AR coefficients and the innovation variance of the AR model. The parameters estimated by this method are called the Yule-Walker estimates. Given the autocovariance function C_k, the Yule-Walker estimates $\hat{a}_i$ of the coefficients a_i of the AR model with order m are obtained as the solutions to

$$
\begin{bmatrix}
C_0 & C_1 & \cdots & C_{m-1} \\
C_1 & C_0 & \cdots & C_{m-2} \\
\vdots & \vdots & \ddots & \vdots \\
C_{m-1} & C_{m-2} & \cdots & C_0
\end{bmatrix}
\begin{bmatrix}
a_1 \\
a_2 \\
\vdots \\
a_m
\end{bmatrix}
=
\begin{bmatrix}
C_1 \\
C_2 \\
\vdots \\
C_m
\end{bmatrix}. \tag{A.27}
$$

The Yule-Walker estimate of the innovation variance is then obtained by

$$
\hat{\sigma}^2 = C_0 - \sum_{i=1}^{m} \hat{a}_i C_i. \tag{A.28}
$$

References

Akaike, H. (1969), "Fitting autoregressive model for prediction," *Annals of the Institute of Statistical Mathematics*, Vol. 21, 243–247.

Akaike, H. (1973), "Information theory and an extension of the maximum likelihood principle," *2nd International Symposium on Information Theory*, B. N. Petrov and F. Caski, eds., Akademiai Kiado, Budapest, 267–281. Also reproduced in *Breakthroughs in Statistics, Volume 1: Foundations and Basic Theory*, S. Kotz and N. L. Johnson, eds., Springer-Verlag, New York, (1992) 610–624.

Akaike, H. (1980), "Likelihood and Bayes procedure," *Bayesian Statistics*, J. M. Bernardo, M. H. de Groot, D. V. Lindley and A. F. M. Smith, eds., University Press, Valencia, Spain, 143–166.

Akaike, H. and T. Nakagawa (1972), *Statistical Analysis and Control of Dynamic System*, Saiensu-sha, Tokyo, (in Japanese).

Kitagawa, G. (1993), *FORTRAN 77 Programing for Time Series Analysis*, Iwanami, Tokyo. Iwanami Publishing Co., (in Japanese).

Ozaki, T. (1988), *Time Series Analysis*, University of the Air Press, Tokyo.

Sakamoto, Y. (1991), *Categorical Data Analysis by AIC*, Kluwer Academic Publishers, Dordrecht.

Sakamoto, Y., Ishiguro, M. and Kitagawa, G. (1986), *Akaike Information Criterion Statistics*, D. Reidel Publishing Company, Dordrecht.

Index